FRONT PAGE PHYSICS

FRONT PAGE PHYSICS

A CENTURY OF PHYSICS IN THE NEWS

Compiled by

AJ Meadows
Department of Information and Library Studies
Loughborough University

and

MM Hancock–Beaulieu
Department of Information Science
City University, London

INSTITUTE OF PHYSICS PUBLISHING
BRISTOL AND PHILADELPHIA

British Library Cataloging in Publication Data

A catalogue record for this book is available from the British Library

ISBN 0-85274-310-6

Library of Congress Cataloguing-in-Publication Data

Front page physics: a century of physics in the news/[compiled by]
  A.J. Meadows and M.M. Hancock–Beaulieu
     p. cm.
   ISBN 0–85274–310–6
    1. Physics––History. 2. Science News––History. I. Meadows, A.
J. (Arthur Jack) II. Hancock–Beaulieu, Micheline.
QC7.F76 1994
500. 2'09'041––dc20

Published by Institute of Physics Publishing,
wholly owned by the Institute of Physics, London
Techno House, Redcliffe Way, Bristol BS1 6NX, UK
US Editorial Office: Institute of Physics Publishing,
The Public Ledger Building, Suite 1035, Philadelphia,
PA 19106, USA

Printed in Great Britain by The Bath Press, Avon

Contents

Acknowledgements

A long view of science reporting requires a media source that has consistently reported on science to a high standard over many years. We are therefore deeply indebted to *The Times* for allowing us to reproduce a considerable number of illustrative extracts used in this book. In a similar way, we are very greatful to the Editors of *The Observatory* for permitting us to use some of the material that has appeared in their publication over the past hundred years. The National Aeronautics and Space Administration (NASA) has been most generous in supplying copies of graphics released to the media.

Finally we would like to record our great indebtedness to the editorial and design staff at Institute of Physics Publishing, both in the UK and the USA, who have made a major contribution to this work at all stages.

INTRODUCTION

Attempts to present developments in science to a wider audience have existed as long as modern science itself. As far back as the end of the seventeenth century, few people could understand Newton's *Principia*, so its basic ideas had to be presented in more popular guises—for example, by Richard Bentley in the 1690s. But the *Principia* was an exception. Until the nineteenth century, educated people supposed that they could join in most discussions of science as and when they wished. As science became increasingly specialised during the nineteenth century, and scientists became increasingly professionalised, this assumption proved harder and harder to maintain. The journal *Nature,* founded in 1869, was initially expected to appeal to everyone with an interest in science. As its career unwound, this expectation receded into the background. Those 'scientific' journals that were widely read during the period, such as *Scientific American* and *English Mechanic,* concerned themselves mainly with applications of science, rather than with developments in basic research. Such journals were obviously only read by people who already had a definite interest in science and engineering. Some of the general interest periodicals of the time did mention science reasonably regularly. Though intellectually influential, these periodicals typically reached a fairly limited audience. From the viewpoint of presenting scientific progress to the general reader, it was the newspapers therefore that possessed the most potential.

The end of the nineteenth century was a time of rapid progress in the newspaper world (due to a range of factors—developments in printing, high literacy levels, etc). By this time, science was too much of a specialist interest to figure greatly in the news. However, scientific matters of immediate concern to readers—for example, the application of x-rays to medicine in the 1890s—now received much wider circulation than would have been possible in earlier times.

At the time, the scientists themselves often provided articles on science to periodicals and newspapers. Other contributors were often not sufficiently interested, or, if they did write on science, only managed to annoy the scientific community. An early editorial in *Nature* was devoted to *'Newspaper science'*. It began:

> 'Whether some knowledge of Science or some love for scientific truth will ever penetrate the masses, may well be questioned when we read such an article as the following, which appeared in the daily paper boasting the largest circulation in the world.'

The condemned article was on Joule's equivalent: as the editorial acknowledged, at least it had the virtue of placing the concept before a wider audience than had ever heard of it before. The scientists themselves were, in any case, making it more difficult for their colleagues to contribute accurate knowledge directly to the papers. It was increasingly being urged that professional scientists should not be involved in the popularisation of their subject. John Tyndall, Michael Faraday's successor at the Royal Institution in London, was a leading populariser of physics in the latter part of the nineteenth century. An eminent contemporary physicist commented magisterially:

> 'Dr Tyndall has, in fact, martyred his scientific authority by deservedly winning distinction in the popular field. One learns too late that he cannot "make the best of both worlds".'

This edict against popularisation applied somewhat less strongly to books than to magazines and newspaper articles. At the time, the distinction between textbook and popularisation was less clear-cut than it tends to be today. A series, such as the International Scientific Series (which contained some hundred titles published in the half-century before the First World War), not only purveyed a picture of scientific progress, but could also be used for teaching purposes. When radio broadcasting developed between the two world wars, its attempts to present science typically followed a similar path, with the popularising and didactic elements intermixed. W H Bragg at the Royal Institution gave radio talks of this genre—in essence a development of the long tradition of mounting popular science lectures, but now to a far larger audience.

Twentieth-century science reporting has therefore faced a problem. On the one hand, ordinary journalists have found it hard to achieve a proper understanding of scientific developments: on the other, scientists have been unwilling to act as journalists. The natural response has been the appearance of journalists specialising in science reporting. These were few in numbers between the wars, but the impact of scientific developments—more especially the atomic bomb—in the Second World War has led to a considerable increase in numbers over the past half-century. Correspondingly, radio and television have recruited specialist science reporters and producers over the same period.

The aim of the present book is to sample the reporting of physics in the newspaper over the past hundred years. Such an aim immediately raises a series of questions. Why physics? Why for a hundred years? Why in newspapers? The answers relate partly to the growth of science reporting and partly to the history of physics.

Some areas of science have always been presented reasonably frequently to a wider audience. An example is medicine, which has been reported in newspapers virtually from their first appearance. Physics, in contrast, has moved from a relatively minor media interest to a major concern over the past century: this gives it a special interest as a media topic. During this time, not only have physics and its applications grown rapidly; the nature of the subject itself has been transformed. The classical physics of the 1890s contrasts strikingly with the modern physics of today. The coverage of specifically a hundred years can be justified in terms of this change, for it is in the 1890s that the origins of modern physics can be discerned.

The reliance on newspapers might also be queried. It is, after all, undisputed that television has dominated the dissemination of science via the media in recent decades. However, if the intention—as here—is to examine the development of reporting over a long time-span, newspapers provide the most reliable and balanced material for study. In any case, a comparison between current mass media descriptions of science indicates similar emphases in terms of subject, even if presentation differs. Consequently, physics reporting in newspapers can act as an examplar of the ways in which media comments on science have changed during the twentieth century.

It was actually not until the latter part of the nineteenth century that physics came to be seen as a unified subject at all. Prior to that the need to study heat along with electricity (say) as a normal concommitant was less apparent. Thus, during a discussion of new science degrees in the University of London in the 1850s, Tyndall suggested that students should be allowed to take degrees limited to what we would now consider to be specific branches of physics, such as heat. The word 'physicist' itself was not coined until the 1840s, whereas the word 'chemist' goes back for centuries. This newness of physics is underlined by the dates of creation of the professional societies. In the UK, the Chemical Society was set up in 1841, but the Physical Society followed only in 1874. Even after the subject matter of physics came to be better defined, physics research tended to divide between the theoretical and the experimental. Theoretical physics was for long seen as a branch of mathematics, even though individual scientists always bridged the gap. The ability and training of theoreticians and experimentalists to work together and to be seen as a unified profession has grown gradually during the twentieth century.

A somewhat similar development occurred in astronomy in the latter part of the nineteenth century. Traditional astronomy was concerned with the positions and movements of celestial bodies, and interpretation of the results had become highly mathematical. The application of such new equipment as the spectroscope to astronomical observation signalled not only the beginning of astrophysics, but also the appearance of astronomical observers who paralleled experimenters in physics laboratories. One of the pioneer astrophysicists reflecting on his early work in the 1860s and 1870s, commented:

> 'Then it was that an astronomical observatory began, for the first time, to take on the appearance of a laboratory. Primary batteries ... together with a battery of Leyden jars; shelves with Bunsen burners, vacuum tubes, and bottles of chemicals ... lined its walls'.

As this implies, the latter part of the nineteenth century saw the beginnings of a convergence between physics and astronomy that has become increasingly close ever since. A similar observation holds for both meteorology and geophysics. Meteorology remained primarily a descriptive activity until the last hundred years, since when atmospheric processes—especially those related to weather forecasting—have been explored in ever-increasing detail. The modern study of geophysics essentially dates back to the early years of the present century, at which stage the observation and interpretation of earthquake waves was put on a firm footing. A concentration on the last hundred years of physics reporting therefore corresponds to a period when the application of physical principles has ranged wider and wider. For this reason, the definition of 'physics' used in collecting the material presented here includes astrophysics, geophysics and meteorology, as well as experimental and theoretical physics. An analysis of media reports can correspondingly cast some light not only on how much physics is being reported, but on what branches of physics attract most popular attention.

Science journalists usually have a number of factors in mind when they are deciding whether, or not, a particular item is suitable for their audience. The following five factors provide typical indicators of what such a journalist seeks.

(1)   The information possesses some immediate relevance to the lives of human beings. An obvious example is the announcement of the existence of the atomic bomb at the end of the Second World War.

(2)   The scientic development may lead to applications of value to society, which deserve reporting. For example, work on radio transmission a hundred years ago was considered well worth publicising widely because of its implications for communication.

(3)   A third category concerns physics that has something useful to say about the human environment. An obvious example here is provided by discussions of astrophysical and geophysical factors in climatic change.

(4)   Any advance that the physics community judges to be of major significance to their field will deserve some mention, however brief, to a wider audience. An example is the continuing reporting of advances in fundamental particle physics, despite the esoteric nature of much of the work.

(5)   Anything which smacks of the bizarre or unusual will be looked at for reporting. For example, the occasional claims that astrology has been supported by experimental studies stand a good chance of being mentioned.

Any such list of the factors that lead to preferential selection makes it obvious that some areas of physics are more likely to be reported than others. For example, astronomy, meteorology and, to a more limited extent, geophysics discuss the immediate human environment—stars, clouds, earthquakes, etc—which ordinary human beings can observe directly. Much laboratory physics, on the contrary, occurs under controlled conditions of which the average reader has little or no experience. Consequently, developments related to the former areas are much more likely to be discussed in the media than are advances in the latter. In the years since the Second World War, perhaps a per cent or two of all the science reported in the media has been mainline physics. In comparison, astronomy and space science have sometimes occupied as much as a third of all the space (or time on radio and television) devoted to a year's reporting of science. This imbalance in reporting is necessarily reflected in the extracts reprinted here.

The factors that lead to preferential selection of particular areas of science for reporting have always been at work, though the relative emphases may change with time. For example, the last fifty years have seen less concern with inventions and inventors than earlier decades. The emphasis may also change with the medium involved. For example, television is obviously concerned with visual presentation, so that photogenic astronomy gains further at the expense of less photogenic laboratory physics. But even within the same medium the emphasis can vary. Thus tabloid newspapers have always been much more concerned with the bizarre than have quality newspapers. In addition, their prime concern is with human interest—something that medicine has much more frequently than physics. Consequently, physics reporting has always been concentrated in the quality newspapers.

Just as factors that affect the selective reporting of science can be listed, so can the ways in which science is reported. Here, too, the emphasis can change with time, medium, etc. For example, two approaches that can be found prior to the First World War are the romantic and the religious. The former approach emphasised the romance and wonder of science and technology. (Indeed, children's books bore titles of the type, 'The romance of ...', until well after the war.) The second approach dwelt on the study of the physical world as providing evidence for the existence of God. (This was particularly true of popular articles on astronomy.) Both styles of reporting disappeared almost completely in the 1920s.

Other styles are still with us—for example, two which might be labelled the 'spine-chilling' and the

Introduction

'gee-whiz'. As an illustration of the former, take a short newspaper item from the 1930s:

> 'Johannesburg has been startled by the relevation that the Earth recently escaped collision with a minor planet by little more than five and a half hours.'

It would have been equally correct to report this as: 'a small lump of rock recently missed the Earth by several hundred thousand miles'. But such a summary somehow lacks the spine-chilling element.

The 'gee-whiz' factor, by way of contrast, is usually intended to produce a sense of awe, or even bewilderment. It is particularly common when describing the very big or the very small. To quote another 1930s newspaper report:

> '[Dr Hubble] said that the Universe was a finite sphere 6,000,000,000 light years in diameter and that it was composed of 500,000,000,000,000 nebulae [and so on]'.

it is most unlikely that the reporter expected readers to comprehend these numbers: they were meant to be impressed.

In these instances, the style of reporting was not falsifying the data—merely presenting it in a particular way to try and evoke a particular response. Much more annoying for scientists have been instances of misreporting, whether deliberate, or due to ignorance and carelessness. One survey of news items dealing with climatic change looked for actual errors in reporting, or misrepresentation of scientists, or of scientific data. It found that about a third of the articles contained errors in the science, and a third had errors in reporting what a scientist has actually said. In addition, a quarter had significant omissions or exaggerations. If this sounds alarming, it is worth remembering that professional science writers usually do rather better; but many science stories still originate from reporters with only a limited knowledge of science. Certainly the errors to be found in science reporting prior to the Second World War—that is, before specialist science reporters became a regular part of the scene—are typically more egregious than those since. One twentieth-century science writer has characterised the approach of Victorian journalists to science reporting in the following way (not so vastly different from tabloid reporting today):

> 'Suppose it's Halley's Comet. Well, first, you have a half-page decoration showing the comet ... If you can work a pretty girl into the decoration, so much the better. If not, get some good nightmare idea like the inhabitants of Mars watching it pass. Then you want ... a scientific opinion which nobody will understand, just to give it class'.

A special category of error is the misprint. In principle, such errors are of no great significance. However, misprints can give enormous pleasure both to the general public and to the scientific community (though the latter may also be highly annoyed). Consequently, the material reprinted here includes a representative selection of misprints.

If the reporting of their research leads to problems for scientists—as it can do—ought they not to keep science journalists at arm's length? Some do, though many consider it a duty to help in the public presentation of science: after all, it is from the public that many scientists receive their financial support. There are other, more immediate reasons for supporting the public understanding of science. One is the role it can play in recruiting new members of the profession. In a recent autobiography, a physicist recalled the impact that a 1932 headline— 'Splitting the atom at the Cavendish Laboratory'—had on the choice of her subsequent career. Fifty years later, she could still accurately recall how and where it appeared on the front page of the newspaper.

More importantly for many scientists, media notice of their research may be of importance for their future funding. Many of the people who have an ultimate say in the provision of funding for science (e.g. politicians, civil servants, heads of universities) may have little, or no knowledge of science. They are inevitably influenced in part by what they see or hear via the media. Some scientists, recognising this influence of the media, naturally try to exploit their own media contacts. (The ultimate limitation on this approach remains the innate suspicion of scientists for colleagues who are seen as 'media scientists', along with the scepticism of experienced science journalists.)

Scientists can be expected to be generally sympathetic to the progress of their subject. Nowadays, science reporters may be less sympathetic. In the early days of the space age, American engineers commented wryly that, if a satellite was launched safely, it was a scientific triumph: if not, it was a technological failure. In fact, the general public does not usually draw a distinction between science and advanced technology.

Introduction

Recent disasters—particularly those affecting the environment, such as the explosion at Chernobyl—are seen as relating to science, as well as technology. Though the attitude of the public remains basically favourable to science, scientific developments are now assessed more cautiously than in the immediate post-war years.

Media presentation of science is now unavoidable. The mirror that the media hold up may not always delight the scientific community. But the ability to see ourselves as others see us can be both illuminating and entertaining—as, perhaps, this book will illustrate.

The 1890s

In literature, the 1890s—the *fin de siècle* period—were seen as a time of change. In retrospect, this is even more fundamentally true of physics. The 1890s were the high noon of classical physics; yet, at the same time, they were the period when experimental results fated to undermine the foundations of classical physics began to accumulate. But this is said with the benefit of hindsight. It was not clear to most physicists, and certainly not to lay people at the time, that fundamental change was on the way.

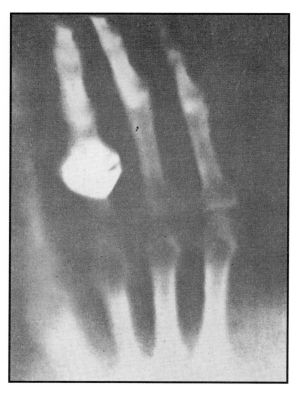

X-ray of the hand of Röntgens wife

What dominated public interest in physics then was electricity and, in particular, new applications of electricity. Readers were fascinated by the exploits of Nikola Tesla, who followed up the invention of the alternating current induction motor in 1888 by introducing the Tesla coil (a means of producing a high voltage at high frequency) in 1891. Alternating current was an especial interest of the period. A gruesome example in 1890 was the first use of the electric chair in the United States to execute a condemned criminal. This was facilitated by Thomas Edison, who provided an alternating current for use in the prison.

Alternating current was also involved in another area of great public interest in 1890s—the production and reception of radio waves. Maxwell's theoretical prediction of electromagnetic waves was taken up by Heinrich Hertz, who, towards the end of the 1880s, first produced radio waves experimentally and examined their properties. Hertz's results were followed up by Oliver Lodge, who was soon able to demonstrate that the waves could be used for signalling. But the big name in this field so far as the public were concerned was Guglielmo Marconi. After some initial work in Italy, Marconi shifted his sphere of operations to the UK. The development work he carried out in the UK during the latter half of the 1890s was basic to the subsequent wide-spread use of radio waves in broadcasting messages. Though the waves were originally called 'Hertzian waves' by physicists, Marconi popularised the term 'radiotelegraphy waves', because of their message-carrying capacity. This then became shortened to our present 'radio waves'.

An equally widely discussed discovery of the decade was at the other end of the electromagnetic spectrum—x-rays. These were first observed by Wilhelm Röntgen in 1895. Within a few months they had been applied medically: to detecting and helping set a broken arm. It was applications such as this that fascinated the general public. From the viewpoint of the history of physics, however, the interest of x-rays is that they bring into focus the breakdown of classical physics. Interest in x-rays led to more detailed studies of the emissions in cathode-ray tubes. There was, for example, a prolonged debate as to the nature of x-rays, whether they were waves or particles. J J Thomson's investigations had shown by 1897 that at least some cathode rays were low-mass, electrically charged particles, soon to be labelled 'electrons'. At the same time, Antoine Becquerel followed up the fluorescent effects of x-rays and, in the process, found that uranium was continually emitting a penetrating radiation. This discovery of what Marie Curie subsequently called 'radioactivity' was the first step on the road towards nuclear physics. By the end of the 1890s, therefore, the basis for a study of sub-atomic physics had been established. But there was little hint as yet to the general reader that major changes in classical physics were in the offing.

One exciting development during the decade linked physics, geology and astronomy. In 1895, William Ramsay isolated helium from terrestrial minerals. It was immediately identified as an element predicted as existing over a quarter of a century before from observations of the Sun's spectrum. The fact that helium was often associated with radioactive materials on Earth was not explained until the next century.

During the 1890s, the growth of astrophysics was aided by the continuing improvement of photographic emulsions. The new science of astrophysics was increasingly concentrating on observations of stars, but public interest continued to concentrate on the solar system. Here photography led to much new knowledge of asteroids. The existence of small, faint bodies orbiting between Mars and Jupiter had been noted originally at the beginning of the nineteenth century, but visual observations only allowed a small number of asteroids to be discovered. In 1891, Max Wolf showed that new asteroids could be detected much more easily using photographic methods. The number discovered soon rose so rapidly that it became difficulty to keep track of them all. But it led in 1898 to the discovery of a new asteroid—Eros—which, unlike other asteroids, could come quite close to the Earth in its orbit. In the twentieth century, this discovery let to a new and more precise method of measuring the Earth–Sun distance (a basic parameter for distance measurement in astronomy).

1890s CHRONOLOGY

1890
- Hollerith introduces an electrically-driven punched-card data system

1894
- Ramsay and Rayleigh discover argon (followed by the other noble gases during the decade)

1895
- The Curie point for magnetic materials is described
- Lilienthal flies the first real glider
- Lorentz describes the forces acting on a moving charged particle
- The Lumière brothers invent the cinematograph
- Perrin provides evidence that cathode rays are negatively charged
- Röntgen discovers x-rays
- Tsiolkovsky publishes his first ideas on space travel
- Wilson develops the first cloud chamber

1896
- Zeeman effect discovered

1897
- J J Thomson discovers the electron

1898
- The Curies discover polonium and radium
- Dewar liquefies hydrogen
- The first viruses are identified

1899
- Rutherford discovers that uranium emits two types of particle

ELECTRIC PROPULSION: WHAT DO YOU CALL IT ?

Sir,—In your issue of the 9th inst. you publish a letter from Lord Bury, asking for suggestions for a short verb to signify moving by electricity.

Might I suggest the adoption of a word from Hindustani and coining another from it ? The word for lightning in Hindustani is " bijli," and is used in common parlance for electricity. If we adopt it and substitute a " y " for the second " i " we should have "bijly," with the first syllable pronounced short.

From this the verb to " bijle " might be coined, making " bijling " in the present, " bijled " in the past.

We should then hear of a ship " bijling " (along) at so many knots.

Sir,—A word is wanted to describe "going ahead by electricity."

Why not " moto " ? I have a very limited knowledge of electrical expressions, but fancy I have heard this word used by electricians. If it is, I suppose it is derived from the Latin word *motus* or the Italian one *moto*. Then let us say we " moto'd at the rate of ten knots up Chiswick Reach."

Sir,—Permit me to amplify my suggestion of " electrized " in your issue of to-day, as follows :—

The verb to electrize, electrizing, electrized ; the noun electrizer ; the person an electrist ; the science electrism ; the force electricity.

Sir,—If a launch or carriage is propelled by electric force, either of two short words will describe the mode of propulsion—"forced" and "sparked."

One is English and forcible, the other is somewhat of an Americanism, but sparkling.

The launch " forced " from Windsor to Marlow ; or, it " sparked " from Marlow to Henley.

Sir,—Comparing the letters appearing in your edition of this morning with my own experiments yesterday, I find I had put down, " We flashed from A to B," " sparked," shocked," " franklined," " battered," " lightened," " poled," " piled," " currented," " electromoted," " circuited," and finally " motored." Of these I liked " flashed " best, but that it invited sarcastic remarks from passengers when the speed was unfortunately not up to the mark. Of the words suggested in *The Times* this morning I think I prefer to " motor." There will be danger, however, of the suffix " ed " in the past tense lengthening the preceding " o " and making it motóred. I prefer " motered," although it is slightly hybrid. How would the Electric Traction Company like "tracted"?

HOW GOOD ARE THE WEATHER FORECASTS ?

Those who have lately consulted day by day the weather forecasts in our columns—and who is now indifferent to the subject ?—cannot fail to have been struck by their remarkable accuracy. But we doubt whether it is generally understood how much meteorology—a science still in some respects in its infancy—has advanced, and how valuable is the assistance which it already lends in the practical business of life. We have before us the last Report, lately published, of the Meteorological Council of the Royal Society, and the results which it attests are such as to convince the most sceptical. Comparing the weather forecasts of 1881 with those of 1889-90, it appears that the percentage of complete successes had risen from 34 to 49, and that the percentage of successes of all kinds, complete and partial, had increased from 76 to 81 per cent. We cannot say that the progress is uniform. The percentage of successes in 1885 and 1887 was higher than it has been since ; probably there were special circumstances to account for this. But in the course of nine years a distinct advance is discernible. It has of late been customary for the department to issue forecasts as to the weather probable in the hay harvest. As to some districts—for example, the Midland Counties—the percentage of correct anticipations was no less than 95. The forecasts for 1889 were five per cent. better than those for 1888, and were the most successful ever published. There is considerable room for improvement in the system of storm warnings for the coast of the United Kingdom originally introduced by ADMIRAL FITZROY. But, on the whole, the results are satisfactory. The department can point with pride to the fact that in 1889 81·2 of its warnings were fully justified, and that only 16·9 of the warnings were not borne out by subsequent weather. Here, however, there are not unequivocal signs of improvement. What explanation is there of the fact that in 1884 the percentage of predictions completely verified by subsequent gales was as high as 66·4 per cent., while in 1889 it was only 47·7 ? It is to be borne in mind that notwithstanding the increase of observations there are large tracts of ocean, of the phenomena originating in which the meteorologist knows absolutely nothing until long after they have expended their energy ; and we notice in the Report the despondent remark that " without questioning that improvements " may be hoped for, as a result of further research, " it is doubtful how far greater success is likely to " be attained, so long as, as at present, no informa- " tion can be obtained concerning atmospheric " changes going on over the Atlantic beyond the " west coast of Ireland." Out of that unknown region may come at any moment a gale which will upset the most careful calculations founded on observations telegraphed from land stations. The communications from the observatory on the summit of Ben Nevis, more than 4,000 feet above the level of the sea, are extremely useful ; and it is possible that Bermuda may become a valuable centre of meteorological observation.

For the time there appears to be a pause in the progress of the art of scientific prediction. Until we know within a few hours what is going on in mid Atlantic and regions nearer our shores, it seems improbable that much greater accuracy will be obtained. And yet there are on sea or shore observers who zealously, so far as their opportunities admit, labour without fee or reward. Hundreds of seamen laboriously collect observations as to the direction and force of winds, the temperature of the sea surface and of the air. All over the country is a small army of volunteer observers. In 1866 there was only one "station of the second order" —that is, a station supplying information from observations taken twice a day by amateurs. At the end of 1889 the number had risen to 97, exclusive of the self-recording observatories and the anemographic stations. But recruits are in request. In many districts from which information would be most desirable there are very few observers, and the result is of course prejudicial. No other science depends so much upon varied co-operation. Every new observer, if only accurate and honest, may, from the first day he sets to work, feel that he is distinctly contributing to the advance of science. The marvel perhaps is that more country gentlemen with time on their hands do not find in the chronicling of accurate observations for the benefit of meteorological science a happy combination of business and pleasure. No large amount of instruments is needed, and the few required are not complicated. With a thermometer, an anemograph, a rain gauge, and a sunshine recorder, the best of work may be done. The friends of idle people are always seeking out a light occupation for them. What can be better than this ?

One thing is very clear—the whole subject of meteorological phenomena is infinitely more complicated than was once supposed. The notion that the knowledge of a few general laws, supplemented by the study of phenomena at many different stations, would soon enable meteorologists to foretell all variations in the weather must be abandoned. We are overwhelmed with observations. But from time to time occurs something to disconcert and humiliate the most skilful meteorologist. Yet, if much remains to be done, much has been done. No small advance has been made when one out of every two forecasts is perfectly correct, and four out of every five are partly so. The agriculturist has good reason to be grateful to those who can tell him nine times out of ten the weather in store for him in harvest ; and the loss of life averted by the system of storm warnings, with all their imperfections, is almost incalculable. This advance has been made chiefly in about a quarter of a century. What will be the condition of things a quarter of a century hence if there be like progress ? With multiplied stations on shore and means of rapidly communicating from mid-ocean the advance of storms, the problem need be no longer handled in the somewhat rough manner now inevitable. Prediction will be possible to an

extent now unknown. A storm totally unforeseen may be as rare as the apparition in the heavens of a comet absolutely unknown to astronomers. To foretell far-off meteorological phenomena may baffle the human intellect. But to be so forewarned as to be forearmed, to be able to take by sea and land precautions sufficient to avert great calamities, seems not only not impossible, but within measurable distance of attainment. What will be the result ? Of this we may be assured— much more than the annual saving of so many lives or so much property. We have all grown up under the empire of other ideas. We have taken it for granted that the storm and the whirlwind must come like a thief in the night without warning. Our habits, our civilization, and even our morals are coloured with the assumption that none can tell whether on any day the sun will shine or rain will fall. Our imaginations have been so under the spell of what may be called the occasional phenomena of Nature, that we shall hardly feel at home in a world in which the tornado is "timed" with the regularity of an express service, and a fall of snow is predicted just as is the arrival of a Cunarder in the Mersey. The world will not lack mysteries, but they will be different from some of those which have perplexed past generations. There will be uncertainties as great as we now know. But the type of instability, the idea of inconstancy, may be, not as now the incalculable tempest, the ever-changing winds, but the nature of man himself, his own mutable ways.

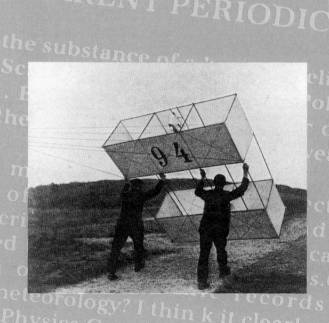

BOX-KITE WITH METEOROLGICAL INSTRUMENTS

1891

ELECTRICITY STANDARDS

The advances made in electric lighting are perhaps best indicated by the fact that there are at present no fewer than 14 companies which have been established with the view of supplying the public with electricity from central stations in the metropolis alone. In every establishment into which the electric current is introduced there must be a meter, in order that the consumer may be correctly charged for the amount used, and these meters, moreover, must, according to the Board of Trade regulations, be standardized and the accuracy of their registering power must be certified. Besides this the central supply stations, of which there are a number established and in course of establishment all over the country, require testing and measuring instruments, which have to be recalibrated and put in accurate working order from time to time. Further, in order to insure uniformity in the electrical instruments of the different makers, it would doubtless be desirable that such instruments should be tested and standardized by some recognized authority. In order to meet these requirements, which constitute a growing want in this comparatively new industry, Mr. Robert Hammond, about a year ago, founded the Electrical Standardizing, Testing and Training Institution. In connexion with this institution, not only has the work we have indicated been carried on, but also that of educating and training students for the profession of electrical engineers. Hitherto the institution has occupied temporary premises in Adam-street, Adelphi, during the construction of a permanent and convenient building. This building, which has now been completed, is situated in Charing Cross-road, and has been appropriately named Faraday-house. It is provided with a perfect set of electrical standardizing and testing instruments for dealing with the work referred to, and there is also a competent staff for carrying out the work which the institution has been formed to undertake.

Faraday-house comprises a number of laboratories, workshops, and class-rooms, the premises being six storeys high, with a basement for engines and heavy tools. The various departments are connected with the public electricity supply mains, so that a constant supply of current in large quantity is always available for the work to be carried on. Another advantage offered is that private rooms, together with the use of every kind of electrical appliance, are, for a reasonable fee, placed at the disposal of experimenters. Inventors or investigators can thus pursue their researches with the strictest privacy and without having to purchase expensive electrical apparatus, and obtain high electric power. The work of training is carried on by a competent staff, under Mr. H. Erat Harrison, B.Sc. Lond. At present there are over 30 students, and ten more are about to enter. The system of training adopted is a combination of the college and workshop systems, or, in other words, the theoretical and the practical. Students, moreover, have the advantage of being in constant touch with the practical business of the generation and distribution of electricity, by visits to the works of various electrical supply companies with which the institution is affiliated. Altogether the institution possesses advantages in connexion with electrical science, of which many will doubtless gladly avail themselves. It is unquestionably a step in the right direction.

DEFECTS IN METALS

THE SCISSOPHONE.—An instrument for detecting flaws in metal castings and forgings, which is called the scissophone, has been invented by Captain de Place, of Paris. The apparatus consists of a small pneumatic tapper worked by the hand, and with which the piece of steel or iron to be tested is tapped all over. Connected with the tapper is a telephone with a microphone interposed in the circuit. Two operators are required, one to apply the tapper and the other to listen through the telephone to the sounds produced. These operators are in separate apartments, so that the direct sounds of the taps may not disturb the listener, whose province it is to detect flaws. The two, however, are in electrical communication, so that the instant the listener hears a false sound he can signal to his colleague to mark the metal at the point of the last tap. In practice the listener sits with the telephone to his ear, and so long as the taps are normal he does nothing. Directly a false sound—which is very distinct from the normal sound—is heard, he at once signals for the spot to be marked. By this means he is able not only to detect a flaw, but to localize it.

STORING ELECTRICITY

The question of successful commercial electrical distribution involves many important points, but next to efficient current-producing power there is, perhaps, not one of higher importance than efficient storage of the current produced. Attention has, therefore, from the first been carefully given to accumulators in which to store the electric current for use as required, just in the same way as gas is stored in a gasholder for distribution to consumers. The result has been the invention of many variously-constructed accumulators, some of which have never survived experiment, while others have lived to do useful work. Among the latter is one which has been quietly making its way in Paris for the past three years, until, owing to its inherent high qualities, it has taken the lead in that city, where it is now in use storing the electric current for about 70,000 lamps. It may be inferred that to do this it exceeds in efficiency all other known batteries. Having recently had the opportunity of inspecting the manufacture and use of these accumulators, we are able to state that it does so, and that it is a marked advance in this class of apparatus. The manufacture is carried on by the Société Anonyme pour le Travail Electrique des Métaux, at their works at St. Ouen, near Paris, where the present output of plates is at the rate of five tons per day. The capacity of the works, however, is a production of ten tons per day. In these accumulators the main objects sought are to increase both the working capacity and the durability, without increasing either their weight or cost. To this end the plates are made of a grid pattern, the square holes being filled in with lead having a very high degree of purity and porosity, which is the secret of the great efficiency attained.

The largest installation in connexion with which these accumulators are used in Paris is that of M. Popp, in which case about 25,000 lamps of 16-candle power each are taking their current from them. There are no fewer than 16 separate Popp distributing stations fitted with these accumulators, which are all charged from one central generating station. The engines and dynamos at this latter station used for charging the batteries cease running at about 4 p.m. daily, the batteries carrying the entire load from that hour until the next morning. We were shown one of the distributing stations, that in the Rue Cambon, from which about 1,500 lamps were being supplied with the current.

SUN AND EARTH

In the natural order of things, astronomy should, no doubt, come first in any review of scientific progress. In astronomy as in biology, during recent years investigators have been dealing with the inside rather than with the outside of their subject. While the "architecture of the heavens," the mathematical relations of the various celestial bodies to each other, continue to form the subject of investigation by observatories and individuals, the internal constitution of these bodies, the physiology, if we may use the term, of suns and planets, of comets and meteorites, is the aspect of astronomy of which we now hear most. The continued improvement of the spectroscope and the constant advances in celestial photography render investigations in this direction more and more fruitful and more and more precise. The latest results and speculations in this department of astronomy have been given to the world during the past year in Professor Norman Lockyer's masterly work, "The Meteoric Hypothesis," noticed in the columns of *The Times* when it was published. That the hypothesis is one which already has led to important results no one will probably deny; but what precisely are the relations of meteors to the stuff out of which the great heavenly bodies have been proximately formed still remains to be proved. But there is every reason to believe that if investigators continue to work along the lines of the hypotheses expounded by Mr. Lockyer, keeping their eyes open to light from all directions, some definite conclusion will in time be reached as to the process by which the orbs of heaven take shape and form. To this end the researches which are being regularly carried on at South Kensington under Mr. Lockyer's superintendence must greatly conduce.

As the sun is the nearest of the so-called fixed stars, he is the object of a crowd of observers, and every year adds to our knowledge of his structure and activity; and through this knowledge, by comparison, we are able to diagnose, to some extent, the nature of the distant suns whose myriad lights constitute the starry heavens. During the past year some of the results of the solar eclipse which took place on January 1, 1889, have been published, and the observations on the corona are of special interest. Thus, from some photographs which have been prepared at the Smithsonian Institution the principal fact deduced is the periodicity of the outer corona in a cycle probably of equal duration with that of the solar spots. The epoch of greatest extension of the equatorial corona appears to coincide very nearly with the epoch of minimum sun-spots. Again, we have Professor Schaeberle, of the Lick Observatory, propounding a mechanical theory of the corona on the basis of the observations made during the eclipse of December 22, 1889. His investigations seem to prove that the corona is caused by light emitted and reflected from streams of matter ejected from the sun by forces which in general act along lines normal to the surface, these forces being most active near the centre of each sun-spot zone. The truth of this theory remains, of course, to be tested. It was to get a clearer look at the sun, unobstructed, as far as possible, by the constituents of our earthly atmosphere, that M. Janssen had himself dragged on a sledge by a score of guides to the summit of Mont Blanc during the past year. His observations confirmed those obtained on a previous occasion at the Grands Mulets; he seems to have definitely proved the absence of oxygen from the sun, at least in the state in which we know it on our earth. Signor Tacchini has continued during the year his constant observations of the solar activity, publishing the results at regular intervals.

The moon has been the object of some interesting observations during the past year. The Brothers Henry, of Paris, have taken some new photographs with their wonderful apparatus, which bring out some features on the surface not before observed; while M. Landerer, by carefully observing the angle of polarization of the rocks of our satellite, comes to the conclusion that her origin and structure are identical with those of the earth.

In other directions we have Koenig's remarkable researches on the physical basis of music, which were brought before the Physical Society by Professor S. P. Thompson. While Koenig's researches cover those of Helmholtz, they are much more fundamental, and upset many of the conclusions reached by the latter. Mr. John Aitken, again, has been continuing his investigations into the dust particles which are everywhere found in the atmosphere, and are probably absent not even from its uppermost layers. On the top of mountains like the Rigi Kulm as few as 200 particles per cubic centimetre were found, and even fewer in certain places in Scotland. The air of Switzerland is indeed remarkably free from dust particles; while in great cities like London and Paris, and some great manufacturing centres, as many as 200,000 per cubic centimetre have been found.

COLOUR PHOTOGRAPHY

At a recent sitting of the Académie des Sciences, Paris, one of its members, Professor Lippmann, explained a new physical method of reproducing the colours of the solar spectrum in the photographic way. Lippmann's process is very simple. Any film employed in photography may be used for it, providing that the sensitive substance is permanently present in it. Those emulsions in which this substance is distributed in the form of fine grains cannot be employed for the purpose. Lippmann used in his experiments bromegelatine plates. The chief feature of the process is that there is under the sensitive film a metallic layer—in Lippmann's method a thin layer of quicksilver—which reflects the rays of light. The further development of the photograph is proceeded with in the usual way. The theory of Lippmann's process is described by him as follows:—The light entering through the objective passes without sensible deflection through the sensitive film of the plate, and effects, with the light reflected from the metallic layer, the division of the film into extremely fine waves, each of which has the exact thickness required for producing the corresponding colour. The vibrations of light are consequently fixed and reproduced almost in the same way as the sound waves in the phonograph. Professor Lippmann submitted to the Academy the results of his experiments in various plates and prints, in which the colours were completely fixed, without being subject to change by the action of light or air.

1892

ELECTRICITY OVER LONG DISTANCES

Sir,—Those who followed the series of articles in your columns in September last upon the Electrical Exhibition of Frankfort will be interested in hearing of the results which have attended the remarkable demonstration of the electrical transmission of power from Lauffen, on the Neckar, to Frankfort, a distance of about 110 English miles.

The transmission was designed for about 100-horse power, which was taken from the rapids in the river at Lauffen by means of turbines, driving a dynamo-electric machine of the "three-phase" pattern, and transmitted to Frankfort by three copper wires, each less than ¼in. in diameter, stretched from pole to pole along the route, beside the railway, through Hanau, Esselbach, and Heilbronn. At Frankfort these wires supplied electric currents to 70,000 glow lamps, or their equivalent, or to a 60-horse power motor, and an additional number of lamps. By means of special transformers at each end of the line the electric pressure was raised to the amount, first, of 12,000 volts, then to 15,000, and later to 30,000 volts, thereby enabling large power to be conveyed through the comparatively thin wires without the great loss in efficiency which would have occurred at a lower voltage.

For three months the result of the elaborate tests applied by the jury of experts under Professor Weber, of Zurich, has been anxiously awaited. Now that the tests are completed, it is gratifying to know that the prophets of evil have been disappointed. To put it briefly, the final result is as follows:—When 113-horse power was taken from the river at Lauffen, the amount received 110 miles away at Frankfort through the wires was about 81-horse power, showing an efficiency, in spite of all possible sources of loss, of 72.1·6 per cent.

With this splendid result to encourage electrical and hydraulic engineers, it will be expected that many schemes for further developments will now be put upon an assured basis. It will obviously be a mere question of means whether, as is proposed by the electricians of Chicago, the coming Exhibition of 1893 will witness there the transmission through wires of 1,000-horse power taken from the Niagara Falls. Already electric transmission has supplanted rope-transmission at Schaffhausen. And we are yet only at the merest beginning of this new branch of engineering development.

I am, Sir, your obedient servant,
SILVANUS P. THOMPSON.
City and Guilds Technical College, Finsbury,
Jan. 25.

ASTEROID PHOTOGRAPHY

circular from the Kiel Central Bureau for Astronomical telegrams announces the discovery of two new asteroids by Charlois, at Nice, on November 29. The right ascension of the first was 57° 43′ 31″, and North Polar distance 77° 36′ 23″, with daily motion of −15′, and 0′ in the two elements respectively. The second lay in R.A. 59° 52′ 41″ and N.P.D. 76° 30′ 50″, with daily motion −13′, and −2′. Both were of the 12th magnitude. The employment of photography has greatly simplified the difficulties of search for new asteroids, and we may fairly assume that none above the 14th magnitude will remain much longer unknown. Longer exposures for the photographs will then be required to pick up those of fainter magnitude, and the question will arise as to how far it will be desirable to carry on this process of extension. The necessity for laborious calculation of orbits to keep up knowledge of such planets as are already known is also greatly lessened by the new method, since the progress of these bodies can be kept under review by repeating photographs of the region in which they move from time to time.

INTERNATIONAL ASTRONOMY

In your issue of February 5, 1892, you published a letter from Professor J. Norman Lockyer and a paper read to the Royal Society by him on " The New Star in Auriga." The paper opened with the following sentences :—

" From a note in The Times of Wednesday, February 3, I learnt that a new star had been discovered in the constellation Auriga, and that photographs had been obtained at Greenwich on Monday night.

" Observations were, therefore, impossible here before last night. This is much to be regretted, and suggests that some organization is needed to further quick transmission of news to observing stations relating to phenomena which may change in a few days, or even hours."

The Astronomer Royal immediately replied (The Times, February 6) that such an organization as that suggested by Mr. Lockyer had been in existence at Kiel for several years, under the name " Centralstelle für Astronomische Telegramme," and had " sent out, on February 2, immediately after the receipt of Dr. Copeland's notification, telegrams to its subscribers announcing the discovery of the new star. It is unfortunate," continues the Astronomer Royal, " that so few English astronomers have hitherto subscribed to this international organization, the necessary expenses of which have to be met by the contributions of the subscribers."

VIEWING MARS

At 10 o'clock to-night the planet Mars will be in opposition with the sun, its distance from the earth being at that hour about 35 millions of miles. This is its point of greatest nearness to the earth since 1877, when the astronomer Hall discovered two satellites. It will also be nearer to the earth than at any future time before 1909.

During July the astronomers of the staff observed Mars for two hours each night. They will continue their observations while it is on the meridian and during the whole of August. The highest magnifying power practicable is 700 diameters, bringing Mars visually within 50,000 miles of the earth. The south polar cap of Mars was an usually large and bright a month ago, but is rapidly diminishing in both respects. This variability has led the astronomers to believe that the cap is composed of snow and ice, as it diminishes as summer comes and increases with the approach of winter. At present there are numerous dark markings within the polar cap, resembling large areas from which the snow had disappeared, and these have not been noticed before in any observatory. A dark region, hitherto observed as single, now appears through the large refractor as double.

The most remarkable phenomena which were observed three weeks ago consisted of three or more prominent bright projections on the south-west limb of Mars. They were visible on several evenings for a short time on each occasion, and will probably again be seen in about two or three weeks' time on the south-west limb. It is hoped that their character will then be more fully determined.

Votaries of other sciences frequently complain at the present day that the fascinations of electrical research and the substantial rewards of applied electrical science are attracting a disproportionate amount of the intelligence of the rising generation. They say that few young men are found to devote themselves, say to astronomy, or biology, while they flock to the study of electrical engineering. We may probably take it that this is more or less the case, and the explanation is not very far to seek. Electricity has laid hold of the imagination of the age. It enters into every department of life, and always with the effect of simplifying, accelerating, and improving the arrangements to which we had been accustomed. The newest torpedo is driven, controlled, and exploded by electricity ; and the electric motor is the destined solution of half our domestic worries. Alike among serious students and among the general public there is strong faith in the possibilities of the science, which offers at once one of the richest and most attractive of modern fields of research, and one of the most promising applications of science to the acquisition of wealth. Some of the sciences now complaining of neglect do not offer the same bread-and-butter attractions, but this, we believe, is a less serious drawback than their want of the scientific outlook that electrical engineering possesses. In some directions we are too visibly reaching, not indeed exhaustion of what is to be known, but the limits of the means of research within our reach. We cannot hope, for example, to see anything much smaller than our present microscopes reveal, nor can we hope for much solid addition to the information given by our best telescopes. But every one feels that we are only on the threshold of knowledge about electricity, whether abstract or applied, and that feeling has determined a flow of young men to the workshops of electrical engineers which cannot be financially justified save by a very rapid development of the comparatively youthful science.

If anything were needed to stimulate enthusiasm or strengthen faith, it was surely supplied last night by the very remarkable lecture with which Mr. Tesla for two hours held a professional audience entranced at the Royal Institution. His beautiful experiments not only open up a new and most promising field of investigation, but also suggest more or less distinctly a revision of many general physical conceptions and a stimulating expansion of our speculative ideas. Mr. Tesla is working on the borderland where light, heat, electricity, chemical affinity, and forms of energy which we cannot confidently identify with any of these, meet and blend. Watching some of his striking experiments, one feels that old lines of demarcation are fading away, and that some new and fruitful generalization cannot be very far off with which we may start upon new voyages of discovery. To take the most obvious and elementary reflection, the spectator involun-tarily asks himself what precisely is meant by electrics and di-electrics, by conducting and insulating bodies. Mr. Tesla establishes an arc between two poles, then passes between them a plate of the best di-electric we know of, and the result is not to check, far less to interrupt, the discharge, but positively to facilitate it. In other experiments of the same kind he shows his high potential currents absolutely disregarding all the devices by which ordinary currents are held in check. It seems to follow that there is no di-electric, no thickness of ebonite, and no space of air that cannot be pierced or bridged by a current of appropriate intensity. It may not seem a remarkable discovery when once it is stated, but nevertheless it is one of the things that people do not realize, and it breaks down quite a number of conceptions which, in the long run, rest upon the tacit assumption that there are hard and fast division lines. Upon this follows the remarkable discovery that as electricity increases in physical power it loses its effect upon the human frame. The lecturer stands in an electrostatic field capable of setting a lamp aglow without wires and feels nothing. He puts one hand upon a terminal from which a brush of violet discharge is crackling and sputtering, in the other he holds a lamp or a vacuum tube, thus making himself the channel for a current at something like 50,000 volts. The vacuum tube glows like the gates of sunrise, but the lecturer feels nothing, though a current of one five-hundredth of the intensity might easily terminate his career. As the sun pours forth rays which the human eye cannot see, so it would seem that the flux of energy in which we live and move finds no response in the human nerve until it is slowed down to the pace of a common dynamo.

1893

RECENT DEVELOPMENTS

Chemistry and Physics have become intimate allies of one of the most interesting and suggestive departments of Astronomy. Chemistry itself, as a science, has reached almost unmanageable dimensions. It has a terminology and a symbolism of its own, into which certain anomalies have been creeping that have produced confusion and misunderstanding. This led to an International Conference of Chemists, held at Geneva at Easter, at which certain very precise rules were adopted as to the methods to be followed in giving names to new combinations. In other respects this conference, it is believed, will have beneficial results on the vast army of chemical-workers all the world over. Noteworthy, also, was the address on Progress in Chemistry and Physics by Professor Ostwald, one of the most eminent of Continental chemists, on recent advances in physics and chemistry. Professor Ostwald has also been continuing his researches and discussions on the theory of solutions. A very beautiful instance of chemical research during the year has been that by Professor Thorpe and Mr. Sutton, resulting in the discovery of the real nature of phosphorous oxide. Equally remarkable was Deslandres' account of his discovery of 14 lines in hydrogen, rendered possible by spectral observations of the sun and stars, resulting in the detection of a striking analogy between these lines and certain harmonics of sound. Lord Rayleigh published during the year the results of his long-continued research on the relative densities of hydrogen and oxygen, 1 to 15·882. Professor H. E. Armstrong's paper on some of the results of his researches into the origin of colour must also be mentioned. The inevitable new element was discovered in Egypt, and was appropriately named Masrium.

Physics is the inadequate name given to a vast and varied department of science. It includes all the multifarious phenomena connected with electricity and magnetism, which in their theories and their applications engage quite as many workers as chemistry itself. We can only say that during the year there has been no slackening in the progress made in both directions.

A paper of considerable importance was given to the Physical Society (one of the most "living" of our scientific societies) by Professor Fitzgerald on electro-magnetic vibrations. Both at the Royal Society and at the Royal Institution, Professor Oliver Lodge, one of our most original and daring investigators, described his delicate researches as to the connexion between ether and matter and the motion of ether near the earth, making us wonder more and more what this intangible something can be. The Electrical Exhibition at the Crystal Palace was a noteworthy event of the year.

Tardy justice has been done to an investigator who was born before his time, by the publication by the Royal Society of Waterston's Theory of Gases, read in abstract to the Society in December, 1845, but pronounced by the referee to whose judgment it was submitted to be "nothing but nonsense, unfit even for reading at the Society." Lord Rayleigh in his preface to the paper points out that Waterston had indicated some of the most recently accepted theories as to gases. Another service done by the Royal Society during the year has been the publication of the report of its Committee on Colour Vision, the practical recommendations of which it is to be hoped will receive attention.

Mr. John Aitken has been continuing his researches on the subject of Dust Particles in the Atmosphere. One conclusion reached by Mr. Aitken is that a great amount of dust increases the day temperature and checks the fall of temperature at night. M. Mascart has been investigating the mass of the atmosphere, and finds it one-sixth greater than is usually calculated. Dr. Assmann, a German physicist, has been observing temperatures at a considerable height in the atmosphere from a captive balloon, and in winter found that the temperature aloft was higher than on the ground, which may be an encouragement to Dr. Nansen to make use of captive balloons in his proposed Polar expedition. Some new ideas on the general circulation of the atmosphere were brought before the Paris Academy by Dr. Pernter.

Professor Roberts-Austen brought before the Institution of Civil Engineers during the year a photographic method of recording high temperatures, devised by himself, which is being adopted in the larger iron and steel works in this country. Results of singular scientific interest have already been obtained by its aid, and it is evident that the introduction of accurate physical measurement in connexion with metallurgy promises to be of great industrial importance.

ELECTRICAL INTERFERENCE

The hearing of evidence before the joint committee presided over by Viscount Cross was continued yesterday. Mr. Langdon, superintendent of the telegraph department of the Midland Railway, gave evidence as to the effect which electric tramways at Walsall had on the wires of the railway company. The only method of rendering railway working safe from electric currents was the introduction of the metallic circuit for both railway telegraphy and tramways, and the dissociation from the earth on both sides. The metallic system was necessary for railway signalling and for the safety of the public. The railway company had 6,000 miles of wire for the block system, and if a third was taken as the amount needing the metallic circuit the company would have to spend £17,000 or £18,000 in the erection of additional wires for the metallic circuit. Capitalizing this sum for maintenance purposes would represent a cost of £57,000. Mr. C. E. Spagnoletti, electrical engineer of the Great Western Railway Company, also recommended as the best system to ensure the safety of the public on railways that there should be an insulated conductor for both the outgoing and incoming currents. The best security for railway signals was to have a metallic circuit. It would not be sufficient for the railways to adopt that system and the electric tramways to keep to the earth system, because the area of tramway disturbance from electric currents was very large. The same held good with reference to currents for electric lighting purposes. The signals on the underground railway in London had been deranged between St. James's-park and Sloane-square, owing to a leak from the wire of one of the electric lighting companies. Mr. T. Swinburne, of the firm of Swinburne and Co., electrical engineers, was called on behalf of the electric traction companies. In his opinion, insulated returns would give the best protection for telephones, telegraphs, and railway companies. He thought that metallic returns ought to be used in all large towns and centres. In the present state of electrical science he would recommend the adoption of the system known as the overhead electric trolly system, which had been adopted at Leeds, and was very greatly used in America.

Mr. Crompton, of Crompton and Co., electrical engineers, expressed himself in favour of the metallic circuit system, and Mr. Bennett, the manager of the New Telephone Company, stated that the company were in favour of metallic returns, believing them to be necessary. The system did not necessarily mean duplicating a number of wires, but a cable containing a number of wires banded together would be preferable. Mr. A. Siemens, of Siemens Brothers, thought that the Budapest conduit system was free from disturbance to telephones. He was in favour of the metallic circuit. Mr. J. H. Greathead, C.E., said the City and South London Electric Railway conductor was metallic, but not absolutely insulated. It passed under two railways, the South-Eastern at London-bridge and the London, Chatham, and Dover at Newington, without in the least affecting the signals or instruments of the railways. On the City and South London Electrical Railway instruments and telephones were in use with insulated returns, and no difficulty was experienced. On the Liverpool Overhead Railway there was a system of automatic electric signals, and no signalmen were employed. The signals had a complete insulated circuit and were worked by the trains themselves. If an insulated return was imposed on the electric railways the effect would be serious. The conductors for the Central London Railway, six miles long, were estimated at £75,000. An insulated return would involve about an equal additional amount, and would add very materially to the cost of working and maintenance. The Budapest conduit system with two conductors would be found to be too expensive for adoption in this country. It would be better to give a power to the telephone companies to lay cables underground than to require the tramways to construct tubes with slots open to the streets. If such a power was given to the telephone companies probably the whole difficulty would disappear. The position of the telephone companies would be improved and electric traction would have a free hand to carry out what was practicable for the improvement of locomotion, especially in cities. Mr. P. J. Parsons, electrical engineer, Paris, speaking with reference to Cincinnati and Boston, showed that the traction companies worked with the return metallic circuit, there being a single wire over the trolley and propelling the car, while the return wire was laid between the rails below the car. In order to get over the effects of stray currents they had in America increased the capacity of the return wires. Mr. Sellon said it was not practically possible to insulate conductors in tramways so as not to interfere with the telephones, unless at very great cost. Major-General Webber said that from his experience of telephonic communication the metallic circuit was indispensable for a good service.

PHYSICS AND GOLF

Prof. P. G. Tait is busily investigating the conditions of flight of a golf-ball, and has been surprised to find how important is the rotation of the ball. A sphere travelling in a resisting medium, and rotating about an axis perpendicular to the line of flight, is deflected from its straight course into a curve : if, for instance, a ball travelling horizontally be rotating about a vertical axis, it will curl to the right or the left according to the direction of the spin ; if it be "topped," and is thus rotating about a horizontal axis in the direction in which it would roll if on the ground, it will fall more quickly than can be explained by gravity alone ; while if the rotation, still about a horizontal axis, is in the opposite direction, the ball will tend to rise. These facts have long been generally known, the case of rotation about a vertical axis being especially well known to golfers practically, and, indeed, to tennis-players, cricketers, and other players of ball-games. But the new fact brought out clearly by Prof. Tait is the *amount* of change thus caused in the flight of a projectile. The range can be enormously increased by increasing the amount of spin about a horizontal axis perpendicular to the line of flight, the spin acting as a levitating agent, which keeps the ball in the air long after it would otherwise have fallen. Thus a golf-ball which, if struck without rotation would fall in 3 seconds at a range of 110 yards, if properly undercut with the same initial velocity, may remain in the air 6 seconds and fall at 180 yards distance. It does not seem impossible that these ideas may prove useful in gunnery or even in aerostation. Their astronomical bearing lies chiefly in the domain of meteoric astronomy ; for new light is undoubtedly thus thrown on the cases of curved paths in meteors. It is only reasonable to suppose that meteors are rotating, and when they meet our atmosphere they should be deflected in consequence. If we suppose a meteor moving in the plane of the ecliptic and rotating, as the planets do, about an axis nearly perpendicular to it in the same direction as that of revolution round the Sun ; then if this strikes our atmosphere on the side away from the Sun, so as to be seen at night, it should have a tendency to curl in towards the Earth, increasing as the atmosphere becomes denser. The matter is perhaps worth careful consideration.

PROGRESS IN PHYSICS

Some of the most interesting work of the year has been done at the Royal Institution, where, with the aid of new and powerful machinery, Professor Dewar has produced liquid air in the large quantities required for purposes of research. Thus armed with the means of studying the behaviour of bodies at very low temperatures, he has carried out, in conjunction with Professor Fleming, an extensive and laborious series of investigations upon the electrical conductivity and resistance of metals at various temperatures down to 200deg. below zero. The results point to the total disappearance of resistance in pure metals at the zero of absolute temperature. The exceedingly perfect vacua used for the protection of liquid air from radiated heat have themselves given rise to investigations into radiation occurring in conditions as far as possible uncomplicated by the presence of convection ; that is to say, in conditions resembling those supposed to obtain in interstellar space. Hydrogen, notwithstanding foolish statements to the contrary, repeated parrot fashion in text-book after text-book, has hitherto resisted all attempts, even with the aid of appliances of unprecedented power, to effect its liquefaction. Until that difficulty is overcome the answer to many questions of the highest interest must remain largely speculative and conjectural.

While the zero of absolute temperature is thus earnestly sought on the surface of the earth, other investigators are endeavouring to measure the temperature of the higher regions of the atmosphere. Experimental balloons carrying automatic recording instruments of great delicacy have been sent up from Paris to a height of 17,000 metres, or over ten miles, and their freight has been recovered. The temperature recorded at that elevation is 51deg. Centigrade below zero.

Lord Kelvin has attacked the problem of crystalline structure, one of the many marvels hidden from common observation by their familiarity. His great command of the most refined methods of mathematical analysis has enabled him to make distinct additions to our knowledge of phenomena which are intimately connected with the inner secrets of the constitution of matter.

An interesting discussion, in which such men as Tait, M'Aulay, Heaviside, and Willard Gibbs have taken part, has been carried on with some vigour upon the basis of the method of quaternions. The value of the method is unquestionable, though oddly enough, it is not taught at Cambridge, yet the ultimate form of its notation and the best mode of using it as a working instrument still remain unsettled.

Professor Rowland, of Baltimore, well known throughout the world by his beautifully-executed diffraction gratings, which rank among the most precious instruments of the astronomer, has completed a work of enormous labour, in the form of a new table of standard wave-lengths, which will supersede the measurements of Engström. In this connexion it may also be noted that great progress has been made during the year in the production of coloured photographs by wave interference.

The recent remarkable employment of inter-ference phenomena for the measurement of objects too small to be seen directly, such as the discs of the minor planets, has during the past year been applied by Professor Michelson, the discoverer of the method, to the solution of a problem of great interest and of practical importance, the determination of the value of the mètre in terms of a natural unit - namely, of the length of a light-wave. He had already found that some 20 spectral lines, which had been supposed to be single, are indeed complex : for example, each component of the yellow sodium lines is itself double, the components being in the ratio of 7 to 10, and their distance being about the one-hundredth of that between the principal components. The red line of cadmium, however, appears to consist of a simple radiation, of which the intensity falls off symmetrically. Making use of this line and an intermediate standard consisting of a bronze bar carrying two plane parallel silvered glass plates, the distance between which can be compared on the one hand with the mètre, and on the other with the length of the red light wave, Professor Michelson found the number of waves of red light in one mètre of air at 15deg. C. and 760mm. to be, from two independent determinations, 1553163·6 and 1553164·6, the difference from the mean being about half a wave length. This unit depends on the properties of the vibrating atoms of the radiating substance and of the luminiferous ether, and is probably, therefore, one of the least changeable quantities in the universe. Professor Michelson says :—

If, therefore, the mètre and all its copies were lost or destroyed they could be replaced by new ones, which would not differ from the originals more than do these among themselves. While such a simultaneous destruction is practically impossible, it is by no means sure that, notwithstanding all the elaborate precautions which have been taken to insure permanency, there may not be slow molecular changes going on in all the standards : changes which it would be impossible to detect except by some such method as that which is here presented.

There is not as yet any immediate prospect of man acquiring the power of flight, but a great amount of solid work has been done during the year in France, in America, and at home, in the investigation of the laws upon which flight depends. It may be said at all events that, whatever mechanical difficulties may remain to be surmounted, there is no theoretical reason why, with motors known to be possible, flight should not be achieved.

In the domain of applied science one of the most noticeable features of the year is the steady extension of the employment of electricity for effecting direct decompositions, hitherto brought about by more circuitous methods. The production of sodium and aluminium, for example, has been immensely cheapened, and the electrolysis of brine threatens to become a formidable rival to the older modes of obtaining alkali and bleaching powder. But the obvious advantages of electrolytic processes will not suffice to insure their adoption on a really extensive scale so long as electricity has to be obtained through the steam-engine. What is indispensable is a more direct utilization of natural forces, and it is to this object that inventive talent may now be most hopefully applied.

BLOWING UP GREENWICH

WE need scarcely inform our readers that the eyes of England, if not of the world, have been centred on the neighbourhood of Greenwich during the last week or two. Notoriety from such a cause is hardly worth the risk run in obtaining it. The explosion of an Anarchist bomb so near our National Observatory, whether as the result of evil design or otherwise, is a little unsettling.

It may be worth while to put on record here the facts as known to those inside the Observatory. On the afternoon of Feb. 15, soon after the close of the usual office-hours, when few persons were on the premises, an explosion, rather louder and more protracted than a rifle-shot, was heard (4^h 51^m 25^s G.M.T.), apparently quite close to the Observatory. Looking from the boundary-railings a little later, Messrs. Thackeray and Hollis, and MacManus the gate-porter, saw a man picked up by the park-constable in the "zigzag," a path in a secluded part of the Observatory Hill. Going to the spot they found the man terribly mutilated, and with his left hand completely severed from his wrist. The man died shortly afterwards. The subsequent investigation, which was carried out in the fullest manner by Government officials, leaves no room for doubt but that the man was one of a band of foreign Anarchists now living in London, and that he had brought an explosive bomb to Greenwich for the purpose of damaging the Observatory and its contents, which exploded as he was in the act of preparing it for its final purpose. But it is not evident why he should have thought that the destruction of scientific instruments could in any way help any quasi-political cause. His intended programme might be described as one of safety, simplicity, and sensation; and it is gratifying to find that the first item of this policy was not practicable.

———

THE only amusing thing in connection with the incident was a letter from a person who wished the Astronomer Royal to use his influence to put a stop to Prof. Dewar's experiments, which he (the correspondent) thought were encouraging Anarchical methods.

RADIO WAVES

In 1884 it was noticed that messages sent through insulated wires buried underground in the Gray's-inn-road were read upon telephone circuits carried on the tops of the houses 80ft. away. In 1885 it was found that currents passing in a telegraph line between Durham and Darlington produced effects in a parallel line 10¼ miles distant. These results were not conclusive, because it could not be proved that they were not due to accidental connexions through the network of wires between those two places. In short, it was doubtful whether they were the result of induction or of conduction. But in 1886 experiments were made on parallel lines 4½ miles apart between Bristol and Gloucester, where there were no intermediate conductors to vitiate the results. Moreover, the circuits used were metallic throughout. Under these circumstances only weak disturbances were detected, though when the experiments were repeated in 1889 in the light of experience more success was obtained. The theoretical possibility of signalling without wires was thus shown, and in 1892 a practical test was made in the Bristol Channel. On Lavernock Point, near Cardiff, a copper wire 1,267 yards long was hung on poles, the circuit being completed by earth. On the sands at low-water mark, 600 yards from this circuit, and parallel to it, two covered copper wires and one bare one were laid down, their ends being buried deep in the sand. On Flat Holm Island, 3·1 miles away, another covered wire 600 yards long was laid down. On the shore an alternating current (which was controlled by a Morse key) with a frequency of 192, a voltage of 150, and of any desirable strength up to 15 ampères was sent through the primary circuit. The signals received on Flat Holm on the secondary circuit produced sound and were read on a pair of telephones. By this means messages were successfully sent. The same method was tried with another island, Steep Holm, 5·35 miles away, but was scarcely successful. Disturbances were indeed perceptible, but the signals could not be read. Such experiments, in which careful precautions were taken to eliminate the effects of earth-currents, proved the possibility of signalling between England and France without any wires being carried across the Channel. But the expense would not be negligeable, for at Lavernock Point a 2-horse power engine was required to get results over three or four miles. Mr. Preece also described experiments made with parallel wires about 1¼ mile apart stretched on poles along the banks of the Caledonian Canal, near Inverness. Here, with the use of ordinary telegraphic apparatus, Morse signalling was easy and speaking by telephone possible. Mr. Preece's critics contend that his results are due to conduction through the earth. He himself maintains they are due to electro-magnetic induction. The rapidly alternating current in the primary circuit throws the surrounding ether into oscillations, and energy is radiated in electric waves. These waves spread out, as do waves of light, and, if they fall on conductors properly placed and sympathetically prepared, are reconverted into an alternate current in the secondary circuit.

POPULAR ASTRONOMY

" I BELIEVE," said Professor Barnard in his lecture last evening in Y.M.C.A. Hall, " that people who ordinarily are supposed to only appreciate such reading-matter in the daily papers as pertains to prize-fights and horse-races will be found to be deeply interested in astronomical subjects, if the same are only placed before them in an easily intelligible form. . . The papers, good as they are, and perfectly wonderful as they are, in getting news that interest the public, do not fully realize yet the interest taken by the masses in astronomical discoveries " (San Franscisco Examiner, 1895, Feb. 10). This paragraph occurs in a very well reported account of the lecture and of an interview with Prof. Barnard ; an account with illustrations, head-lines, and all the other devices for securing the attention of those who read, not exactly as they run, but as they travel by express train. This particular paper, in fact, seems to have taken the hint, and realized for once how attractive a three-column article could be made out of the occasion. But is not Prof. Barnard encouraging rather false hopes ? Part of the attraction in the case of horse-racing is certainly the continuity of the news ; those who appreciate such reading-matter are rewarded by finding, for a large part of the year, something of fresh interest every day : they open their daily paper with that expectation. Now Astronomy has nothing to lose by the dissemination of results when these are obtained ; but to encourage the daily papers to look with any regularity for astronomical news of a popular character would seem rather dangerous to its best interests. News of this kind is scarce. The occasions of an eclipse or other unusual events are, of course, to be excepted, for here the weather supplies the " sporting " interest, and everyone is keenly interested in weather. But ordinarily the tendency is, perhaps, already to publish results too soon instead of working them out thoroughly, and any encouragement of news-gathering would probably increase this tendency, and is therefore to be deprecated.

HYDROELECTRICITY

A great amount of public interest has been aroused in the north of Scotland in connexion with the formation of a British aluminium company to utilize the water-power at the Falls of Foyers, on Loch Ness, which form one of the great attractions for tourists during the summer. It is proposed to bring the material from Ireland to Foyers, and there manufacture the metal with electricity as the motive power. The works will give employment to 100 men in a district where at present there is little or no population ; and the buildings, the company state, will be of an ornamental character to harmonise with the beauty of the surroundings. Public feeling was somewhat allayed by an official statement made at a meeting on Friday to the effect that it is not proposed to do away with the falls, and that the flow of water will only be affected in very dry seasons. As tourists from all parts of the world visit the falls, the subject is one of national interest, and it is understood that the Secretary for Scotland has ordered an official report on the proposed operations of the company.

(UN) STANDARD TIME

" There is a Church of England clergyman in Victoria who absolutely declines to recognize the standard time adopted by Parliament. He rests upon Nature and the God of Nature, and rings his bell for service twenty minutes after other churches are in full swing. The following was his preliminary address to his congregation just before delivering his sermon on the first eventful Sunday morning :—' My brethren,— During the past week an extraordinary attempt has been made by Parliament to alter the time of day. I wish to announce that I intend to take no notice myself of such alteration—nor will the Sun do so either. The Sun will rise and set as heretofore regardless of the unauthorized mandate of men in temporary authority. I do not know of sufficient cause for such alteration, and I repeat that I shall adhere to the only true time, and let those who are weak enough to adopt this false Parliamentary time do so if they wish. The text for this evening's sermon will be found in the nineteenth verse of the hundred-and-fourth Psalm—' The sun knoweth his going down.' ' "

SUNLIGHT AND SUNSPOTS

A paper by Mr. D. M. B. GEMMILL was read on the "Zodiacal Light," the faint glow seen in the sky some time after sunset under favourable circumstances, and supposed to be due to some sort of solar appendage. The writer said it was visible in this country much oftener than was supposed, even in the neighbourhood of towns. The President recommended the phenomenon to the attention of members, stating that there was a surprising scarcity of information with regard to it. He also recommended the study of the form of the Milky Way as another field for observation by members who did not possess telescopes.

A paper by Mr. V. HALE-WHITE, describing a series of observations of sun-spots which he had made to test the truth of Wilson's theory, which supposes the spots to be cavities, was followed by some discussion. Mr. Hale-White said that when a spot had been seen near the edge of the sun, the umbra, or dark central portion, thus foreshortened had invariably been shown as a black line, which would seem to show that it could not be at any great depth, and the appearances generally were inconsistent with the theory of a simple cavity with sloping sides.

The REV. A. L. CORTIE said he believed in the truth of the Wilson theory. A drawing which he had made showing a spot as a notch on the sun's limb was verified by a photograph taken the same day at Greenwich. Some spots looked lighter than others; these were surface spots, and their spectrum showed reversal of the hydrogen lines. The speaker also referred to the *faculæ* or bright markings on the sun, which Professor Hale supposed to be due to outbursts of calcium vapour.

Remarks were also made by Mr. A. ADAMS, Mr. HOLMES, Mr. KENNEDY, and Mr. SEABROKE. Several observations were noted of cases where, when a spot was on the sun's edge, the umbra was seen projecting "like a rock" against the sky. This might be an optical illusion.

The PRESIDENT stated that he was inclined to believe that the spots were cavities, but very shallow, hardly saucer shaped, but rather in form like a shallow plate. He agreed with a remark made by Father Cortie that there was an interesting field for study in the similarity of the outbursts taking place at different parts of the sun at the same time. At one time the sun would be all over little spots for a week, then for a time only good-sized spots would appear, and so on. We had scarcely begun to touch upon the problems the study of sun-spots opened to us.

LIGHT AND PAINTINGS

AN experiment of the gravest importance to artists and owners of valuable pictures is being conducted at South Kensington. Under Capt. Abney's superintendence a gallery has been lighted so as to shut out the most important actinic rays which cause paintings to fade, and yet to show pictures in their true colours ! This marvellous result has been attained as follows : daylight is admitted through skylight windows of *coloured* glass, the panes being alternately green-blue and yellow, so that actinic rays are excluded. At the same time the mixture of these colours produces the effect of white light to the eye ; and a committee of artists has decided that pictures are seen sensibly in their true colours. The gallery ("Raphael Cartoon" Gallery) has been open to the public some little time, and it is curious to note how completely unconscious the visitors are of anything strange in the lighting.

ELECTRICITY FOR CHILDREN

Professor J. A. Fleming, F.R.S., began by discussing the chemical action of the current. Some liquids, he said, can be decomposed into their constituents by means of an electric current ; such liquids are called electrolytes, and the products of the electrolysis ions, these terms having been coined by Whowell at Faraday's request. Professor Fleming then showed the electrolysis of a solution of sugar of lead into metallic lead and acetic acid, and pointed out that the deposition of lead upon the electrode became irregular if the process were hurried too much. This simple experiment he declared to involve the whole art of electro-plating, which he further illustrated by employing electricity to deposit copper on a carbon rod from a solution of sulphate of copper. He then proceeded to discuss the theory of electrolytic action, taking as an example the decomposition of chloride of silver. He supposed that one electrode was positively charged, and the other negatively ; that each molecule of salt was not continuously composed of the same two atoms of silver and chlorine ; and that each atom liberated from union with another carried with it a charge of electricity. He then employed a happy analogy drawn from a ball-room, possessing on one side a peculiar attraction for the ladies and on the other for the gentlemen. No pair of partners were to dance together for more than a few turns, but were to separate and either find new partners or yield to the attractions of one of the sides. The room would thus at any moment be full of pairs dancing and of individuals gradually making their way to one side or the other according to their sex. In the same way the vessel in which electrolysis was going on might be viewed as full of molecules of chloride of silver, and of disconnected atoms of chlorine and silver passing to one electrode or the other according to their polarity. Professor Fleming then stated the law that a given quantity of electricity was always able to liberate the same amount of ions from a particular solution, and, after explaining how an ampère may be defined from this fact, went on to give an experimental proof of its truth by showing that, if a certain tube could be filled with oxygen from the decomposition of water by a certain current in 20 seconds, it could be filled in 10 seconds if twice the current was employed. He then described the curious change which was undergone by two platinum plates used as electrodes, in the decomposition of water, for example. They were no longer the same electrically, but could give off a current as if they were made of zinc and copper respectively. The reason was that one was impregnated with hydrogen and the other with oxygen, and thus their electrical behaviour was altered. The secondary battery had been developed from this fact. Planté was the first to follow it up, and he found the best results could be obtained from lead plates. When these were used as electrodes in acidulated water, oxide of lead was formed on one and spongy lead on the other by the action of the current, and Faure, in 1880 or earlier, noticing this fact, conceived the idea of employing ready-made lead oxide on his plates instead of making it by the current. Secondary batteries were now made both on Faure's and Planté's principles since each had its advantages. Professor Fleming then passed on to consider the heating effects of the current. He showed that a conductor through which electricity was travelling became heated, but to different degrees according as the material of which it was composed was of higher or lower electrical resistance. Thus a current passing through a chain made up of alternated links of iron and copper rendered the iron red hot, but left the copper unaffected to the eye. The heating effects of the current were also illustrated by passing it through a platinum spiral immersed in a little water, which soon began to boil, and its practical applications exhibited in the form of electric ovens, kettles, flat-irons, cigar-lighters, &c. It was proved experimentally that a conductor, when heated by the passage of a current, became elongated ; and a Cardew voltmeter was shown, the action of which depended on this fact. The lecturer concluded by lighting a candle by means of a small piece of red hot platinum wire and a match, thus explaining the principle of electric fuses.

HELIUM AND ASTRONOMY

the unexpected discovery by Runge and Paschen that the principal lines of the new gas, including the bright yellow one, are double appeared for a moment to throw some doubt upon the reality of the certain identity of the clevite line with that in the sun, as the helium line had been regarded as a single line. It was soon proved, however, to confirm strongly the identity of the new substance with that in the sun, since new observations showed to Hale in the United States and to Huggins in this country that the solar line was a really double line precisely similar to that in Runge's photographs.

But it is especially desirable here to call attention to the importance of this discovery as affording an explanation of a number of lines hitherto of unknown origin in the sun's surroundings and in certain stars. In particular, a number of the Orion stars are distinguished by certain dark lines which appear to be bright in the spectrum of the nebula of Orion. These lines are now found to be due to the new gas from clevite.

Vogel has recently made an important advance in our knowledge of the distribution of star types by showing that these characteristic lines of the Orion stars are by no means confined to the Orion quarter of the heavens. In addition to four similar stars already pointed out by Scheiner, Vogel has found some 25 white stars which contain these lines of clevite gas.

Each year of astronomical progress brings to the front with increased emphasis the importance of the far-reaching method of astronomical investigation by means of the spectroscopic determination of the motion of the heavenly bodies in the line of sight, the fundamental principle of which method was first suggested by Doppler in 1842, although it only became a practical method of investigation in the hands of Huggins in 1864.

We have to record an application of no little interest of this method by Keeler to the rotation of Saturn and his rings, which gives us the first direct proof by actual observation of the truth of the view which had come to be held from theory of the nature of the rings, though it may be pointed out that Seeliger's photometric observations of the unvarying brilliancy of the outer rings under all angles of illumination from 0° to 30° require that the rings should be assumed to consist of discrete satellites. Mathematical investigation had shown that a continuous solid or fluid ring could not exist in the circumstances in which the actual ring is placed. This led to the early suggestion by Cassini of an hypothesis which became firmly established by the publication of Clerk Maxwell's classical researches on the subject, that the rings of Saturn are composed of an immense number of comparatively small bodies revolving round the planet in approximately circular orbits. It is obvious that if the rings were continuous, rotating as a whole, the outer edge must rotate

faster than the inner one, whereas if they consist of a number of small independent satellites then lines across the spectrum in Keeler's photographs showed not only that the inner edge moved faster than the outer one, but enabled him to determine with considerable accuracy the relative speed of the outer and inner edges of the ring, giving results closely in accordance with what had been worked out from theory. These observations by Keeler on the rotation of the ring, and also on that of the ball itself, have since been repeated at the Lick Observatory by Professor Campbell. They confirm the accepted view of the rotation of the planet and the theory that the rings consist of a number of small bodies.

In a similar manner the period of rotation of Jupiter determined by observations of markings on the disc has been confirmed by the spectroscope.

The past like the previous year has been one of sensational discovery in chemistry. The discovery of argon announced in 1894 was officially registered in all necessary detail at a meeting of the Royal Society on January 31; and it then appeared that, when everything was told, little could be said of it from the chemical point of view. We are still uncertain whether argon consists of one or several substances; on this and other points concerning it there has been much speculation, but no facts justifying any positive conclusion. At the annual meeting of the Chemical Society on March 24 Professor Ramsay announced the discovery on the earth of a substance known since 1868 to be present in the sun. Helium, as it was named by Mr. Lockyer while still some 92 millions of miles beyond our touch, was stumbled upon—for no other words express the facts—in a most remarkable manner, and, had argon not been discovered, would probably have remained unrecognized. In 1889 Dr. Hillebrand, of the United States Geological Survey, in the course of examination of certain rare uranium minerals (uraninites), discovered that when boiled with weak acids they gave off gas which he, for thoroughly good reasons, concluded to be nitrogen. Mr. Miers, of the Mineral Department of the British Museum—on whose recent appointment to the chair of mineralogy the University of Oxford has much reason to congratulate itself—drew Professor Ramsay's attention to this work as remarkable, and as perhaps of special interest in connexion with argon. On examining the gas given off by clevite, Professor Ramsay found that it was one new to chemists, showing in its spectrum a strong yellow line, which was subsequently shown to be identical with the helium line in the solar spectrum. Helium is apparently as inactive chemically as argon; it is little more than twice as dense as hydrogen, and has a very low refractive index. It appears to be widely distributed, and has been found in the gas given off from various mineral waters—among others in that from Bath water—by Lord Rayleigh.

LISTENING TO SOUND

FIREBALL OVER MADRID

TYPEWRITTEN MESSAGES

WE heard with surprise and admiration that the fame of Baltimore and Prof. Rowland bring to the University annually some 20 or 30 students in *advanced* physics. Considering the number of other universities and institutions in the Eastern States, this is a very large number. No doubt they will still come from afar to hear the *ipsissima verba* of famous men, whatever advances may be made in our methods of communication at a distance—a reflection suggested by the recollection of an apparatus, which Prof. Rowland himself has just completed, for type-writing at a distance of a few hundreds or even thousands of miles. By this apparatus the following small scheme is rendered practicable: eight persons can sit down in Baltimore to eight separate sets of keys and type eight independent letters or poems; and over one and the same wire the results will be flashed to New York, and there appear type-written on eight separate pieces of paper. The simultaneous transmission of eight messages over the same wire is arranged by using different and incommensurable periods between the signals.

The lecture dealt mainly with the phenomena of sound. The sensation of loudness was shown to be due to the amount of pressure on the drum-head of the ear, the greater the pressure the louder being the sound. These pressures were communicated to the inner ear and stimulated the auditory nerve, which was more affected by a loud than by a faint sound. As regards the delicacy of the perception of the ear for varying pressures, the lecturer pointed out that the amount of movement of the diaphragm of the telephone under the influence of a feeble sound was almost inconceivably small, and only to be measured in millionths of an inch. Yet the pressure produced by such a movement was sufficiently great to act on the ear, which, in this respect, was an organ at least as delicate as the eye. Pitch was the next character of sound discussed, and the scale and intervals were explained. Experiments were made, illustrating the well-known fact that pitch depended on the frequency of the vibrations falling on the ear, the greater the number of vibrations striking the ear in a given time the higher the pitch of the tone. A metal disc, perforated with holes and driven at high speed by an electric motor, was caused to sing out tones and even to execute a piece of music when a blast of air was driven against the holes. This led to experiments with the compound siren of Helmholtz, by which the ratio of the frequencies of vibrations in musical intervals, such as the third, fifth, &c., was demonstrated. Then the lower limits of pitch down to tones of about 30 vibrations a second were illustrated by vibrating reeds in Appunn's apparatus, and the curious rough effects of intervals and chords in the lower ranges of the scale were pointed out. Passing to the upper limits of pitch, the lecturer produced from steel bars sounds having vibration frequencies of about 30,000 a second, and intervals and chords were shown by a splendid series of heavy tuning-forks kindly lent by Lord Kelvin. Vibrations of a frequency of about 30,000 a second formed the upper limit of hearing, and those of about 30 the lower. Between these two extremes there was no break, and Dr. M'Kendrick pointed out the analogy in this respect between the senses of hearing and vision, between the range of audible sounds and the spectrum-band. He next exhibited a diagram showing the compass as regards pitch of the chief instruments of music, and noted the enormous range of a complete organ from 16 or 20 up to 4,000 vibrations a second. He showed that the highest tones were produced by instruments in the following order—organ, piccolo, clarinet, harp, flute, piano, &c., and the lowest by the organ, contra-bassoon, bass tuba, piano, ophicleide, &c. He also compared the compass of the human voice with that of the instruments, and showed that it occupied about the middle of the range from the lowest to the highest tone of the organ. It was interesting to notice that, in the upper part of the scale of pitch audible to the ear, there were two or three octaves of sounds not directly used in music, which might, however, be heard in the music of nature—in the hissing of waterfalls, for instance. This led to consideration of how the ear dealt with pitch. Presumably we had in the ear a mechanism adapted for the reception of tones of any pitch, at least from 30 to 30,000 vibrations a second, on the principle that of two bodies tuned to vibrate at the same rate, motion in the one would excite motion in the other. The microscope revealed such a mechanism in the ear. Sympathetic vibration was illustrated by tuning-forks, and the lecture closed with experiments with the phonograph, in which the instrument was driven at slow and fast speeds over records of scales from various instruments. It was thus shown that, the faster the cylinder of the phonograph revolved and the greater the number of impulses consequently given to the glass disc in a given time, the higher was the pitch of the tone given out.

1897

RADIO DEVELOPS

He had, he said, spent 47 years of his life in the study of electricity. Not a day passed that he did not come across something new and interesting, but he ventured to say that the system of telegraphy he was there to explain was the greatest and most important discovery that had yet been made in this branch of science. They knew that the universe was filled with a homogeneous, continuous, elastic medium which transmitted heat, light, electricity, and other forms of energy from one point of space to another. This medium was ether, not air; and the discovery of its real existence was one of the greatest scientific events of the Victorian era. What ether was they did not know. But this agency was utilized in the new means of telegraphy. Mr. Marconi had produced an instrument which he had no hesitation in describing as the most delicate electrical instrument they possessed. The distance to which signals could be sent by Mr. Marconi's system was remarkable. On Salisbury Plain a distance of four miles had been covered. In the Bristol Channel the distance had been extended to over eight miles. The system was now to be put in operation officially between Sark and the other Channel Islands, and in a short time a telegraph office would be opened there and messages would be received and transmitted without the aid of any communicating wires. He believed that this new system would in the near future prove of great commercial and naval and military value. Even if it turned out that it was impossible to communicate over very long distances, could they estimate the value of the system as a means of communication between ship and ship, or ship and shore? Mr. Marconi briefly addressed the audience.

LOW TEMPERATURES

Professor Dewar said it was a mistake to suppose that the establishment of the various cryogenic laboratories that now existed in Europe was due to a desire merely to attain very low temperatures. Rather the desire was to investigate the conditions of gases at all temperatures and thus carry on the work of Faraday, Regnault, and Dr. Andrews, of Belfast. The laws of Boyle and Charles, connecting together the volume, pressure, and temperature of a gas, broke down as the liquid state was reached and indeed before, and for them Van der Waals, working on the careful measurements made by Andrews, proposed a new formula which connected the gaseous and liquid states. From this it was possible to calculate the critical temperature of a gas, above which no amount of pressure could liquefy it, and the work of Cailletet and others consisted in the experimental verification of the results thus obtained. Van der Waals also predicted that there would be found one isothermal line which would express the pressure, temperature, and volume relations of all gases, and Amagat's work had confirmed this prediction. Referring to the difficulty of getting to very low temperatures, the lecturer said that if we had liquid hydrogen tomorrow we should still be a long way from the zero of absolute temperature and should have no conceivable way of reaching it. No liquefied gas could be utilized to produce a fall of temperature more than one-third to one-half its absolute critical range. Hence the lowest temperature that could be obtained by means of liquid hydrogen would still leave us some 20 deg. above the absolute zero. In the course of the experiments which occupied the rest of the lecture, Professor Dewar illustrated the application of extreme cold to the purposes of analysis by showing how with its aid certain constituents, such as ethylene and marsh-gas, could be separated out in a liquid form from a mixture like coal-gas. He also showed the extreme contraction produced by cold in gaseous and other bodies by cooling one end of a tube full of ethylene, the other end being immersed in mercury. When the ethylene became liquid the mercury had risen in the tube substantially as high as in the barometer, proving the vapour pressure to be practically as small as in the Torricellian vacuum.

LODGE vs MARCONI

It appears that many persons suppose that the method of signalling across space by means of Hertz waves received by a Branly tube of filings is a new discovery made by Signor Marconi, who has recently been engaged in improving some of the details.

It is well known to physicists, and perhaps the public may be willing to share the information, that I myself showed what was essentially the same plan of signalling in 1894. My apparatus acted very vigorously across the college quadrangle, a distance of 60 yards, and I estimated that there would be some response up to a limit of half a mile. Some of the hearers of Mr. Preece's recent lecture at the Royal Institution seem to have understood his reference to these previous trials to signify that I had asserted or prophesied that more powerful apparatus would always be limited to some such distance; whereas my statement was a scientific one, concerning the small and early apparatus which, with the help of my assistant, Mr. E. E. Robinson, I had at the time devised and constructed. My apparatus was substantially the same as that now used by Signor Marconi—there was a row of sparking spheres; the sparks were taken under oil sometimes, as suggested by M. Sarasin; there were iron and brass filings in a high vacuum and likewise in hydrogen; there was also my own coherer with a single contact, which is more sensitive, but less manageable, than a filings tube; and the restoration to sensitiveness was effected by an electrically-worked hammer. Signor Marconi uses nickel and silver filings in a lower vacuum, and by employing greater power he has obtained signals over much greater distances; moreover, instructed primarily by Professor Righi, and aided in his trials by the British Post Office, he has worked hard to develop the method into a commercial success. For all this full credit is due—I do not suppose that Signor Marconi himself claims any more—but much of the language indulged in during the past few months by writers of popular articles on the subject about "Marconi waves," "important discoveries," and "brilliant novelties" has been more than usually absurd. The only "important discovery" about the matter was made in 1888 by Hertz; and on that is based the emitter of the waves; the receiver depends on cohesion under electrical influence, which was noticed long ago by Lord Rayleigh and has been re-observed in other forms by other experimenters, including the writer in 1890.

OLIVER LODGE.

X-RAYS AND OTHER ADVANCES

The chief interests in pure as distinguished from applied electrical science during the year 1896 have centred round the important discovery which Professor W. K. Röntgen, of Wurzburg, announced at the beginning of the year. Electricians have long been familiar with the general nature of the effects taking place in the interior of a high vacuum tube, and which the investigations of Mr. Crookes did so much to elucidate 20 years ago. Lenard, guided by Hertz, had succeeded in showing that cathode rays, or at least some radiant effect, could be transmitted through an aluminium window sealed into a vacuum tube into the space outside the vacuum tube to a short distance, but it remained for Professor Röntgen to discover a kind of radiation proceeding from the tube and propagated out to considerable distances from the tube walls, to which the appropriate name of the X radiation has been given.

Although a torrent of scientific literature has been poured out since the announcement of that discovery and having these rays for its subject, and although an army of investigators have attacked the subject, it can hardly be said that very much has been added to the original facts discovered by Röntgen, although much has been done in improving and applying the knowledge. Broadly speaking, the facts appear to show that inside the vacuum tube from the cathode or negative pole a projection takes place of molecules or atoms of the residual gas which are negatively electrified. This was Crookes's original hypothesis, and in spite of some divergence of opinion, chiefly coming from German sources, it appears to have stood well the criticism to which it has been subjected. Where these electrified particles of the residual gas strike the tube wall, or an object placed within the tube or the anode or positive pole, they give rise to the new X radiation. It appears that it is preferable, but not essential, that the object struck should be the anode or a plate connected with the anode. As a matter of practical experience, it has been found best to employ a cup-shaped cathode, and direct the bombardment from it against a plate of platinum placed obliquely, and forming the anode of the tube. The tube itself is best made of German glass. It seems beyond question that the origin of the X radiation is the surface of the object struck, whether the glass walls or the metallic plate. This X radiation causes fluorescence in certain salts, notably in the barium and potassium platinocyanides and the tungstate of calcium. It affects a photographic plate, and it passes freely through some bodies, but is stopped by others. It appears also to be slightly reflected by some metallic plates. In these things it resembles certain kinds of visible light. It is, however, not polarizable. It is not refracted, neither is there any good evidence of interference or diffraction, all of which are characteristics of ethereal wave motion as we know it. It has been stated that the wave length of X rays has been measured by means of diffraction by MM. Calmette and Huillier, but, as this has not yet been confirmed, it will be desirable to suspend judgment on those experiments at present. So far the evidence is not strongly in favour of the view that the X radiation is a wave motion at all, although it has the characteristics of some known ethereal undulatory rays.

What chiefly drew attention to the new radiation was the fact that, owing to the transparency of flesh and opacity of bone, the feat of photographing the bones of the skeleton of a living animal or human being became possible, and an undoubted aid was thereby rendered to surgery in the exact determination of the nature of certain injuries and the location of metallic objects in the tissues. The so-called new photography has passed already into the region of a scientific commonplace. The new radiation has also certain properties in discharging electrification, in which it resembles known forms of light. It has been stated that non-conductors, such as paraffin wax, are rendered conductors when the X rays fall upon them. This fact has, however, not been universally admitted, and certain interesting experiments of Mr. Tesla seem to contradict it entirely. On the whole, we must say that much yet remains to be done before we can pronounce positively on the real character of the new radiation. Its discovery remains, however, one of the notable scientific achievements of the year. An early hypothesis that the X rays were longitudinal vibrations in the ether does not appear to have been strengthened by subsequent work, and it was pointed out that certain of the effects to account for which this hypothesis was erected could equally be ascribed to transverse vibrations of very high frequency. The X radiation has been looked for in other sources of light, such as the sun and the electric arc, but there is no evidence that these sources of light yield any of the peculiar rays proceeding from the place where cathode rays in a high vacuum tube strike an object therein contained or the walls of the tube itself. As regards the nature of cathode rays an experiment by M. Perrin seems to point strongly to the fact that they are, in part at least, composed of streams of negatively electrified particles. One of the most remarkable facts in connexion with this subject is the discovery by Becquerel that certain uranium salts emit, after exposure to sunlight, radiations which, like the Röntgen rays, can pass through plates of aluminium or sheets of cardboard and affect a photographic plate. He found that the potassium platinum compound of uranium emitted these rays for 15 days after it had been exposed to sunlight. It is possible that a connexion may exist between this phenomenon and the increased duration and intensity of phosphorescence of many bodies at low temperatures as discovered by Dewar. Edison and Bleekrode have also shown that increased penetrating power of the X radiation results from low temperatures. As regards relative opacity and transparency of bodies to the X radiation, it appears that opacity is directly proportioned to atomic weight, and that in any homologous chemical series increase of atomic weight of the variable results in increased opacity. It is not at all improbable that the so-called X radiation is a complex thing, and that much research will be necessary before we thoroughly understand the method of its propagation.

1898

RADIATION AND OTHER ACTIVITIES

In electric traction enterprise England is still greatly behind other countries. At the present time there are 950 electric tramcar lines in the United States, having a total length of 14,000 miles of track. In the United Kingdom there are 80 miles actually running and 75 miles under construction. For the ensuing Session applications have been lodged with the Light Railway Commissioners for powers to construct 125 miles of railway, to be worked electrically. In Leeds, Dublin, Bristol, and other places electric tramlines are in operation, and in other towns horse-car lines are being converted to electrical lines. The interpretation which has been placed upon the purchase clauses of the Tramways Act has, however, operated to retard private enterprise and throw the responsibility for progress into the hands of the municipal authorities. In Budapest the last horse tram-car has been lately taken off the lines, and the whole 70 miles of urban tramline in that city are now worked electrically. During the last year a pioneer band of electric cabs have made their appearance in the metropolis. These cabs are driven by electric motors geared to the wheels, and furnished with currents by a storage battery carried behind the cab. They are easily managed, and likely to become very popular.

In the region of pure science, one of the most interesting discoveries is that announced quite at the end of 1896 by Dr. P. Zeeman. He discovered that if sodium or lithium salts are introduced into a flame placed between the poles of a powerful electro-magnet and the light examined by a spectroscope the spectral lines of the metals are seen broadened out on exciting the magnet. Peculiar changes are also seen in the nature of the light near the edges of these broadened lines. In some cases a multiplication of the spectral lines has been observed. The interest in the effect arises from the further evidence thus furnished of a connexion between magnetism and light. The refinement of modern spectroscopy could not be better shown than by the elaborate investigation carried out by Humphreys of the Johns Hopkins University into the variation of wave-lengths under pressure. Only extremely high dispersion and the greatest instrumental delicacy could display the phenomena at all. The shift of similar lines in different spectra seems to follow the analogy of the periodic law of Mendeleeff, being a function of the atomic weight. By photographing the electric spark upon a film moving with extreme rapidity Professor Schuster, of Owens College, has investigated the velocity with which metallic particles are projected from the terminals. The lag of the metallic particles is shown by the inclination or curvature of the image—the air lines showing none—and its amount is thus susceptible of computation. It appears that the *maximum* velocity of the particles is over 6,000ft. per second, or about three times that of a rifle bullet.

Persevering research continues to be undertaken to elucidate the mystery still surrounding the Röntgen or X rays. It has even been considered necessary to establish a Röntgen Society to bring together investigators in this department. M. Henri Becquerel has found that various salts of uranium after six months' enclosure in a radiation-tight box still possess the power to give out rays capable of passing through glass and black paper. These uranic rays also possess the faculty of endowing gases exposed to them with the power of discharging electrified bodies, and they appear to be capable of reflection and refraction.

The nature of the radiation commonly called the Cathode rays, proceeding from the negative terminal in a high vacuum tube, has received in the last year considerable attention. Professor E. Wiedemann has lately found that certain bodies, such as a mixture of calcic and manganese sulphates, if exposed to Cathode rays, afterwards possess the property of becoming luminous when heated to much lower temperatures than would ordinarily suffice. This effect he calls thermo-luminescence. Interesting contributions to our knowledge of Cathode ray phenomena have been made during the past year by Messrs. Campbell Swinton, Deslandres, M'Clelland, and others. The view that these Cathode rays consist of a torrent of electrified atoms or molecules proceeding from the negative pole still receives support from experimental work. Professor J. J. Thomson has, however, suggested that the actual carriers of the electric charge may not be atoms in the chemical sense but some subdivision of them into yet more elementary forms. Every step in a knowledge of the complex phenomena of the electric discharge in vacuo shows that it is to the study of this effect we must chiefly look for an enlargement of our knowledge of the ultimate nature of matter.

In magnetic science Lombardi has made a valuable contribution to the knowledge of dia-magnetic susceptibility and has shown that this is independent of magnetizing force. Cantone has continued his monumental work on the magneto-elastic properties of nickel. One of the most interesting of these is the enormous increase in magnetic susceptibility which nickel experiences when under slight torsion and small magnetizing force. Much study has been given to the di-electric quality of insulators in the past year. Hopkinson and Wilson have studied particularly ice and glass. They have shown that the di-electric constant of glass is much affected by temperature when the electromotive force changes are effected slowly, but very little if the electromotive force changes are very rapid. Dewar and Fleming have completed a long series of researches on di-electric constants at very low temperatures. They have shown that the abnormally large di-electric constants of certain organic and inorganic liquids disappear when these are cooled to the temperature of liquid air. They have also shown reason to believe that Maxwell's law connecting refractive index and di-electric constant is fulfilled exactly by all bodies at very low temperatures.

Lampa has measured the di-electric constants of many bodies with electromotive force reversals of very high frequency, and his results, as well as similar ones by P. Drude, are of great interest. The value of all the above experiments on di-electric capacity rests on the fact that they are of crucial importance in testing the theory which we owe to Maxwell that all electric and magnetic phenomena are really due to actions taking place in the ether, but these are modified by the presence of matter in the same place.

SENDING ELECTRICAL MESSAGES

Beginning with Professor Ewing's well-known experiment whereby the molecular process which takes place in iron during magnetization is represented by the motions of a number of minute magnetic needles under the influence of a powerful magnet, the lecturer proceeded to explain the construction of a Morse key and the method of connecting it up to telegraphic instruments in actual working. The use of the earth for the return circuit was discussed, and it was pointed out that a closed circuit was really as necessary for the flow of a fluid as for the electric current. People were sometimes puzzled as to why the electric light in their houses required two wires whereas the gas had only one pipe. But in truth there was a return circuit in the case of gas just as much as with electricity, because for every cubic foot that flowed along the pipes from the gasometer an equal quantity of air returned through the general atmosphere. With respect to insulation, the real insulator of an aerial line was the air, even aqueous vapour possessing insulating properties. But when aqueous vapour was deposited on solid surfaces as moisture it became a conductor, and therefore the earthenware insulators on telegraph posts had deep channels cut in them, into which the wire was bound so as to be kept dry as by a sort of umbrella arrangement. Telegraph wires were expensive to put up and maintain, and one difficulty in telegraphy was that one wire could carry only a limited number of messages. The lecturer went on to explain the devices by which the carrying capacity of a single wire might be increased. Taking, first, the duplex system of working, he showed how it was possible to signal simultaneously in opposite directions along one wire by the use of differentially-wound instruments and an artificial cable that could take half the current direct to earth at each station. This did not mean that two currents were flowing in the same wire in different directions at the same time, for, when one sender only was at work, one current only would be passing; while if the senders at both ends were in use there would be no current whatever in the line wire. In the latter case, though the current sent from each station would only affect the receiver at the same station, still signals would be received at each end from the other, because the course of a current sent from one station would depend on the condition of the line wire, which was under the control of the other station. By the duplex system, again, two messages could be sent in the same direction along the same wire. A current in a wire at a given instant could have only one definite strength and one definite direction; but either a change of strength or a change of direction could give a signal. Hence by using a reversing key and a strengthening key at one end signals could be produced at the other respectively on a polarized receiver and an unpolarized receiver arranged to be affected only by strong currents. The combination of the duplex and diplex systems gave the quadruplex, by which four independent senders with corresponding receivers could be operated simultaneously over one line, two of each at each end. The lecturer next described the Wheatstone automatic transmitter, explaining how holes were punched in a continuous strip of paper to represent the various letters, and how the paper was then fed into machines which automatically transmitted the message to a number of places in the ordinary dot and dash alphabet. Political speeches were telegraphed in this way, being divided into parts which were punched out by different operators and then put on the transmitter. Over short lines 400 words a minute could be sent by such apparatus, and the surprising character of this achievement might be realized from the fact that on the average each word required 17 or 18 signals. It was not, however, possible to work at this speed over long lines and especially over underground lines, owing to line retardation, which was quite a distinct matter from instrumental retardation. The effects of this line retardation were illustrated by means of Muirhead's artificial cable. With the equivalent of 480 miles of cable in the circuit, no results at all were obtainable with instruments that responded immediately on a short line, while 180 miles were sufficient to cause an interval of perhaps two seconds between the sending and the reception of the signal. In conclusion, the Hughes type-printing telegraph was described and an explanation given of the devices by which the necessary synchronism was maintained between the instruments at each end of the line. The Hughes machine was largely used on the Continent and to a less extent in this country, and the tape machines seen in clubs, &c., were a development of the same principle.

The lecture was copiously illustrated with diagrams and with examples of the actual apparatus, some of which were lent by the Post Office. The next lecture will deal with telegraphy over ocean cables.

SOUND AND HEIGHT

Yesterday a series of experiments in aerial research were conducted in the grounds of Shaw-house, near Newbury, in the presence of several thousands of spectators. The experiments were carried out under the direction of the Rev. J. M. Bacon, Dr. R. Lachlan, Mr. J. N. Maskelyne, and others, with the advice and assistance of Lord Kelvin, Lord Rayleigh, and other scientists. The balloon was in charge of Mr. Percival Spencer and his brother, and was filled with 40,000 cubic feet of gas. During a preliminary captive ascent Mr. Bacon succeeded in taking some excellent photographs from the car with a kinematograph. The main object of the experiments was to discover in what measure the intensity of sound is influenced by altitude, by the presence of clouds, &c. The weather proved favourable for the observations, and the ascent was successfully made at 20 minutes past 5 o'clock, the balloon drifting steadily in a north-westerly direction. The car was occupied by Mr. Spencer, the Rev. J. M. Bacon, Dr. Lachlan, and Mr. Henry Eyre. As soon as the balloon had had a fair start the series of experiments commenced. The first experiment in acoustics was with the voice, followed by five tests with musical instruments, these being succeeded by the discharge of rifles and blasts of the siren from an engine. Then came a rifle volley, followed by a roll of musketry, succeeded in turn by discharges of cotton-powder, 4oz. being used in each charge. After this came three further discharges of cotton-powder, with 8oz. in each charge. When the balloon had travelled a considerable distance there were two explosions of cotton-powder with double charges, the final experiment being a comparison between a discharge of 4oz. of gunpowder and 4oz. of cotton-powder. The aeronauts had with them a receiving instrument, and by noting the altitude and the sounds which reached them took the angular distance. The balloon made an excellent descent at ten minutes to 7 o'clock at North Denford. All the experiments proved highly successful.

1899

VARIABLE TELESCOPE

THE *Daily Mail* is showing a praiseworthy interest in the big telescope for the Paris Exposition next year; but I fear the variety of its information on different dates may confuse its readers. On November 2 last it gave a creditable approximation to the proper design—mirror, horizontal telescope, and screen were all represented, though the clockwork for the mirror was not shown; on the screen was a rather impossible mountain in the Moon; and the instrument was "to bring the Moon within two miles." But on March 30 a quite different picture was given, showing a large equatorial of the ordinary pattern, with a small Eiffel Tower as pier. Letterpress with the picture stated that the magnifying-power was to be 6000—rather a drop from the 120,000 of November 2. The length of the whole instrument is reduced from 197 feet to 185 feet; but even then one cannot deny the truth of the comments that "a man beside it will look a very small and insignificant object," and "its great length will reach high into the air." But by April 26 new information is to hand; no reference is made to the story of March 30; indeed, since it is now remarked that "a telescope of such proportions would fall down of its own weight if constructed on the ordinary upright principle," it would seem wiser to omit any reference to the previous information, though, as above remarked, readers of both articles are likely to be confused a little. The design of April 26 is the correct one, being copied (without acknowledgment) from a French authority; or possibly from the *English Mechanic* of April 14.

NAME OF NEW ASTEROID

IT is after all just as well that the new planet 433 has been named Eros instead of gloomy Pluto. I have not seen any account of the reasons for the name: perhaps there are none capable of being stated. But it is worthy of note that such people as found amusement in making allusions to the tender passion on the occasions of the Transit of Venus will not be robbed of their occupation when the solar parallax is in danger from the new planet. Indeed there is a good deal more meaning in the "Opposition of Eros" than in the old phrase.

NEW ASTEROID

The discovery of an aberrant asteroid revived drooping interest in a company so humorous and so uniformly circumstanced as to offer a very slight prospect of repaying the heavy cost of their computative maintenance. A new vein was however struck with No. 433, which left its "trail" on a sensitive plate exposed by Herr Witt at the Urania Observatory, Berlin, on August 13. The travelling object was found to be in unusually rapid motion, and to pursue a most exceptional orbit. Its mean distance is actually less than that of Mars; and at favourable oppositions, once in 30 years, it approaches within 15 million miles of the earth, being thus our nearest planetary neighbour. By its means, accordingly, increased accuracy can be given to determinations of the solar parallax; and "DQ," as this unique body has been temporarily called, is expected to prove useful and instructive in other ways as well. At favourable oppositions it shines as a seventh magnitude star, though estimated to be no more than 20 miles in diameter; and it has a year of 645 days. Future research will decide whether it can claim intra-Martian associates, or should be regarded merely as a straggler from the familiar trans-Martian band. The latter alternative seems the more probable.

STOKES LECTURES

We can imagine no greater intellectual treat than to be one of the audience at his (Stokes's) professorial lectures, to admire the clearness of expression, the way in which difficulties are faced and removed, to see how everything becomes clearer after passing through his mind, and to enjoy his enthusiasm for his subject, an enthusiasm which is undeniable after fifty years' lecturing, and which led him last year, when a veteran of 79, to forget the time and continue lecturing uninterruptedly for three hours. In his lectures on light we marvel at the experimental skill with which the most difficult experiments on optics are successfully performed with the simplest apparatus; indeed, it has been said that if you give Stokes the Sun and three-quarters of an hour there is not an experiment in optics which he cannot perform.

FLUID MECHANICS

Mr. Boys said that his lectures would essentially be concerned with fluids in motion, but for a proper understanding of that subject it was necessary for him to make some introductory remarks about fluids at rest. He used the term fluids with definite intent, and included under it not only liquids, but also gases and vapours. The chief and most noticeable feature of a liquid was that it always stood level, no matter what was the shape of the vessel in which it was contained. Hence, if the vessel took the form of a long coiled tube, the liquid would always stand at the same height at the two extremities. Two instruments were pointed to as depending on this property of liquids of finding their own level. The first was the spirit level which was so sensitive if well made, that it could detect a difference of level amounting to less than one inch in a mile, and the second was the artificial horizon used by astronomers to determine the altitude of a star. Another important property of liquids was that their pressure became greater at greater depths. This was illustrated by means of an indiarubber tube filled with quicksilver, which was shown to swell at one end as the other was raised up. That this pressure had nothing to do with the shape of the containing vessel but depended solely on the depth of the liquid from the top surface to the bottom was proved by experiments in which it was shown that the same depth of water was required to force open a certain valve, whatever might be the shape and capacity of the vessel at the bottom of which the valve was situated. A fluid in a U-shaped tube stood at the same level in the two legs, the two balancing columns being equal, but if two fluids of different densities were used a longer column of the lighter one was necessary to balance the heavier. Thus in the case of the fluids air and quicksilver about 30in. of the latter was able to balance many miles of the former, as was seen in the barometer; in other words, the pressure of the air in which we lived was the same as we should be subjected to at the bottom of a bath of quicksilver 30in. deep. This pressure of the air was illustrated by two well-known experiments. In the first, that of the Magdeburg hemispheres, it was seen that the two halves of a brass ball could be easily pulled asunder if it were full of air, but not if it were exhausted; and in the second a closed tin vessel containing steam was shown to collapse when the steam was condensed, and a partial vacuum formed within it by means of cold water. Turning to the subject of buoyancy, the lecturer showed that a lighter fluid like alcohol would float on a heavier like water just as an egg would sink in fresh water but float in salt. This fact was taken advantage of in the testing of precious stones. A solution of boro-tungstate of cadmium could be made so dense that heavy gems like diamonds, rubies, and sapphires would, if genuine, sink in it, though imitations or other stones outwardly resembling them would float. In a similar device used by geologists the minerals composing a rock were made to sort themselves by being dropped into a column of liquid heavy at the bottom and gradually becoming lighter towards the top. Wood would float in water because it was lighter, and similarly a stone hung by a string seemed to lose part of its weight when immersed in water. That the same thing was true of air was shown by an experiment in which a glass bulb full of air was arranged in a closed vessel to balance exactly a glass weight hanging at the other end of the beam; when the vessel was exhausted the glass bulb was seen to weigh more, through being deprived of the buoyant effect of the surrounding air. In the same way a balloon rose if filled with a gas lighter than air, or if the air it contained was made lighter by being heated. The hydraulic press was next explained and the tremendous power which it could be made to exert was shown to depend on the fact that pressure in a liquid was the same in all directions and all parts throughout its extent. It was not therefore surprising that hydraulic power pipes were laid all over London and used to drive lifts, engines, &c. The hydraulic engine, however, had usually the disadvantage that it used as much water when it was doing work as when it was not, but in the form devised by Mr. Rigg, the lecturer believed that for the first time the problem was solved of making a motor whose consumption of water was proportionate to the work being done. The property possessed by liquids of expanding when heated and occupying more space was next pointed out, and by means of a model the action of the hot water supply system of a house was explained as depending on the superior lightness of hot water as compared with cold, whereby it was made to rise. The action of the siphon was then discussed, and the instrument shown to be applicable both to liquids and gases. Finally, a description was given of the Harcourt standard pentane lamp, which depended on the action of two siphons working with gases and which, by providing a light that was always steady and constant, deserved to be called a Godsend to the gas examiners who daily tested the quality of the London gas supply.

AMATEURS AND EARTHQUAKES

THE best way to eradicate our mistrust of earthquakes and snakes and similar terrors of childhood is, no doubt, to keep them as pets. There seems no reason why the horizontal pendulum for studying earthquakes should not become nearly as popular as the barometer and thermometer for studying the weather, except that in England we get more weather than earthquakes. The late Mr. Latimer Clark wrote his little book on the transit-instrument "for country gentlemen," in the hope that the instrument might become as popular with them as "the stereoscope or the camera." He was, in this matter, a little too sanguine, but no doubt there were some who profited by his advice. And there must be other country gentlemen who would find it fascinating to read in their own homes the echoes of upheavals in distant Japan. There need not be any calculations as in the case of the transit-instrument. The earthquake-instrument wants winding up regularly, and that is all. Days may pass without any record of interest, but "Milne in his own seismological observatory at Shide, Isle of Wight, obtains records of distant earthquakes at the average rate of about six per month"; so that there is really quite enough excitement for leisure moments. If anyone is considering the desirability of setting up a horizontal pendulum, I may mention that the complete outfit costs something like £30, besides the working expenses of the photographic records.

The 1900s

From the viewpoint of physics reporting, the first decade of the twentieth century presents something of a paradox. In retrospect, it is obvious that several of the most important developments in modern physics began then; yet they went mostly unreported in the popular press. This was the decade of Planck's quantum hypothesis, and of Einstein's *annus mirabilis*—1905. In that year, Einstein explained the photoelectric effect on the basis of the quantum hypothesis, demonstrating that light must sometimes act as though it consists of particles. In the same year, he showed that Brownian motion can be interpreted as direct evidence for the existence of molecules in motion. In 1905, too, he laid out the basis for the special theory of relativity. Yet all these breakthroughs mostly passed the newspapers by.

Popular interest concentrated much more on the experimental discoveries. Not only was more being uncovered about the complexities of radioactivity; in addition, the observations began to suggest possible explanations of what was happening in radioactive processes.

Fleming gives the Christmas lectures at the Royal Institution

At the beginning of the period, Rutherford and Soddy discovered that uranium breaks down into some other substance (actually found later to be radium). Following up this work led them to a general description of radioactivity as the decay of one element into another. Soddy further defined isotopes as atoms which are chemically identical, but which have different weights. Consequently, radioactivity could now be seen as the breakdown of atoms through a series of steps, decaying from one isotope to another.

By the early part of the decade, it was recognised that radioactive substances produced three types of emission—the α, β and γ-ray—and it was known that the first two of these were charged particles. Rutherford subsequently showed that α-particles were in some way associated with helium. Along with his colleagues, Geiger and Marsden, he used these α-particles to bombard gold foil, and was extremely surprised to find that some particles were actually reflected back by the foil. 'It was', he said, 'almost as incredible as if you fired a 15 inch shell at a piece of tissue paper and it came back and hit you'. Trying to think how this observation could be explained ultimately led Rutherford, at the end of the decade, to the idea of an atom with a small, dense, positively charged nucleus. At the same time, J J Thomson and others demonstrated that the number of negatively charged electrons in an atom was limited. The scene was now set for the theoretical explanation of atomic structure that followed early in the next decade.

Experiments on radioactivity figured significantly in the popular press. Part of the reason was the intrinsic interest—rather like that of a detective story—in disentangling what was happening. Another part was the potential importance of radioactive processes for energy generation. As the humorous magazine *Punch* explained:

Radium very expensive, the source of perpetual motion,
Take but a pinch of the same you'll find it according to experts,
Equal for luminous ends to a couple of million candles

A final attraction of radioactivity was the applications of it that now began to appear. In the middle of the decade, Boltwood in the USA suggested that the final decay product of uranium was lead, and that a comparison of the amounts of uranium and lead present in a rock would allow the age of the rock to be calculated. His first estimates using this method were sufficient to reveal that the Earth must be much older than was then supposed. Previous estimates had stemmed from work by Kelvin, who had supposed that the Earth's heat derived purely from the slow cooling down of an originally molten Earth. It was now realised that Kelvin's results must be wrong, because no account had been taken of radioactive heating. At almost the same time, increasingly sophisticated studies of the way earthquake waves travelled through the Earth showed that the Earth possessed a fluid (and so, presumably, hot) central core.

Yet the great media interest of the period related to another area of applied physics—the development of radio. In the latter part of the 1890s Marconi in Britain had concentrated on extending the range of radio signals. In 1901, he succeeded in transmitting a morse signal across the Atlantic from Cornwall to Newfoundland. In the same year, Braun in Germany demonstrated that radio waves could be picked up using a crystal detector (as compared with the 'coherer' previously in use).

Marconi and Braun shared the Nobel prize for physics in 1909. But Braun's introduction of what would now be labelled a semi-conductor receiver was already being overshadowed by the development of the vacuum tube, or valve. The diode was invented early in the decade by Fleming in England, and was very quickly followed by the triode, developed mainly by De Forest in the USA. Over the next few decades, radio detection was to be dominated by the vacuum tube. At the same time, new methods of generating radio waves were being explored. Thus, Marconi had relied on a spark generator in 1901, but an arc generator was successfully put into use in the following year.

This newspaper interest in radio can be seen as the culmination of the Victorian fascination with electricity. The growth of the various parts of the electrical industry represented the first major industrial application of new advances in physics: as such, it played a major role in drawing public attention to physics research. However, it was not the only such application. For example, colour photography—for which the French physicist, Lippmann, received a Nobel prize in 1908—also attracted attention from the press over a number of years. Though the origins of flight were more empirical than theory-based, the discussion of aerodynamics became far more intensive once powered flight was under way. The flights by the Wright brothers received little public attention initially. Consequently, it was only in the latter part of the decade that the mechanics of flight were much mentioned in the press.

The 1900s were not a decade of spectacular advance in astronomy so far as the general public were concerned. The major excitement was the discovery of two new moons of Jupiter (another illustration of the value of photography). A curious result obtained by the Dutch astronomer, Kapteyn, also attracted some discussion. His detailed examination of the way in which the stars move seemed to show that there are two distinct streams of stars observable in the sky. (It was not until some decades later that this odd observation could be properly interpreted in terms of the nature of our Galaxy.) Early in the decade, Lebedev in Russia showed experimentally that light can exert a pressure. This led to discussion of the possibility that sunlight might affect the shape of comets—a matter of some import as the time approached for the return of Halley's comet in 1910.

1900s CHRONOLOGY

1900

- Drude demonstrates that metallic conductivity can be explained by moving electrons

- Hilbert provides a list of the main mathematical problems that need to be solved in the twentieth century

- Mendel's work on genetics is rediscovered

- Planck enunciates the quantum theory

- First amino acid is discovered

1901

- De Vries introduces the idea of genetic mutations

- Harvard classification of stellar spectra attains its final form

- Marconi receives a radio message across the Atlantic

1903

- First powered flight by the Wright brothers

- Rutherford and Soddy identify radioactivity as transmutation of elements

1904

- Fleming patents the first vacuum tube

1905

- Einstein explains the photoelectric effect, demonstrates the cause of Brownian motion, and describes the special theory of relativity

1906

- Oldham shows from a study of earthquake waves that the Earth has a core

- Rutherford identifies α-particles as helium nuclei

- Schwarzschild develops the idea of radiative transfer in stellar atmospheres

1907

- Weiss proposes the domain theory of ferromagnetism

1908

- Geiger and Rutherford begin to develop the Geiger counter

- Hale discovers that sunspots have strong magnetic fields

- Onnes liquefies helium

1909

- Blériot flies across the English Channel

- Mohorovičić identifies the boundary between the Earth's crust and mantle

- Peary reaches the North Pole

1900

PROGRESS IN PHYSICS

To physicists undoubtedly one of the most interesting events of the year just closed has been the celebration of the centenary of the epoch-making electrical discoveries of Volta. In June a meeting of telegraphists and representative electricians assembled at Como and Milan, and various conferences and festivities were held in honour of the great scientist. A second congress was opened on September 18 by their Majesties the King and Queen of Italy, under the presidency of Professor P. Blaserna, of Rome. The members of the British Association, meeting at Dover, under the presidency of Sir Michael Foster, at the same time, placed themselves in sympathy with these events at Como by the despatch of appropriate messages, and assembled in the Town-hall on September 18 to hear an address by Professor Fleming on " The Centenary of the Electric Current."

Amongst the electrical events of the year which have attracted public attention a prominent position must be given to the striking advances made in telegraphy across space without connecting wires. On Christmas Eve, 1898, an apparatus was placed on board the East Goodwin lightship, establishing communication between it and the South Foreland lighthouse. Evidence has recently been given that property to the value of £52,000 has been saved from loss by the establishment of this means of communication between the lightship and the shore. In spite, however, of these facts, the Brethren of the Trinity House move slowly in the matter. The Admiralty, always more prompt, took advantage of the naval manœuvres in 1899 to test this method thoroughly. On one occasion the Europa communicated with the flagship through the Juno at a total distance of some 55 miles. Remarkable demonstrations were also given near New York during the cup races between the Shamrock and the Columbia. Apparatus of the same kind has recently been despatched by the War Office to South Africa.

In the old established telegraphy with wires there have not been wanting important improvements in 1899. Between Budapest and Vienna the Pollak-Virag system of high-speed telegraphy has been tried, sending 1,500 words a minute; and in the United States preliminary trials of the synchronograph of Messrs. Crehore and Squier have shown that it is possible to transmit with it 6,000 words a minute over short distances. Rumours have been circulated of revolutionizing inventions in transmitting visual images by wire, said to have been made by Jan Szczepanik, but a popular disclosure of the methods has not served to establish their credibility.

In electric lighting engineers had their curiosity and interest greatly excited in the early part of 1899 by the announcement that Dr. W. Nernst, an eminent German chemist, had invented a new incandescent electric lamp having an efficiency more than double that of the present carbon filament lamp. Dr. Nernst's invention is based upon the fact that mixtures of oxides of the rare earths with magnesia, when moulded into rods, produce a material which, though non-conductive when cold, yet becomes a conductor when initially heated, and is then susceptible of being raised to a very high temperature in open air without destruction. So far, however, beyond a few public exhibitions and some resulting company promotion, the public have not been able to make practical acquaintance with the new lamp, or judge of its qualities as a rival to the existing carbon filament lamp. Amongst other small but interesting electrical novelties in the past year must be included the Wehnelt current interrupter, a simple device whereby a platinum wire and a lead plate immersed in dilute acid are made the means of rapidly intermitting a continuous electric current. The device employed in connexion with an induction coil becomes a means of greatly increasing its power of producing Röntgen rays when the discharge is sent through a vacuum bulb.

In the more purely scientific region of investigation a high position must be accorded to work conducted by Professor J. J. Thomson in 1899 in further analysing the nature of electric discharge *in vacuo*. It has long been recognized as the result of the work of Crookes, Goldstein, Perrin, Lenard, and many others that in the electric discharge through rarified gases we have a phenomenon which essentially consists in the conveyance of electric charges by moving matter. The question is, are these material conveyances molecules, atoms, or something smaller than chemical atoms? By reasoning and experiments of a remarkably ingenious character Professor J. J. Thomson has been able to show that these electrical conveyers are probably masses of matter of about one-thousandth part of the mass of a chemical atom of hydrogen. According to Professor Thomson's views, chemical atoms are built up of smaller masses called corpuscules or electrons, which may be detached from their association with each other by the electrical operations taking place in the vacuum tube. These corpuscules when free constitute a fourth state of matter, to employ a term originated by Sir W. Crookes, and they are so minute that they appear to be capable of penetrating through thin sheets of certain metals. In close connexion with this subject we may mention that the phenomenon discovered a short time back by Professor Zeeman, in which the spectral lines of a light-giving body are split up into sets of triple lines when the radiator is placed in a strong magnetic field, has continued since to receive attention, and the interpretation of the results will undoubtedly lead to a knowledge of events taking place in the microcosm of the atom. Mr. C. E. S. Phillips has added to our knowledge in this region of research by the discovery of a curious luminous effect arising in a vacuum tube provided with iron electrodes, when these are magnetized subsequently to the passage of an electric discharge.

A ninth Saturnian satellite was photographically discovered by Professor W. H. Pickering in April. The superposition of plates taken at Arequipa with the 24in. Bruce telescope, on August 16, 17, and 18, 1898, disclosed it as a progressively shifting grey dot of such extreme faintness that no visual observations of the object it represents have yet been secured. Viewed from Saturn, it would appear no brighter than a sixth-magnitude star; nevertheless, if no more reflective of light than Titan, its diameter cannot be much less than 200 miles. The period of " Phœbe," as this most remote member of the Saturnian family is to be called, must be of many months, but is still undetermined.

A fairly bright comet, picked up on March 3 at the Lowe observatory, California, by Professor Swift—the Messier of the 19th century—developed in May a tail nine degrees long, shown in the Lick photographs to be made up of separate twisted strands of luminous matter. A gradually-widening fissure detected in the nucleus by Professors Perrine and Barnard perhaps indicated the beginning of a process analogous to that which led to the disruption of Biela's comet in 1846. But Swift's is an " irrevocable traveller"; along a track virtually parabolic it has retreated finally from view. A much fainter comet was discovered by M. Giacobini at Nice on September 29; and known comets—Tuttle's of 1858, Tempel's of 1873, and Holmes's of 1892—made anticipated returns.

Stellar chemistry advanced notably in 1899. Dr. Gill's re-identification, with the McClean spectrographic apparatus, of several groups of oxygen-lines found by its donor in the spectra of some bright southern stars was capped by Sir William and Lady Huggins's recognition of a different set of lines due to ultra-violet absorption by the same substance, in the atmospheres of Rigel and Beta Lyræ. Their detection of nitrogen-lines in the spectra of Rigel and Bellatrix quickly ensued and was confirmed by M. Bélopolsky's spectrographs of P Cygni, Janson's "Nova" of 1600. All these stars belong to the "helium" type; they appear to be at the very highest stage of incandescence; hence the copious presence in them of one or both ingredients of common air is a point of curious interest in sidereal physics.

Among the numerous spectroscopic binaries lately discovered at Lick by Professor Campbell, two particularly arrest attention. The Polestar has a radial movement affected by fluctuations showing it to revolve certainly round one dark companion in a period of four days, and probably round another in the course of many months. The primary of this obscure ternary system emits at least four times the radiance of Sirius, and is about 100 times more massive than the sun. The close duplicity of Capella, independently detected by Mr. Newall at Cambridge, is a still greater surprise. This star is a gigantic globe of solar constitution, while its newly-found satellite gives a distinct, non-solar spectrum. Their mutual circulation is attested by the relative shifting of the two sets of lines.

The Crossley reflector has been employed most effectively by Professor Keeler in picturing nebulæ, the work for which Dr. Common's record result with it in 1883 proved its exquisite adaptation. No fewer than 120,000 nebulæ are computed to lie within its grasp in the pure air of Mount Hamilton; delicate details, seen by Lord Rosse in the Lyra annular nebula, have at last been photographically verified; and a spiral structure has been demonstrated to be no exception, but the rule, for non-gaseous nebulæ. Of their great exemplar in Andromeda a legible spectrographic impression was secured by Dr. Scheiner at Potsdam in January. He found its light to show the peculiarities of inconceivably attenuated sunbeams and concluded this prodigious object to be in effect a cluster of solar stars. [7]

Professor S. I. Bailey's photographic investigations of 85 variable components of the globular cluster "Messier Five" have made evident their approximate conformity to a period of half-a-day, as well as a striking similarity in their mode and amount of light-change. The perturbed operation of a primitive law is, he thinks, indicated. On the other hand, the still more numerous variables in the great cluster Omega Centauri tend in no way towards cyclical congruity.

QUESTIONS AND ANSWERS

Q.: EXPLAIN the statement that the Sun never sets on the British Empire.

A.: The Sun sets in the west; now the British Empire lies in the north, south, and east.

———

A.: WHAT is the derivation of the name *theodolite*?

B.: Oh! I suppose it comes from θεωρέω, ὁδὸς, and λίθος, to observe stones in the road, doesn't it?

A.: Oh, no, it doesn't. It is called after a Frenchman named Theodolite, who invented it.

COLOUR PHOTOGRAPHY

he began with a preliminary description of the characters of wave-motion in general, drawing his illustrations from waves in water and sound-waves in air. Of the latter he showed a number of photographs, taken by Professor R. W. Woods, in which the sound waves produced by the crack of an electric spark were photographed by the light from another spark, so as to afford visual representations of phenomena such as reflection from surfaces of various forms, retardation in passing through a dense medium, &c. Turning to light waves, he explained the mechanism of refraction, showing how a prism decomposed white light into a series of coloured rays or spectrum, which could again be recombined by the aid of a lens, and how the colour of an object depended on what rays of the spectrum it absorbed or stopped, and what it reflected or transmitted. He then pointed out the application of this fact to the problem of colour photography, and concluded his lecture with a discussion of the measurement of lights and shadows and of colour, mentioning the difficulties in the way of such measurements and describing several of the instruments which had been designed to effect them.

1901

THREE BODIES PLAGIARIZED

CASES are not infrequent where the editor of a literary journal is victimized by some one copying out a poem from the works of an eminent man and sending it in as his own, but I had not heard until recently of any scientific crime of this kind. The curious will find in the *Comptes Rendus* for 1894 (cxix. p. 451) a paper by *Vernier*, on the problem of three bodies, which is almost a literal transcript of a paper by *Siacchi* published twenty years earlier in the same journal (*C. R.* lxxviii. p. 110). The plagiarist no doubt thought *Siacchi* might be dead, or might not see the second paper; but he was very much alive, and protested, with the result that Vernier has been heard of no more. But there the two papers stand in print, to show what human nature is capable of. Query, when we get the new subject-index, will it be necessary to compare all papers submitted with their predecessors? The life of a referee will not be a very happy one; but on the whole we shall gain by avoiding mistakes of this kind.

INTERNATIONAL COOPERATION IN ASTRONOMY

it should be remembered that astronomical co-operation had not always been successful. The band of astronomers who at the end of the 18th century divided the sky among them and maintained a constant watch to detect a minor planet met with failure, for by the irony of fate the first discovery of the kind fell to another who was not engaged in any special search. Failure of a more real kind had over-taken enterprises to chart the stars or to map the moon. Yet a little more energy on the part of some one might have led to a satisfactory issue, especially if there had been supervision, a sort of foreman of the works. It did not seem unlikely that this general supervision was best performed by one not actually engaged in the work itself. One of our great London schoolmasters had declared that a nominally idle man should be at the head of all enterprises—that "he never knew any good come to any work where there was not a man with his hands in his pockets looking after it." The danger of attempting too much was illustrated by the Eros campaign. The 18 observatories had their hands full with the astrographic chart when there came a special occasion, which would not be repeated for 30 years, to determine the solar parallax. The 18 observatories could not refuse to take photographs, and this they had done, but the work of reducing and measuring them must throw back the chart work. A different kind of danger to which co-operation was liable arose from the necessity of binding the associating individuals by certain rules, which might have the effect of checking that originality which was almost vital in scientific work. We must not shut our eyes to the fact that astronomical work was terribly liable to

settle down into routine, and we had already many small observatories where only routine work was carried on. The astrographic chart illustrated both the dark and the bright side. The danger of crippling individuality had been averted in an unexpected, almost a comical, manner. Rules were drawn up after many conferences, and the participating workers went off to their observatories with a copy in their pocket—and did not observe them. Such as they found convenient they adhered to closely, but when they found that a rule would not work they did not hesitate to prefer their own experience. This was the salvation of the scheme, for the rules which were broken were those which experience proved non-essential, and which ought never to have been made. Thus, when those who had actually carried out a considerable portion of the work met last July, they found that they had arrived at practically the same conclusion by a diversity of routes. The comedy of this result had a very serious significance, and there were reasons why we might heartily congratu-late ourselves that the time was not yet come when astronomers were prepared to lose their individuality in a co-operative scheme of work; and still more that such schemes could be found where such loss of individuality was unnecessary.

A new gauge for small pressures was described by Messrs. E. MORLEY and C. F. BRUSH, who also gave an account of experiments on the "Transmission of Heat through Water Vapour," conducted with the aid of their gauges. The authors have determined the rate of trans-mission of heat through water vapour at pressures from that of saturation at 0 deg. to less than a millionth of an atmosphere. At low pressures water vapour transmits heat more rapidly than air, but not so rapidly as hydrogen. The superiority over air is a *maximum* at 20 or 30 millionths of an atmosphere, and is not far from 30 per cent. At 60 or 80 millionths air and water vapour transmit heat at the same rate; at higher pressures water vapour transmits heat less rapidly than air at the same pressures. Statements more precise could not be made, because the form and dimensions of the apparatus used modify slightly the curves which represent the relations between pressure and rate of transmitting heat, and the place of the intersection of the curves is therefore un-certain.

SCIENCE AND EDUCATION

Of science I would gladly speak if I could do so with anything but the enthusiasm of ignorance, for it is at once a most practical and attractive subject. On the one hand it is a fairyland of the marvellous whence at any moment may emerge some dream of wonder, or it may at any moment open out an unlimited vista of possibility, while on the other hand it appears to offer the only bedrock, the foundation on which most technical education can be based. A sound and adequate scientific training seems to be a necessary qualification of our captains and lieutenants of industry. But against one sublime department of science I would beg to enter a respectful warning. I do so with some apprehension, because I have distributed a prize in that department of science to-night. It is dangerous, in my judgment, to study astronomy, for astronomy kills ambition. What mind, after contemplating the eternal procession of unnumbered worlds, perhaps with their infinite generations of life, their various splendours, their history, their endless rolls of celebrities, their separate myriads of heroes, of achievements, can return without a disheartening sense of the pettiness and futility and transitory fame of his own narrow universe?

THE NATIONAL PHYSICAL LABORATORY

Mr. GLAZEBROOK remarked that the idea of a physical laboratory, in which problems bearing at once on science and industry might be solved, was comparatively new ; perhaps the first was the Physikalisch-Technische Reichsanstalt, founded in Berlin, by the joint labours of Werner von Siemens and von Helmholtz, during the years 1883-87. It was less than ten years ago that Dr. Lodge outlined the scheme of work for such an institution in England, and in 1895 the late Sir Douglas Galton called attention to the question. A petition to Lord Salisbury followed, and as a consequence a Treasury Committee with Lord Rayleigh in the chair was appointed to consider the desirability of establishing a national physical laboratory. This committee examined over 30 witnesses and then reported unanimously, " That a public institution should be founded for standardizing and verifying instruments, for testing materials, and for the determination of physical constants." It was now realized at any rate by the more enlightened of our leaders of industry that science could help them. This fact, however, had been grasped by too few in England, though our rivals in Germany and America knew it well ; and the first aim of the laboratory was to bring its truth home to all, to assist in promoting a union which was certainly necessary if England was to maintain her supremacy in trade and manufacture, to make the forces of science available for the nation, to break down by every possible means the barrier between theory and practice, and to point out plainly the plan which must be followed unless we were prepared to see our rivals take our place. The effect of the close connexion between science and industry on German trade might be illustrated, if illustration were wanted, by the history of the aniline dye manufacture and artificial indigo, and by the German scientific apparatus industry, the growth of which had been expressly attributed to the influence of the Reichsanstalt. Mr. Glazebrook proceeded to describe the means at disposal for realizing the aims of the laboratory.

THE EFFECTS OF CLIMATE

meteorology was often considered a dull and statistical science, but, properly looked at, the heavy columns of figures, the marshalling of which was the chief work of the meteorologist, were found to occupy some very interesting ground. Popular interest in science could rarely be aroused unless there was something to show, and the atmosphere when in its most desirable state was invisible. The effects of atmospheric phenomena, however, were often very striking ; this was true not only of such exceptional and catastrophic occurrences as whirlwinds, avalanches, and floods, or such curious but unimportant incidents as showers of so-called blood or sulphur, but also of the normal effects of climate, which were too familiar to us to attract much notice at home. After explaining that climatology is as much a department of geography as of meteorology, the lecturer illustrated the results of typical climatic conditions by means of the electric lantern with a large number of photographs which he had taken during visits to various parts of Europe and Canada. They dealt with the results of radiation, prevailing winds, extreme cold, and especially with precipitation. Among the most interesting effects of climate referred to were the adaptations of industrial processes or of dwellings in order to take advantage of climatic conditions or to mitigate their unpleasant features. Such, for instance, were the arcaded footways of southern towns, where shade was the first desideratum, the temporary roofs placed over the lemon gardens in northern Italy on a threatening of frost, the covered bridges of the Tyrol constructed to prevent the accumulation of snow on the wooden roadway, and the huge waterspouts which disfigured the finest buildings in St. Petersburg, but were required to carry off the great volume of water resulting from the sudden melting of the winter's snow on the roofs in spring. Special attention was given to the devices used by farmers in drying hay and other crops in wet countries—the stakes fashioned like hat-stands on which the hay was hung in northern Tyrol, the vast frames, resembling the supports for wires on a telephone exchange, used for the same purpose in southern Tyrol, the hurdles arranged like fences across the direction of the prevailing wind in Norway. A photograph of the last-mentioned device taken in British Columbia showed how similar methods might be employed in similar climates even when widely separated. In every part of the world the climate depended on the position of a place in latitude, its distance from the ocean, and, above all, on the configuration of the land, and the climate in turn dominated the life of plants and animals and the culture of mankind.

MARCONI IN NEWFOUNDLAND

It is not given to every man to " stagger humanity " in either a baleful or beneficent sense, but Mr. Marconi, the distinguished young scientist who has just compassed the marvellous achievement of communicating across the Atlantic Ocean by wireless telegraphy, deserves the credit of having startled the world in a manner that will work wonders for the public good.

In the history of modern scientific development four great epoch-marking events may be recorded—the perfecting of the electric light, the laying of the Atlantic cable, the inventing of the telephone, and the discovery of the Röntgen rays. To these may be added a fifth—that represented by Mr. Marconi's exploit last week—and this is the most wonderful of them all, for no more extraordinary conception has ever yet had its birth from the mind of the most imaginative novelist than this seemingly incredible feat of transmitting intelligible signals over 2,000 miles of space without wire or cable or any other visible or tangible agency whatever.

In view of the importance of the discovery and of the subsequent legal complication resulting from the attempt of the Anglo-American Telegraph Company to prevent his continuing his work, I have availed myself of the opportunity of Mr. Marconi's presence here to obtain from him a statement of the whole matter, which he has been kind enough to revise in its technical aspects, and which will doubtless interest your readers.

The story of his early experiments and gradual success in the matter of wireless telegraphy is an oft-told tale. I take up the story where, having accomplished everything that was possible within the limited ocean areas which separate the British Isles from the European Continent, this scientific Alexander turned in search of new worlds to conquer. When his experiments between England and France were shown to be a complete success and the British Admiralty had set its seal of approval upon the system by installing it upon several warships he induced his company to establish a large power plant at Poldhu, Cornwall, for still more important tests. As compared with his ordinary stations located around the British coasts, and used mainly for marine purposes, this Poldhu depôt possessed powers about one hundred times greater. It consists of the buildings wherein the generators and instruments are stored, and 20 masts, each 210ft. high, supporting aerial wires and so connected as to form a gigantic conductor by means of which the volume of electricity produced in the generator is forced through the wires and projected into space, causing the oscillations of ether waves which radiate through the atmosphere, travelling outwards in every direction to the uttermost ends of the earth. Poldhu station was built some eight months ago, but was partly blown down in the big gale which swept the English Channel in September, and is not yet completely restored.

But it proved sufficiently effective to enable Mr. Marconi, in October and November, to experiment—having great things in view—with his most distant station, that at Crookhaven, on the west coast of Ireland, 225 miles away. While he was receiving messages at the latter post he observed that the strength of the signals which were recorded was very great ; so much so, indeed, that his long experience in the subject enabled him to conclude that they would manifest an observable activity at a tenfold distance.

As it was imperative that, if such a result was obtainable, the facts should be determined as soon as possible, Mr. Marconi determined to take ship at the earliest convenient date for Newfoundland.

He shrewdly reasoned that, if he stated his purpose beforehand and failed, the world would discredit his invention in its more modest form, while if he failed now he could try again later, and if he succeeded his success would be all the greater from coming upon the world so unexpectedly. Fortunately for him, a legitimate reason for his visiting Newfoundland existed in the fact that he was interested in the possibility of signalling the Cunard liners traversing the ocean beyond the Grand Banks, 400 miles from our coast, and he announced his intention of undertaking experiments looking to that object—a sufficiently ambitious one, in the minds of the public, seeing that it represented the doubling of his previous achievements. He left England on November 26, in the Allan liner Sardinian, and arrived here on Friday, December 6. He was accompanied by his two assistants, Messrs. Kemp and Paget, and they brought with them all requisite apparatus, and also two balloons and six kites, for the purpose of elevating their aerial wires so that the signals diffused through the ether might be caught in passing and recorded on his instruments located in some suitable building at the base of the wire.

On arriving, and before beginning operations, he waited upon the Governor, Sir Cavendish Boyle, the Premier, Sir Robert Bond, and the members of the Ministry, who promised him the hearty co-operation of the Government and placed the resources of every department at his disposal, to facilitate his work. They also offered him the temporary use of such lands as he might require for the erection of depôts at Cape Race or elsewhere, if he determined to establish the marine stations, which it was then understood he contemplated. He decided to begin his work on Signal-hill, a lofty eminence overlooking the port and forming the natural bulwark which protects it from the fury of the Atlantic surges. This hill is crowned by a plateau two acres in extent, affording an ample area for manipulating the kites and balloons, and on a crag above the harbour is the new Cabot Memorial Tower designed as a signal station. In the vicinity is an old military barrack now used as a hospital, and in part of this a Marconi receiving apparatus was set up. It consisted of a very sensitive coherer, an electric battery, and a telephone. Mr. Marconi for the purposes immediately in view did not require a permanent station here, his equipment of balloons and kites being regarded as adequate, seeing that he could always fall back upon the erection of a high mast or staff, the appliance with which the public have become most familiar in connexion with his work. Accordingly, on Monday, December 9, he and his assistants began their labours on Signal-hill. By Wednesday they had inflated their balloon and it made its first ascent, only to break clear after a short while and drift away to parts unknown. It was a balloon 14ft. in diameter and with a cubic capacity of about 1,400ft., charged with hydrogen gas and designed to uphold the "aerial," which weighed about 10lb., besides a stay to hold the sphere comparatively motionless. On Thursday he succeeded in elevating one of his kites to a height of 400ft., where it kept the "aerial" in position, and by this means he was fortunate in being able to receive the trans-oceanic signals which have annihilated space.

In order that your readers may understand how it was done, I must explain that Mr. Marconi, before leaving England, prepared a series of instructions for the guidance of the chief electrician at Poldhu station, Cornwall. Mr. Marconi was kind enough to show me his press-copied duplicates of these instructions in his official letter book. They intimated that after his arrival in St. Johns he would wire a certain date (as "Tuesday, December 10th") to the Poldhu operator, who was, every day after receipt of this cablegram, enjoined to make the signal "S" (which letter is represented in the Morse code by three dots—thus . . .) at regular intervals daily, during the hours of 3 to 6 p.m. Greenwich time, until ordered to desist. The hours given were, in local time at St. Johns, approximately 11 30 a.m. to 2 30 p.m., the most convenient period for working here, and Mr. Marconi's idea was that, if his theories as to the force of the electric waves were correct, he would be able to receive distinct and effective signals at his base here.

On Thursday, at 12 30 p.m., the great idea was transmuted into an assured fact. Mr. Marconi at his instrument in the old hospital received the expected signals. In view of the importance of all that was at stake, he decided not to trust to the automatic recorder, but to use a telephone attached to the coherer instead.

For ordinary work a recorder is used—an instrument in connexion with which a tape is worked, the dots and dashes being printed thereon. The human ear is, however, infinitely more sensitive than the recorder, and, therefore, the inventor chose it as his medium for the reception of the all-important signals. They were quite audible to himself and his assistant, Mr. Kemp, who also noted them. The two observers were at the instrument at different times, and the signals clicked along at irregular intervals during three minutes. They were again noted at 1 10 p.m. and at 2 20. They came as a frequent repetition of the conventional formula which he had instructed to be sent, were distinct and unmistakable, and were received about 25 times altogether. On Friday, again, at 1 30 p.m., they were repeated, but were less distinct, and not renewed after a brief period, Mr. Marconi's conclusion being that the kite, which worked badly, prevented the records from being made with the same distinctness and regularity as on Thursday. On Saturday a further attempt was made to obtain a repetition of the signals, but it was unsuccessful owing to adverse weather conditions, it being found impossible to elevate the kites, of which no fewer than three were smashed by falling to the earth. He thereupon decided to discontinue the experiments until he could erect a pole 200ft. high on Signal-hill or at some other point which possessed the requisite facilities, so that he would thus be rendered independent of wind or weather. Another advantage with the pole would be that he could use a heavier wire, as with a kite elevated 400ft. the wire must necessarily be slight, while that suspended from a mast, being stouter, would form a better conductor for the electric waves.

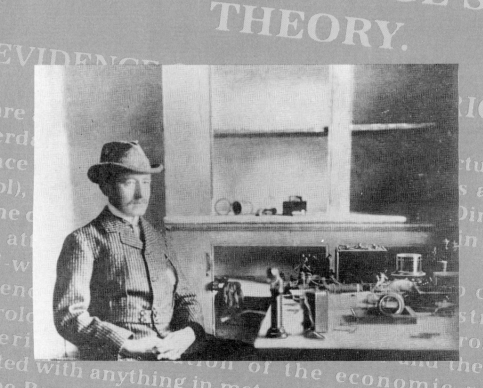

MARCONI IN NEWFOUNDLAND FOR THE FIRST TRANSATLANTIC RADIO SIGNALLING

THE MYSTERY OF RADIUM

M. Curie, a French physicist of the highest reputation and attainments, has made a communication to the Academy of Sciences which would have been received with absolute incredulity had it been offered on less unimpeachable authority. He finds that a substance of comparatively recent discovery, to which the name of Radium has been given, and in the isolation of which he has had the indefatigable and invaluable assistance of Mme. Curie, possesses the extraordinary property of continuously emitting heat, without combustion, without chemical change of any kind, and without any change in its molecular structure, which remains spectroscopically identical after many months of continuous emission of heat. He finds, further, that Radium maintains its own temperature at a point 1·5deg. Centigrade, or 2·7deg. on our ordinary scale, above its surroundings. To put the matter in another way, the actual quantity of heat evolved is such that the pure Radium salt would melt more than its own weight of ice every hour. Or, again, half a pound of the Radium salt would evolve in one hour heat equal to that produced by the burning of one-third of a cubic foot of hydrogen gas ; and this evolution of heat goes on continuously for indefinite periods, leaving the salt at the end of months of activity just as potent as at the beginning.

Radium has excited the keenest interest by its power of throwing off rays, vibrations, emanations, or whatever we may call them, which, when received upon a sensitive screen of barium platinocyanide or zinc sulphide, cause it to glow with a phosphorescent light. Sir William Crookes, who has investigated this subject with the most brilliant results, gave a very beautiful demonstration at the meeting of the Royal Society last week. Viewed through a magnifying glass, the sensitive screen is seen to be the object of a veritable bombardment by particles of infinite minuteness, which, themselves invisible, make known their arrival on the screen by flashes of light, just as a shell coming from the blue announces itself by an explosion. According to the nature of the receiving screen, they show a mass of discrete scintillations or a diffused glow, very much as falling rain will cause a general wetness on one surface while it runs off another in drops. Though working with only a few milli-grammes of the Radium salt, Sir William Crookes found that so extraordinary is the output of these emanations that every vessel with which they came in contact, and even the fingers of the operator, acquired temporarily the power of exciting phosphorescence on the sensitive screen. Yet, notwithstanding their infinite number and the ceaselessness of their emission, the mass of the radiating body appears to suffer no diminution.

Remarkable as are these photogenic properties of Radium, it is obvious that M. Curie has introduced us to forces of a totally different order of magnitude. Phosphorescence occurs in nature, as, for example, in the glow-worm and in certain bacteria, in conditions of energy which is absolutely infinitesimal as compared with what we have to expend to produce light. Hence the phosphorescence of a sensitive screen under the influence of Radium emanations does not necessarily take us beyond a region in which light is an accident of processes infinitely minute, though not on that account less worthy of investigation. But heat sufficient to raise the mercury in the thermometer by 2·7deg. is a different thing altogether, and when the output of this heat is maintained indefinitely without any visible compensation to the heat-giving body, we are in presence of a physical effect which is not only appreciable, but considerable. That effect must have a cause, for we are not to suppose that we have at last hit upon perpetual motion. Investigation of that cause is full of promise for the physicist. Apparently we have in Radium a substance having the power to gather up and convert into heat some form of ambient energy with which we are not yet acquainted. Other substances, mostly of high atomic weight, possess its radiant properties to a less well-marked extent, and research may prove that transparency to the unknown form of energy is merely a question of degree. M. Becquerel gave a powerful initial impulse to the study of this subject by his discovery that uranium continuously emits some temperature at a point 1·5deg. Centigrade, or 2·7deg. on our ordinary scale, above its surroundings. To put the matter in another way, the actual quantity of heat evolved is such that the pure Radium salt would melt more than its own weight of ice every hour. Or, again, half a pound of the Radium salt would evolve in one hour heat equal to that produced by the burning of one-third of a cubic foot of hydrogen gas ; and this evolution of heat goes on continuously for indefinite periods, leaving the salt at the end of months of activity just as potent as at the beginning.

Radium has excited the keenest interest by its power of throwing off rays, vibrations, emanations, or whatever we may call them, which, when received upon a sensitive screen of barium platinocyanide or zinc sulphide, cause it to glow with a phosphorescent light. Sir William Crookes, who has investigated this subject with the most brilliant results, gave a very beautiful demonstration at the meeting of the Royal Society last week. Viewed through a magnifying glass, the sensitive screen is seen to be the object of a veritable bombardment by particles of infinite minuteness, which, themselves invisible, make known their arrival on the screen by flashes of light, just as a shell coming from the blue announces itself by an explosion. According to the nature of the receiving screen, they show a mass of discrete scintillations or a diffused glow, very much as falling rain will cause a general wetness on one surface while it runs off another in drops. Though working with only a few milli-grammes of the Radium salt, Sir William Crookes found that so extraordinary is the output of these emanations that every vessel with which they came in contact, and even the fingers of the operator, acquired temporarily the power of exciting phosphorescence on the sensitive screen. Yet, notwithstanding their infinite number and the ceaselessness of their emission, the mass of the radiating body appears to suffer no diminution.

Remarkable as are these photogenic properties of Radium, it is obvious that M. Curie has introduced us to forces of a totally different order of magnitude. Phosphorescence occurs in nature, as, for example, in the glow-worm and in certain bacteria, in conditions of energy which is absolutely infinitesimal as compared with what we have to expend to produce light. Hence the phosphorescence of a sensitive screen under the influence of Radium emanations does not necessarily take us beyond a region in which light is an accident of processes infinitely minute, though not on that account less worthy of investigation. But heat sufficient to raise the mercury in the thermometer by 2·7deg. is a different thing altogether, and when the output of this heat is maintained indefinitely without any visible compensation to the heat-giving body, we are in presence of a physical effect which is not only appreciable but considerable. That effect must have a cause, for we are not to suppose that we have at last hit upon perpetual motion. Investigation of that cause is full of promise for the physicist. Apparently we have in Radium a substance having the power to gather up and convert into heat some form of ambient energy with which we are not yet acquainted. Other substances, mostly of high atomic weight, possess its radiant properties to a less well-marked extent, and research may prove that transparency to the unknown form of energy is merely a question of degree. M. Becquerel gave a powerful initial impulse to the study of this subject by his discovery that uranium continuously emits some kind of rays or emanations capable of affecting sensitive plates. It is calculated by M. and Mme. Curie that Radium is 500,000 times as powerful as uranium.

The physiological action of these Radium emanations is very powerful, though time is required for its development. A small tube containing Radium, if kept in contact with the skin for some hours, or even if carried in the waistcoat pocket, produces an open sore, by destroying the epidermis and the true skin beneath. Its effects do not, however, appear to extend to the subjacent tissues, and the sore remains superficial. On the other hand, Radium emanations act powerfully upon the nerve substance, and cause the death of living things whose nerve centres do not lie deep enough to be shielded from their influence.

MARIE AND PIERRE CURIE

A NEW FORM OF ELECTRIC RADIATOR

A new description of electric radiator has lately been introduced and will be manufactured by the Electric Ordnance and Accessories Company, of Birmingham. It is the invention of Mr. E. G. Rivers, chief engineer to H.M. Office of Works. An example of the heater has been fitted in a room at the Office of Works at Storey's-gate. The heater consists of a series of panels of composite construction. Between the pairs of enamelled iron plates a layer of very finely powdered gas retort carbon is held, the powder being retained in position by a bordering of asbestos cardboard. Three copper conductors are laid between the plates and make contact with the powdered carbon, and, in this way, two circuits are formed, the current passing through the carbon. The iron plates are bolted together so as to compress the carbon between them. The electrical resistance of the finely-powdered carbon affords an equal and moderate generation of heat, the sections of the radiator being brought to about 190deg. Fahrenheit. Though it is not claimed that the apparatus will compete in regard to economy with steam or hot water for large installations, its cleanliness and convenience render it desirable for domestic purposes.

1904

A NEW APPROACH TO FLIGHT

The attempts in recent years to solve the problem of aerial navigation have for the most part been made with navigable balloons. There has, however, lately been constructed, from the designs of Señor Alvares, a Brazilian, by Messrs. C. G. Spencer and Sons, of Highbury, a new aeroplane flying machine which does away with the gas vessel and its many risks. The main characteristic of this new invention is the simplicity of its structure. It consists of two swing-like aeroplanes having a superficial area of 400 square feet ; these are attached to two outstretching and slightly curved arms and fixed to a bamboo framework, in shape like a cigar. In the front of this framework is fitted a two-horse power motor which drives two two-bladed tractors—each of them 5ft. in diameter—which are placed one on each side of the frame and level with the motor. Beneath the framework and behind the engine is a seat for the aeronaut. At the back of the machine are three rudders, which are worked from the front by means of ropes. Two of the rudders are triangular and are constructed to move horizontally, for the purpose of controlling the upward and downward motion of the machine, while the other, the largest of the three, which is rectangular, is fitted perpendicularly and is intended to guide the machine to the left or right. The weight of the machine is 150lb. without the aeronaut. It does not appear that the invention has any power of raising itself from the ground, as it is stated that during the next few days it is to be taken up by a balloon, at the Crystal Palace, to an altitude of 5,000ft., when it will be released for the purpose of testing its actual power of flight.

THE SO-CALLED WEATHER PLANT

As far back as 1888 a good deal of public attention was drawn to statements that Professor Nowack, of Vienna, had discovered a plant capable of forecasting the weather, and you yourself, in November, 1888, published a letter from a Mr. Radeke, in which it was stated that the weather plant was an " electro-magnetic plant," that it was able to forecast local weather 48 hours beforehand, and to foretell earthquakes from three to eight days before they took place. The following year Mr. Nowack came to England with an introduction from his Imperial Highness the late Crown Prince Rudolph of Austria, and in consequence of his Majesty, then his Royal Highness the Prince of Wales, interesting himself in the matter, a very careful investigation into the powers of the weather plant was carried out at Kew.

It appears that the plant is a well-known tropical weed *Abrus Precatorius*. Its seeds are used in India by goldsmiths as weights, and it is said that the word carat is derived from the name of the seed, *rati*. The plant is very sensitive to heat and to light, and very considerable movements of its leaflets are caused by changes of temperature or variations of light. For a period of apparently about a month careful observations of the plant were made, and Mr. Nowack prepared, from the behaviour of the plant and the movements of its leaves, a series of forecasts, which were actually compared with the reports of the actual weather. A number of prospective weather charts were also made by the same gentleman, and these were compared with charts subsequently drawn up from the actual records of the Meteorological Office. In neither the case of the forecasts nor of the weather charts was there any apparent correspondence.

The result of this very exhaustive series of tests was to convince the authorities at Kew that any connexion between the behaviour of the plant and the future weather, either local or distant, was entirely imaginary, and that the movements of its leaves were due to local variations of temperature and light.

LUNATIC DEATH

Tuesday's eclipse of the Moon resulted in at least the death of one foreigner, and it is appalling to contemplate the number of fatalities that might occur in the foreign community of Kobe if the practice of gazing at the Moon is persisted in. On Tuesday evening a foreigner went into his garden with the object of looking at the Moon. Having cleared away the stones and snakes in his immediate vicinity, he lay down in a blanket and got up with a cold. From the *post-mortem* examination it was elucidated that the deceased died through laboured respiration and shock, and at the inquest Dr. Frederick Augustus Guppy testified that craning his neck to look at the Moon was the direct cause of the man's death, the deceased having never had occasion to look in that direction since he came to Kobe.

FORMS OF IRON

Whenever, in these reports, the carburization of iron has been dealt with, the question of the allotropy of iron has assumed a prominent place. It is well known that the allotropic theory has formed the subject of a more than animated discussion between Sir William Roberts-Austen, who supported Osmond's views, and other metallurgists, of whom Professor Arnold has been the most prominent. It would be impossible in the space now available to give an intelligible account of this much-disputed problem; but those who have followed the subject are interested to know that the theory is receiving distinguished support amongst Continental men of science. Professor Bakhuis-Roozeboom, of Amsterdam, has made an elaborate study of curves given in diagrams attached to Sir William Roberts-Austen's former report. His conclusions are of great scientific interest, and were obtained by applying the phase doctrine of Gibbs. He found that it was necessary in this system to consider the following phases:—Carbon and iron (which may exist in three phases, α, β, and γ); liquid and solid solution of carbon in γ (austenite), iron and cementite. These substances are constituents of steel, the knowledge of which we owe to the microscopists, amongst whom the names of Sorby and Andrews stand out with distinguished prominence in this field. At the Paris International Physical Congress of 1900, Professor Warburg, of the University of Berlin, appealed to Professor Van t'Hoff to give an opinion respecting Osmond's views as to the transformation of carburized iron. In response Van t'Hoff said :—" β iron is capable of retaining variable quantities of carbon in the form of a homogeneous solid solution, known as martensite, while the property is not possessed by α modification of iron."   In addition to this support given to the importance of allotropy in relation to the properties of steel, the report cites the evidence of E. Heyn, of the Imperial Laboratory of Charlottenburg, contained in a communication to Stahl und Eisen, 1899, Nos. 1 and 16.

To the average worker in iron, the " practical man " who received his education before the era of technical schools and scientific training, it may seem a small matter, whether Osmond and Roberts-Austen or Arnold and his followers had the best of the argument ; and it must be confessed that a good deal yet remains to be done before many of the researches of the laboratory or the speculations of natural philosophers are brought within the scope of the foundry or the workshop. We are moving in this direction, however, when we learn that " allotropic changes in the iron itself are the all-important factors in determining the relations between iron and carbon, which are involved in the characteristic capacity of steel for being hardened and tempered. If iron had existed only in the α form it could never have been hardened by quenching ; on the other hand, if β and γ iron were alone known, steel could never have been tempered or annealed." The prevalent belief, therefore, of nearly two centuries, which claimed attention for the carbon alone, has had to be abandoned. The " practical man " will, however, be apt to ask, not whether it is " carbon alone " that makes hardening and tempering possible, but whether these extremely serviceable functions can be performed in any other way than by carbon. Carbon, however, is so plentiful, and so easily alloys with iron, that the latter question is of smaller industrial importance than that of the influence of some other alloys on steel. But here the subject is so vast, owing to minuteness of the percentage of alloys that will produce great results, and the complex interaction of the various elements of each other, that it cannot be hoped finality will ever be reached.

NEW APPROACHES

To remove air from a flask: fill the flask with water, tip the water out, and put the cork in quick.

The probable cause of earthquakes may be attributed to bad drainage and neglect of sewage.

Gravity was discovered by Isaac Walton. It is chiefly noticeable in the autumn, when the apples are falling from the trees.

MARCONI STATION IN CANADA AND NEWFOUNDLAND

MARCONI STATIONS IN CANADA AND NEW-FOUNDLAND.—Marconi's Wireless Telegraph Company has received an intimation from the Marconi Wireless Telegraph Company of Canada that the installation of the Marconi-equipped wireless telegraph stations at Sable Island and Halifax (Camperdown station) under contract with the Canadian Government, has just been completed, and these two stations are now in full operation. Marconi wireless stations at Whittle Rocks and Cape Rich are in course of erection and equipment, and will, it is expected, be in working order by July 31, by which date it is hoped that the effective range of other Marconi stations at Cape Ray, Heath Point, and Fame Point will be increased to such an extent as to provide a complete inter-station service between the following stations on the Canadian and Newfoundland coasts:—Fame Point, Heath Point, Cape Ray, Whittle Rocks, Cape Rich, Point Amour, Bell Isle, Cape Race, Halifax, Sable Island. This service has been established by the Canadian Government primarily as an aid to navigation in Canadian waters, and with a view to establishing an efficient substitute for the land line service along the shores of the St. Lawrence, which is subject to frequent delays caused by bad weather and the breaking of the cables by ice floes.

SATELLITE COMPETITION

AT a lecture at the Royal Institution the other day the following table was put upon the screen as a nucleus for remarks on the discovery of new satellites:—

Date.	M.	J.	S.	U.	N.	
To 1685......	...	4	5	....	...	Foreign.
1787–9	...	...	2	2	...	England.
1848	...	...	1	...	...	America.
1846–51......	...	...	...	2	1	England.
1877	2	...	...	....	...	America.
1892	...	1	...	...	...	America.
1899	...	...	1	...	...	America.
1904	...	1	...	...	...	America.

Score:—England 7. America 6.

It was remarked by the lecturer that since the slide had been prepared the addition of a 7th satellite to Jupiter by an American discoverer had brought the score "all-even," and that England was on the point of being passed in the race.

1905

"Cosmical Physics" is a department of science which is rapidly growing in favour. As a subsection of the British Association, for instance, it is scarcely three years old, owing its birth to the fusion of the existing subsection of astronomy with non-existent sections for meteorology, seismology, and magnetism. But once born it has justified its existence by growing rapidly and healthily, and has apparently fascinated the readers of papers and addresses in the main section. Dr. Arrhenius devoted his address to emphasizing the intimate relationships of all the sciences above mentioned among themselves and with others. For instance, meteorology was closely concerned with the amount of carbonic acid gas in the atmosphere, which now remained nearly constant; but had the quantity always been constant? If it had formerly been much greater, how had plants lived? There was some evidence—not very much, unfortunately, but still some—that certain forms of plant life could exist in an atmosphere strongly charged with carbonic acid gas, perhaps in an atmosphere of that gas alone. More light was needed on this question, so that physiological chemistry must come to the aid of meteorology. Not the least interesting feature of the address was the genial personality of the great Swedish philosopher himself. Spectacles in one hand and a few notes in the other, to which he only occasionally referred, he spoke fluently in a language not his own for nearly an hour, seeming to be totally unmoved by interruptions arising from the constant accession of new auditors. Before the end of his address the room, which was well filled at the beginning, was closely packed with an eager audience.

In the current issue of *St. Bartholomew's Parish Messenger* (Naylor Street) the Rev. John Howard instructs his parishioners in the inner meaning of the "Song of Sixpence." He says:—"Perhaps many who often repeat 'Sing a Song of Sixpence' have never heard this explanation of its meaning:—The four-and-twenty blackbirds represent the four-and-twenty hours. The bottom of the pie is the world, while the top crust is the sky which overreaches it. The opening of the pie is the day-dawn, when the birds begin to sing; and surely such a sight is fit for a King. The King, who is represented as sitting in his parlour counting his money, is the Sun; while the gold pieces which slip through his fingers as he counts them are the golden sunshine. The Queen, who sits in the dark kitchen (? pantry), is the Moon; and the honey with which she regales herself is the moonlight. The industrious maid, who is in the garden at work before the King—the Sun—has risen, is day-dawn, and the clothes she hangs out are the clouds; while the bird, who so tragically ends the song by 'nipping off her nose,' is the hour of sunset. So we have the whole of the day, if not in a nutshell, in a pie." Our old friend the Sun Myth once more! His uses are inexhaustible."

GREEN WICH

L.C.C. POWER STATION

(ASTRONOMER ROYAL)
I WAS HERE FIRST,
I MEAN TIME
YOU MEAN TROUBLE
YOU'VE UPSET LONDON
NOW YOU WANT TO
UPSET THE WORLD

ZEEMAN ON MAGNETO-OPTICS

Professor Zeeman gave a general review of the experimental researches on the relation between magnetism and light which had occupied him during the last few years. His observation, made in 1896, of a slight widening of the spectral lines of sodium under the influence of a magnetic field was, he said, the origin of his work, which he carried on in the light of the theory of electromagnetic and optical phenomena developed by H. A. Lorentz. In accordance with this theory he found that in a strong magnetic field certain spectral lines were divided into three components, when the magnetic force was at right angles to the direction of propagation of the light, and further, that the middle one of these components was plane-polarized in a direction different from that of the two outer ones. When the magnetic force was parallel to the direction of the propagation of the light, the lines split up into two components, each circularly polarized but in opposite directions. From these phenomena it could be inferred that in a luminous gas all vibrations arose from the negative electrons, and the value deduced for the ratio of the charge to the mass of the electron was of the same order as that obtained from the study of the cathode rays. Professor Zeeman next considered the rotation of the plane of polarization close to an absorption band, and then the double refraction and resolution of the absorption lines. Finally, he discussed the behaviour of different spectral lines in the magnetic field. In many metallic spectra a number of lines occurred which were closely related and formed so-called series. It was found that all lines of the same series were split up in the same manner—*e.g.*, all were resolved into triplets, or sextets, or nonets ; moreover, not only was the general type of subdivision the same, but even the amount of separation, measured in oscillation frequency. A second law was that the corresponding series of different elements also showed the same type of resolution and the same amount of separation. The conclusion seemed to be that all the lines of a series were emitted by one oscillating system, and that, therefore, there were as many series in the spectrum of a substance as there were oscillating systems in its atom ; and that the oscillating mechanism was the same in different elements. He thought there could be no doubt that spectrum analysis and especially the magnetization of the spectral lines would give a clue to the inner structure of the atom.

CROOKES AND RADIUM

Sir W. Crookes, as a chemist, had asked himself what was the cause of the heat of radium, and he had traced it to collisions and bombardment. It might be that to a certain extent the heat of radium was superficial, and that when locked up in a rock mass there was no opportunity for emission to go on. He might describe a simple experiment. He took two brass boxes, equal in superficial area and cubical contents. Into one he placed 50 milligrams of bromide of radium, packed tightly, and in the other box a similar quantity of some inert substance, like silica. In a piece of ice, carefully flattened and placed in conditions, which physicists would easily understand, to ensure a constant freezing temperature, two holes were bored of an equal depth into which the brass boxes exactly fitted. The interstices were plugged with pounded ice. Now the box containing the 50 milligrams of radium ought, with lapse of time, to have sunk considerably, but as a matter of fact he could not detect a difference of level amounting to the 1-1,000in. He might add that he had prepared an account of this experiment, and submitted it to the Royal Society, but that he had been entreated privately to withdraw it, as the results were not such as would be accepted by a learned society.

SPECIALIZATION

THE *Proceedings of the Royal Society* have not only been enlarged in size and type but have been divided into two series, one for Mathematical and Physical Sciences, and the other for Biological Sciences. The departure is not entirely new, as the *Philosophical Transactions* (which are now to be reserved for very long researches, perhaps each occupying a separate volume, similar to that on the Krakatoa Eruption) have been so divided for many years. The division of the *Proceedings* is a concession to those who desire to exclude from their libraries literature relating to sciences apart from their own. Their point of view is only too obvious. But at the same time one cannot help regretting that one of the main objects of the Royal Society, which we may take to be the unification of the Sciences, should in this way be placed second to a consideration of convenience. If a physicist does not care to read any of the biological papers, two questions are suggested : firstly, is he not unnecessarily narrow ? should he not be able, for instance, to read with interest the papers on epidemics like malaria and sleeping sickness, which should appeal to every educated human being ? secondly, are the bulk of the papers printed by the Royal Society of the right type if they command so little sympathy outside a narrow circle of specialists ? Papers concerning one branch of science alone would surely be best placed in the proceedings of a special society : there will be no lack of others which concern two or more branches, and may even cross that line of division between physics and biology which has been emphasized by the step recently taken. It is indeed noteworthy that the line has already been crossed by one of the Secretaries. Professor Larmor contributes to Vol. 76 of Series B (Biological Sciences) a "Note on the Mechanics of the Ascent of Sap in Trees " (p. 460), and has got over an obvious difficulty of indexing by giving a cross-reference to the paper in the A series. This expedient may suffice for the present : but it will be a sad thing for both physics and biology if the gap between them is to become wider in the future.

IONS IN SOLUTION

The lecturer pointed out that in order to explain why, during the electrical decomposition of solutions, the products appeared only at the terminals, the intermediate portion of the liquid remaining unaltered, it was necessary to suppose that the opposite parts of the salt in solution moved under the influence of the current in opposite directions. These moving parts, or ions as they were called by Faraday, since they moved under an electric force, must carry electric charges, and it was the movement of these charges through the liquid—the positive charges in one direction and the negative charges in the other—which constituted the electric current. Sir Oliver Lodge was the first to observe directly the movement of the ions, and the lecturer showed another method which depended on the use of two solutions—one coloured, of potassium permanganate, the other, colourless, of potassium chloride. Methods used by Professor Orme Masson and by Messrs. Denison and Steele were also explained. The lecturer said that the evidence went to show that the ions existed in a solution practically independently of each other, whether or not a current was passing. When there was no current, their electric charges could be demonstrated by their coagulative action on solutions of substances like albumen, and such coagulation had been satisfactorily explained by the effect of a definite electric charge brought into action by the chance conjunction of a number of electrical ions.

NEW RADIOACTIVE ELEMENT

The *American Journal of Science* gives details of the discovery by Professor B. B. Boltwood, of Yale University, of a new source of radium. Evidence has been obtained of the existence in the uranium metals of a new radio-active element which emits both alpha and beta radiations, which produces no emanation, and which resembles thorium in its chemical properties. It is, without doubt, a disintegration product of uranium and is in all probability the immediate parent of radium. The name "ionium," derived from "ion," is proposed for this new substance, which would seem to be appropriate, because of the ionizing action which it possesses, in common with the other elements which emit alpha rays.

RAYS OF POSITIVE ELECTRICITY

Professor Thomson explained that if a perforated cathode were employed in a vacuum discharge tube, certain rays were found to stream through the apertures to the back of the cathode. These rays were noticed about 20 years ago by Goldstein, by whom they were termed *kanalstrahlen*. They showed a reddish colour as they passed through the gas, and they could be deflected by an intense magnetic or electric field, though not so strongly as the cathode rays, and in the opposite direction, a circumstance which indicated that they were positively charged. The lecturer said they seemed to afford opportunity for the study of positive electricity, of which very little was known, and therefore he had been engaged during the last 12 months in experimenting on them. To detect them he used the luminiscence produced by them as they fell on willemite, which was the most sensitive substance he had discovered for the purpose. The willemite was powdered and attached to a screen, which formed the end of the discharge tube, by means of water glass. A fine pencil of the rays passing through a very narrow tube such as was used for hypodermic injection syringes was directed on the screen, and the behaviour of the phosphorescent patch was studied as the rays were acted upon by electric and magnetic forces. The phosphorescence being too faint to be directly photographed, its position was mapped out by means of Indian ink painted on the outside of the tube. When the tube contained air at not very low pressure the phosphorescent patch was spread out into a band with straight edges, but the boundaries became curved when the tube contained hydrogen. At lower pressures there was no continuous band, and the phosphorescence was split up into two distinct patches. Determinations for each of these patches of the ratio of the charge to the mass gave a *maximum* value of 10⁴ (corresponding to the hydrogen atom) for one patch and half that value for the other. These two values were obtained whatever the gas or gases in the tube, whereas with the higher pressures, though the *maximum* value was the same, the lower values were different for different gases. Though further experiment was required to settle the question definitely, these results indicated that with very intense fields gases could give off positive particles which were the same for all gases.

TELEGRAPHING PHOTOGRAPHS

At the office of the *Daily Mirror* last night a demonstration was given of the process of transmitting photographs by electricity devised by Professor Korn, of Munich, which has been adopted by that journal for use between London and Paris. The transmitting apparatus consists of a glass cylinder on which is rolled the photograph that is to be transmitted, and which is enclosed in a dark box. Passing through a small hole in this box, a concentrated beam of light from a Nernst electric lamp strikes a reflecting prism mounted in the interior of the cylinder, and as this light has to pass through the photographic film, its intensity, when it reaches the prism, is varied according to the lights and shadows of the picture. From this prism the light is reflected upon a selenium cell which is included in the same circuit as the receiver, and as the electrical resistance of selenium varies with the intensity of the light falling upon it, the strength of the current in the line is also varied according to the lights and shadows of the photograph. The cylinder which carries the picture is made to revolve spirally, so that each portion of its surface is successively exposed to the beam of light, with the result that a series of electric currents of varying strength are sent through the line to the receiving instrument. To translate these currents into a photograph at the receiver a second cylinder, carrying a sensitized film, is made to revolve synchronously with the cylinder at the sending end and with a similar spiral motion, and on this cylinder falls a beam of light the intensity of which is regulated by a shutter, whose movements are governed by the strength of the currents received. Hence each portion of the sensitized film is successively exposed to a degree which varies with the intensity of the light that has reached the reflecting prism in the sending apparatus, and thus the film, when developed, exhibits the same gradations of light and shade as the original, or rather they are reversed, since a positive at the sending station is received as a negative, and *vice versa*. The fineness of the reproduction of course, varies with the number of impulses used in transmitting a given area of photograph, much as the character of a process plate varies with the fineness of the screen used in reproduction, but a photograph measuring 13 by 24 cm. and sent in 12 minutes has a fineness corresponding to a distinctly coarse process plate, though even when sent in six minutes it is perfectly recognizable. Professor Korn, however, hopes to improve both the speed and the quality of the transmission, and also to succeed in working his method over long submarine cables, which at present cannot be done.

WIRELESS RECEPTION

The *New York Herald* (Paris edition) publishes the following telegram of yesterday's date from Newport, Rhode Island :—

"Lloyd Ronney, a boy of 14, whose wireless apparatus, his own invention, is conceded to be one of the best on the coast, caught messages last night from the battleships. His success is verified by the naval station authorities, who also complain that wireless apparatus on sound steamers interfere with the receipt of messages at the station."

HALLEY'S COMET

a lecture on Halley's Comet, giving the latest results obtained from the investigations made respecting its approaching return in 1910. He traced the history of the returns of this comet at intervals of some 75 years, back for several centuries, and described the work which is being undertaken in the study of the perturbations of this body in order to say with certainty whether or not the comet of 1066 was Halley's. The next return of Halley's Comet might be looked for, he said, on May 16, 1910. In February and March of that year it would be nearly stationary in the constellation Pisces—an evening star, but not very bright. It might be faintly visible to the naked eye in February, though, perhaps, a telescope would be required in order to see it. It would be in conjunction with the sun early in April, becoming then a morning star, visible shortly before sunrise. It would be best seen after its next conjunction with the sun, early in June, passing near Pollux on June 9, when it would be at its *maximum* brightness. During the following week or fortnight it ought to be fairly well placed in the evening sky, to the north-west. Its tail would probably be about 30deg. or 40deg. in length, so that, when the comet itself was below the horizon, we might hope to see something of its tail. Dr. Holetschek, of Vienna, was of opinion that there was a slight chance that during October or November next this comet might be photographed, and it was exceedingly desirable that this should be done, as it would enable the mean motion to be obtained at the next return with very great accuracy. It was thought that the light pressure exerted by the sun had a sensible influence on the motion of comets—in other words, that the sun's gravitational action was modified by the action of light; and the only way to verify this by way of Halley's Comet was to observe it over a very long period during its time of visibility.

STELLAR MOTIONS

Mr. A. S. Eddington, Royal Observatory, Greenwich, spoke on the results of his researches into the motions of the stars. Formerly, he explained, the motions of the stars were thought to be quite haphazard, and anything of a systematic nature in connexion with them was supposed to be due to the motion of the sun. The theory advanced by Professor Kapteyn two years ago virtually divided the universe into two distinct systems as the motions showed tendencies to congregate in two favoured directions. His own work, dealing with different data, confirmed the results arrived at by Professor Kapteyn, and seemed to show that these two systems were completely mixed, as it were—that the two drifts were two independent systems of stars which happened to be crossing, and that this world was more or less in the midst of them.

DO NOT ATTACH TOO MUCH WEIGHT

[Seven sextillions], of course, is a tremendous number of tons for any moving mass to weigh, but there is a time twice each year when the Earth actually weighs nothing at all. In October last this Earth gradually began to lose weight like some huge giant dying of a decline, until at a certain moment of time it weighed only an ounce, then half an ounce, then a quarter, and finally, just for about the fifty-thousandth part of a second, it weighed absolutely nothing whatever—not even so much as a soap bubble which a baby might blow away.

To realize this you must remember that the Earth does not travel round the Sun in an exact circle, so that the distance from the Sun is always varying, which, of course, alters the power of attraction or, in other words, the Earth's weight. But in October and April of each year the Earth is at an exact average distance during the fraction of a second, at which time, as I have said before, it weighs nothing. How short a space of time this is may be judged from the fact that the Earth moves at the rate of eighteen and a half miles every second.

But to weigh the Earth in the manner suggested would be a very costly matter, and so it is found to be more satisfactory to employ mathematics, when we shall arrive at the same result, assuming, of course, that we are correct in our deductions.

THE NEW COMET

The new comet discovered at the beginning of this month by Mr. Morehouse in Wisconsin promises to be of considerable interest. It has now been sufficiently observed at European observatories to enable the elements of its orbit to be deduced, which has been done by Dr. Kobold of Kiel. The result shows that the comet is rapidly approaching both the sun and the earth, and consequently that the brightness will go on increasing for about two months from the present time. Now, as it is already on the verge of naked-eye visibility, there is little reason to doubt that it will become tolerably conspicuous to the naked eye, while it is not impossible that it may be a really brilliant object. Though it is still at twice the earth's distance from the sun, photographs show evidence of great activity in the tail, which is multiple and fan-shaped, like those of many recent comets. The outer streamers are faint and fairly straight; the central portion is bright and curiously twisted, showing at once the intensity and the apparently capricious nature of the force which carries it out from the nucleus in a direction away from the sun.

Throughout September the comet will be within 20° of the Pole, and consequently above the horizon the whole night. At the end of the month it will be on the meridian at 9 p.m. After this it will steadily approach the sun, and will be lost in his rays about the middle of December, reappearing a month or two later in the morning sky. The date of its nearest approach to the sun is given by Dr. Kobold as December 24, when it will be inside the earth's orbit, at 11-12 of our distance from the sun. Its distance at the present time is about twice as great. The motion round the sun is retrograde, or in the opposite direction to that of the planets. This fact makes it likely, but not certain, that the comet belongs to the non-periodic class, for members of this move indifferently in either direction, while only two of the periodic class (Halley's and Tempel's, the latter being the comet associated with the November meteors) move in a retrograde manner.

PHOTOGRAPHING THE HUMAN VOICE

At yesterday's sitting of the Academy of Sciences, M. Poincaré gave particulars of a discovery by M. Devaux Charbonnel of a method of photographing the sounds of the human voice with sufficient precision to enable the record to be read. Vowels and consonants are combined with a Blondel oscillograph. The latter, which is extremely sensitive, impresses the sounds upon a photographic plate in the form of curves characteristic of each category. With a little practice it is possible to decipher these characters.

AN ASTRONOMICAL CLOCK

An astronomical clock, the invention of Dr. Herman Bumpus, the director, which shows all the movements of the earth, has been placed on view at the American Museum of Natural History. The museum authorities state that no similar device has before been exhibited for the instruction of the public.

A stereopticon representing the sun is placed at a distance of 10ft. from a globe of paper-composition, 4ft. in diameter, which is illuminated from the lens in such a manner that only half of the sphere shows the light as the globe slowly revolves and changes its poise by means of mechanism connected with a small steeple-clock. The globe derives its movement from the working of the clock, and the shadow of a wire placed at the back of the lens, which corresponds to the meridian of New York, and which is thrown upon the sphere, shows the time of day with mathematical accuracy.

WIRELESS TELEGRAPHY AND ULTRA-VIOLET LIGHT

The well-known experiments of Professor Zickler with ultra-violet light have recently been extended by J. Köhler, at Oggersheim. He makes use of the fact that if the two poles of an electrical machine are connected respectively to two sharp points, the electricity generated will pass from one point to the other noiselessly and practically invisibly. On alternately illuminating the points with ultra-violet light, a telephone in the circuit can be made to emit sounds that correspond to the illumination. The transmitter is a rectangular wooden box, in which a stroboscope disc of paste board is arranged so as to be readily set rotating by hand. A sheet metal tube in front of the box is used as a reflector. The receiver comprises an influence machine and a small induction coil with the spark gap to be acted upon by the ultra-violet light. Magnesium light was used for producing ultra-violet rays. The transmitter and receiver were installed at a short distance apart, and the light rays reached the receiver after traversing a round hole in the disc. When the disc was rotated a sound whose pitch depended on the speed of rotation was heard in the telephone. By alternately illuminating and screening the receiver, Mr. Köhler succeeded in telegraphing first letters, then words, and finally whole sentences.

THE GROWTH OF APPLIED PHYSICS

During the past year, in the territory of physics, exact methods of measurement have advanced, mathematical investigations of known phenomena have continued, and repetitions of familiar experiments have been in evidence; but notwithstanding our State-aided laboratories, our municipal technical schools, and our modernized Universities, the harvest of original observations is meagre and disappointing. It has been suggested that the savant is passing away, and that he is giving place to the specialist. The specialists in industrial chemistry and in industrial physics have certainly left their impress upon the work of 1907. As examples may be cited the advance in the preparation of material, and in the firing and glazing of pottery for insulators, the improvements in manufacture which have resulted in a more uniform and more trustworthy cement and concrete, the progress of aluminium and steel manufacture, and the establishment of new types of incandescent lamps. Electrical plant for coal cutting, pumping, haulage, screening, picking, winding, and washing at the pit head, and electrical equipment for shipbuilding. yards and for elevators and telferage of all kinds continues to extend. The economy and convenience of electrical haulage was demonstrated in the later stages of the construction of the dock at Keyham—a masterpiece of engineering, now successfully completed. Electric welding was for a considerable time attended with difficulties in regard to annealing and the methods of carrying out the work; the process has recently been improved, and it has consequently increased in favour among cable manufacturers and others, and it promises to be of great service in repairs and for rail joints. Meanwhile an electro-thermic process for the smelting of iron ores has received well-merited encouragement from the Canadian Government. Their tests have shown that charcoal can be used as a reducing agent in the electric furnace, without briquetting, and they have revealed valuable information as to the applicability of electric smelting to various kinds of ore.

MARTIANS VIEW THE EARTH

Professor Highell, of Bannerpole, has observed on Terra a brightening of the tiny spot known as Albion, suggesting a fall of snow. In a general way there is nothing remarkable in a whitish appearance of this spot, which is well within the north-polar cap of our sister planet. The unusual feature is the date, which is more than 30 days after the Sun has crossed the equator of Terra northwards; so that the inhabitants (for whose existence Professor Highell so staunchly contends, in spite of the scepticism of some of his colleagues) of the northern hemisphere might be fairly expecting the advent of summer. If we accept the observation and the interpretation of our distinguished astronomer, we can well imagine the pleasurable surprise of our hypothetical brethren at this unusual phenomenon, which must be the stranger since the neighbourhood of Albion has, during the past winter season, shown few signs of cool weather. There is reason to believe that the climate of our sister planet, from its being nearer the Sun than ourselves, and girt with a dense atmosphere, is unpleasantly warm, with the possible exception of the winter months; and a not inconsiderable body of opinion declares that no beings like ourselves could exist thereon. Professor Highell has, however, suggested that during the summer months the inhabitants gradually withdraw to the polar caps, in search of a suitable climate; and has ingeniously inferred that the population must have reached a high level of civilization, since they can carry out such strategic movements on a vast scale in good order, and can live together amicably in the restricted area represented by the summer polar cap.

THE SEARCH FOR PLANETS OUTSIDE NEPTUNE

Ever since the discovery of Neptune in 1846 from the disturbances which it produced in the motion of Uranus, astronomers have speculated on the possible existence of still more distant planets. There have been two methods of investigation, one based on a study of the motion of Uranus and Neptune, by which Professor Todd was led, about 1880, to search in the neighbourhood of Tau Leonis; the search, however, was fruitless. Another method consists in studying the grouping of cometary orbits. Each of the four giant planets has a family of associated comets, whose aphelia lie near the orbit of the controlling planet, and whose periods are about half that of the planet. Thus, for example, Encke's comet is associated with Jupiter, Tuttle's with Saturn, the comet of the November meteors with Uranus, and Halley's comet with Neptune. Professor Forbes, of Glasgow, has in the last month published a paper in which he has collected a large amount of evidence from the grouping of cometary orbits, of the existence of a planet at 100 times the earth's distance from the sun, and with a period of 1,000 years. In particular, he supposes that the comet of 1556, whose return was expected about 1850, was split up about 1700 into three different comets, which appeared in 1843, 1880, 1882; this would necessitate a very close approach to the supposed planet, since the elements of the comet's orbit were entirely changed. Combining this with the behaviour of some other comets, he has given an orbit for the planet, of which the most surprising feature is the inclination, which he makes 53° to the ecliptic. From the position assigned, this planet is now quite out of reach of European observers, but search is being made by photography at the Cape Observatory.

Professor W. H. Pickering, of Harvard College Observatory, has recently been led in some manner similar to that used by Professor Forbes to suspect the existence of another trans-Neptunine planet; the region indicated is in the constellation Gemini, Right Ascension 7h. 47m., North Declination 21°. This region is far removed from that given by Professor Forbes, so that the planets indicated by the two astronomers must be distinct from each other. Search is being made by photography for the Pickering planet, both at Harvard, and Arequipa, and also by Dr. Metcalf, at Taunton, Mass., U.S.A.

Some vague rumours have been published in this country that the planet has actually been found, but of this there is no authentic confirmation. It is probable that they are only based on Professor Pickering's announcement that he suspected a planet there; the details of his paper have not yet reached this country, but, in any case, deductions of this kind can only be classed as speculations, until they have been visually verified.

A NEW SATELLITE OF JUPITER

This tiny object was discovered at the Royal Observatory, Greenwich, just a year ago. It was for some time doubtful whether it was a true satellite or only a minor planet that made a close approach to Jupiter; but, after three months' observations were available, it was shown by Mr. Cowell and Mr. Crommelin, in a paper read before the Royal Astronomical Society, that it was almost certainly a satellite. The object is so feeble that it can only be detected on long exposure photographs, taken in a perfectly dark sky. Consequently nothing can be done when Jupiter is near the sun in the heavens, as was the case from April to October.

Since the latter date the weather conditions have been unfavourable, and it was not till the early morning of Sunday last that it was possible to obtain a satisfactory photograph. On this the object has been detected, its satellite character being now placed beyond doubt. Its Right Ascension was 0m. 22s. less than Jupiter's, and it was 14' north of him. This position accords very closely with that deduced from the Cowell-Crommelin elements, so that these are probably near the truth. These indicate a very remarkable orbit, much more distant from its primary than any other known satellite, very eccentric and highly inclined, and with retrograde motion, or in the opposite direction to that of all the other Jovian satellites. This last feature is, however, shared by Phœbe, the outermost satellite of Saturn. It was nearest to Jupiter last August, the distance being ten millions of miles; the greatest distance, 20 millions of miles, will be attained next September. The eccentricity is thus one-third, and the inclination to Jupiter's orbit is 30°, the period about two years and two months. Owing to the great distance of the satellite, the sun will very greatly disturb its motion round Jupiter, and the stability of the orbit will present an interesting problem.

The object is of the 17th magnitude, implying a diameter of perhaps 35 miles; as seen from Jupiter it would only appear as a faint telescopic star of the ninth magnitude.

MR EDISON'S NEW STORAGE BATTERY

Mr. Edison's long-promised new storage battery, with the details of which he has been occupied for some time, is about to be practically demonstrated. For a long time the inventor has been engaged in perfecting such a device for automobiles, and this has been the subject of frequent premature announcements in the Press. The new battery, however, is for use on street car systems, and will be put on trial shortly by Mr. Whitridge, the Federal Receiver of the Third Avenue Street Railway. The statement of the inventor, explaining the new battery, says that the elements are the same as in the battery which he practically perfected years ago—namely, nickel with an alkali reaction, but with some improvements suggested by continued experimenting, until, he is now convinced, he has the battery needed. He is satisfied he can put a car in service that will run a whole day without recharging. He expressed the opinion that the use of the new battery would revolutionize surface car traffic. He predicts that street car lines will employ none but cars equipped with the new batteries when he has demonstrated their commercial value ; the tracks will be without either overhead or underground wires, or rails for the transmission of current, and comparatively cheap stations only will be necessary where the storage batteries may be charged after they have exhausted their store of current.

NEW X-RAY APPARATUS

new form of Röntgen-ray apparatus from America, where it has been adopted to a considerable extent both by hospitals and by private practitioners. It is the invention of Mr. H. Clyde Snook, of Philadelphia, and the advantages claimed for it are efficiency and simplicity of working, as compared with the usual arrangement of induction coil, with its interrupter and other troublesome accessories. The current is obtained from a motor-alternator, which can be driven from any suitable electric power supply, and is raised to a high potential by being passed through a step-up transformer. The special feature of the machine is that this current is rectified, for use in the Röntgen bulb, by means of a mechanical commutating device which is carried on a prolongation of the shaft of the motor-alternator, perfect regularity and synchronism being thus secured in the commutation. The particular machine now on view in Messrs. Newton's show-room takes 22½ ampères at 200 volts, and its output can be regulated from 1 to 60 milliampères with ease and certainty simply by moving a small crank, which puts in or cuts out resistance.

RADIO COMMUNICATION WITH MARS

If life, says the professor, really exists on Mars they have been trying for years to get into conversation with us, and perhaps wonder what manner of stupid things we are not to respond. Following this assumption, I thought we might feel their presence if we could get high enough away from the noises and ether waves that surround us up into the rarefied regions of our atmosphere with nothing to disturb communication.

To intercept any ether waves that may be radiating from the planets, I shall be shut into a metal box made of aluminium for lightness, and fitted with an apparatus to drive out carbonic acid gas and supply oxygen, and with air-pressure to prevent sickness. In that way I can ascend much higher than balloons heretofore have gone. We shall take a wireless receiving instrument, and try to record electrical waves from Mars, Venus, or any other planets. We cannot presume to send messages to Mars on this coming trip, but shall only try to receive them. The question has been raised as to how we should get our ground connection necessary for telegraphy. There are several answers to that. We shall use what is practicable. If it is established that a 1000-ft. wire, hanging from the car with the surrounding atmosphere forms a grounding we shall use that.

1910s

In terms of the popular press, the decade beginning in 1910 was bracketed by two spectacular astronomical events. In 1910, Halley's comet made all the headlines. In 1919, the headlines were about Einstein's theory of general relativity.

Solar eclipse photograph taken in Brazil. (The stars circled were used to confirm Einstein's predictions concerning the bending of starlight by the Sun.)

For more than two centuries past, Halley's comet has attracted more popular attention than any other comet. It was the first comet identified as 'periodic'—that is, one that returned to the Sun's vicinity at regular intervals (about 76 years for Halley's comet). In addition, it was usually bright enough to be readily seen with the naked eye by untrained observers. For the return of 1910, there was an additional interest: it was forecast that the Earth would pass through the tail of the comet. By this time, spectroscopic observations of comet tails had been made, and it was known that they contained gases that were poisonous to human beings. This gave rise to widespread fears that the life on Earth might be affected by such a contact. The spread of communication meant that rumour of this possibility was spread around the Earth. Equally, the press had a field day in reporting the suicides, and even riots, that could be linked to the fear of Halley's comet.

Reports of the comet also reveal, in passing, the low level of scientific knowledge then common amongst reporters and editors. More than once, readers were told that the comet's tail was so many feet long. What had happened was that astronomers had described the tail as (say) 10′ long. They meant '10 arc min', but the press interpreted the symbol as meaning 'ten feet'.

At the end of the decade, the solar eclipse of 1919 proved a good deal more important for the development of physics. Einstein published his ideas on general relativity in the middle of the First World War, so knowledge of them outside Germany grew rather slowly. However, it was realised, especially by Eddington in England, that the next solar eclipse in 1919 provided an excellent chance for testing the theory. Einstein predicted that light passing close to the limb of the Sun, as seen from the Earth, would be deflected by a specified amount. Because of the Sun's brightness, the small displacements involved could only be measured during a solar eclipse, when the Sun's light was blocked out.

The eclipse of 1919 seemed especially suitable because the Sun then occupied a region of the sky that had many background stars. But little could be done to organise observing expeditions whilst the war was on. Fortunately, the armistice was signed just in time for two expeditions to be dispatched—one to West Africa and the other to Brazil. Analysis of the photographs taken by these two expeditions appeared to provide strong support for Einstein's theory. Immediately, newspapers were full of such headlines as 'Newtonian ideas overthrown'.

Developments in atomic and nuclear physics during this decade were of major importance, but received only occasional mention in newspapers. Rutherford's new picture of the atom, consisting of a small electrically positive nucleus surrounded by a cloud of electrically negative electrons, did receive some attention. So, too, did Bohr's ideas on how this atomic structure could be related to the absorption and emission of radiation via the quantum theory. Newspapers often reported, alongside these, hostile comments on the new ideas by some off the older generation of physicists.

The most important reason for such paucity of comment was almost certainly the impact of the First World War. In the first place, the space available for reporting science was reduced, and what remained was often used to discuss war-related science. Secondly, the amount of civilian science under way diminished appreciably, and so did access to what was happening elsewhere (especially in Germany).

An obvious example of war-related science was the study of aerodynamics. Though only a small amount of the space devoted to flying actually referred to scientific matters, this often came into discussions of stability, whether in relation to aircraft manoeuvrability or to accidents. Alongside this, flying led to a renewed concern with meteorology. In order to advise pilots on flying conditions, it became necessary to explore the vertical structure of the atmosphere in more detail, especially in terms of changes in wind direction and strength with height. These studies fed back into further theoretical developments in meteorology.

Another change produced by the pressures of wartime was the introduction of daylight saving—called 'British Summer Time' in the UK. This was meant as a practical measure, more especially to help with extending daylight hours for agricultural work. However, it also led to a considerable public debate which extended to time standards in general.

At the beginning of the First World War, a number of German scholars—including some physicists—issued an open letter declaring their support for the German state. The proclamation was badly received by scientists in the Allied countries. Later in the war, representatives from these latter countries came together to examine how international science might be organised once the war was over. The antagonism they felt towards German scientists was such that it was decided to exclude the German scientific community altogether from participation in the post-war international scene.

1910s CHRONOLOGY

1910
- Ehrlich initiates the use of chemically developed drugs

1911
- Amundsen reaches the South Pole
- Onnes discovers superconductivity
- Rutherford presents the idea of a nuclear atom

1912
- The Braggs use crystals to measure the wavelengths of x-rays
- Hess discovers cosmic radiation
- Wegener proposes the theory of continental drift

1913
- Bohr completes his theory of a 'planetary' atom
- The Hertzsprung–Russell diagram for the study of stellar evolution is finalised
- Millikan completes his measurements of the electronic charge
- Moseley develops the idea of atomic number, based on x-ray spectra

1914
- First World War starts
- Franck and Hertz confirm Bohr's atomic model experimentally

1915
- The first white dwarf star is discovered

1917
- Einstein puts forward the general theory of relativity

1918
- First World War ends
- Aston builds the first mass spectrograph
- Shapley describes the layout of the Milky Way galaxy

1919
- Observations made at the total solar eclipse support the general theory of relativity
- Rutherford observes artificial atomic fission

1910s

RUSSIA AWAITS THE COMET

In Southern Russia there is much popular terror at the advent of Halley's comet, and prayers are being offered in the country churches and monasteries for the salvation of Russia from the threatened cataclysm. During the last week a paragraph vaguely purporting to represent the views of certain distinguished Russian astronomers has been going the round of the provincial journals to the effect that the visitation of the comet will have the most disastrous consequences.

The superstitious fears of the peasantry are being exploited by persons who are collecting money, falsely representing that it is for the solemnization of masses and special services during the dreaded month of May.

COMET CAUSES SUICIDE

Landowner feared being " killed by a star."

BERLIN, Wednesday.

A Budapest message to the *Frankfurter Zeitung* reports the suicide of Adam Toma, a landowner in the Szozona district, through fear of Halley's Comet. He said he preferred suicide to being killed by a star.

HALLEY'S COMET AND BEER

In an essay on Halley's Comet a Bavarian schoolboy wrote: "In this country the Comet has already caused a rise in the price of beer, but it may cause even greater misfortunes in other countries."

HALLEY'S COMET AND EARLY AVIATION

AERIAL NAVIGATION

You were good enough some weeks ago to give publicity to an extract from a poem by Erasmus Darwin in which that versatile writer foretells the advent of aerial navigation. At the time of writing I was unable to give the reference to the original, as I was then far away from any library; but this omission has since been supplied by one of your correspondents. On my return to town I turned at once to the "Botanic Garden" to check the reference, and, to my great surprise, I came across the following paragraph appended as a note :—

As the specific levity of air is too great for the support of great burthens by balloons, there seems no probable method of flying conveniently but by the power of steam, or some other explosive material, which another half century may probably discover.

Not only did Darwin prophesy aerial navigation, but he actually anticipated in the above passage the use of petrol.

HALLEY'S COMET APPROACHES

The time is now approaching when this famous comet may be expected to form a spectacle of general interest. It has been steadily brightening since its discovery last September, when it was an excessively faint object of magnitude 15½. A photograph taken at Greenwich Observatory on January 30 showed a fairly defined nucleus surrounded by a large, diffused coma, and a very faint tail.

The comet will continue to be a morning star till May 18; in the early hours of May 19 it will transit the sun's disc, which it will enter at 2.22 a.m., six degrees below the western point of the disc, and leave at 3.22 a.m. exactly at the eastern point. The phenomenon will not be visible in Western Europe, but will be observable in Asia, Australia, and the Pacific. In particular we may hope for observations at Kodaikanal, near Madras, where Mr. Evershed proposes to take photographs both with the ordinary photo-heliograph and with the spectro-heliograph in ultra-violet light. It must be understood that it is hopeless to expect to see the outer envelopes of the comet against the splendour of the sun; it is only the nucleus that we may hope to see, and, if the attempt is successful, it should add considerably to our knowledge of the constitution of comets. There can be little doubt that Halley's comet has a considerable amount of solid matter in it, for a mere bunch of vapour could not hold together for even a short time, much less for two thousand years. Moreover, at each return it has poured out an immense tail, and it could not possibly recover to itself the gases spread out over millions of miles of space. In other word, the tail at each return is a new one, and there must be some storehouse in the comet where the gas is occluded, to be emitted and repelled when the comet approaches the sun. There is no reasonable doubt that the storehouse is a swarm of meteoric masses, like the large masses that can be seen at the Natural History Museum, but perhaps still millions of miles of space. In other word, the tail at each return is a new one, and there must be some storehouse in the comet where the gas is occluded, to be emitted and repelled when the comet approaches the sun. There is no reasonable doubt that the storehouse is a swarm of meteoric masses, like the large masses that can be seen at the Natural History Museum, but perhaps still larger, since these must have suffered loss in their passage through the atmosphere. They are found on analysis to contain just such gases as the spectroscope reveals in the tails of comets.

The Earth will in all probability pass through the tail of the comet about the time of the transit; in fact the only doubt is whether the tail will be long enough to reach us; a length of 15,000,000 miles will be necessary for this purpose. It might seem that a passage through the tail would enable us to settle all questions as to its nature, but in reality it is easier to study it from outside than from inside, since it is too attenuated to make its presence in any way manifest while we are crossing it; and its light will be too diffused to render spectroscopic observations easy. Still it is desirable to study the night sky on the nights of May 18 and May 19, to see whether the tail can be discerned; when the Earth passed through the tail of the comet of 1861 a kind of auroral glare was said to be visible; in that case the transit was not anticipated, but it was afterwards inferred that it must have taken place.

NEW LIGHT ON MARS

Professor Lowell said he had the honour to present to the society the corroboration in a most striking manner at the last opposition of what previous observations of Mars had taught of the planet, and evidence of the discovery of certain canals on the planet, not simply new in the sense of never before having been seen, but in the signification that they have never previously existed.

The observations at Flagstaff began in 1894. The two new canals appeared on September 30, 1909. It was evident from investigation among the records that they had never been seen before, but as they had discovered about 550 canals since 1894, they were not surprised, though they tried to treat each with courtesy at the appearance of another. What constituted a far more interesting question was the coming into existence of such curious markings. They had to exclude all the contingencies which might have made the event possible. The first was that they might have been there but escaped notice. That explanation was negatived by the fact that the canals were the most conspicuous features on their part of the disc. They also took into consideration the seasonal changes which affected all the features of Mars, and proved that the canals were new by reference to the drawings made at the identical period of the planet's year. The proof obtained of the existence of the canals was only possible through the long series of observations conducted at Flagstaff.

The Chairman, at the close of the lecture, asked Professor Lowell whether he maintained that these canals were really engineering works for the transport of water from the Poles to the Equator.

Professor Lowell, answering this and other questions, said that the mere fact that these canals were a thousand or more miles long and 30 wide had led many people to the erroneous supposition that they must have been dug. There was no reason to suppose that they were dug canals like the Suez and Panama Canals. They were clearly strips of vegetation fertilized by water from the Polar caps, but as Mars was in fluid equilibrium, there was no force tending to make a particle of water move from one part of the planet to another. The only explanation of the motion of the water was that it had been artificially helped.

The Chairman.—By power?

Professor Lowell.—By power on the planet. If, as was the case with New York, they on this planet had sometimes to go hundreds of miles for a water supply, they could think what must happen where all the water was concentrated in the Pole cap. The water returned to the Pole aerially.

SCIENTISTS GET UP FIRST THING
THIS MORNING TO STARGAZE

The discovery of a blue tap on the south pole of Saturn, which was made by the group of scientists on Mount Wilson this morning at 1 o'clock, has created great consternation among the visitors to the convention and is the cause of much speculation today.

Rising shortly after midnight the astronomers went to the observatory telescope and viewed the planet Saturn. They succeeded by mechanical adjustments in bringing the body up from a scale of 4 to 10 until it was in a scale of 7 to 10. Thus they drew it 3 notches closer than it had ever been observed before. The definition was so perfect that the colour of the statification between the pole and equator which has heretofore appeared to be a dull yellow, appeared to be a very bright yellow this morning. The tap on the south pole of Saturn which appeared blue this morning was always merely a dark blot heretofore. Ten points is considered a perfect measure in observations of this sort, but the scientists succeeded today in bringing the planet 3 points closer to the telescope, which in itself is considered a remarkable accomplishment.

In making their report to the convention the committee on the study of sun rotation by means of spectroscopes asserted that they had been unable to determine whether the Sun revolved faster at the centre than at the outside rim on account of the fact that they had never succeeded in segregating the molecules in the atmosphere sufficiently to get a distinct measurement. A full vote decided that the committee should be continued for a series of 3 years more. Prof. M. H. Deslandres explained how M. Berropt of France had used the Slit Method in his one line process of sunlight measurement and the same scheme may be employed by this committee.

AN ASTRONOMER'S DAY

This is Sunday, and I came over hoping to have a quiet morning in my office. It is very hot. The north door is out of order and won't lock. I hoped I was here all alone, but I soon found the hall filled with a crowd of people who came in an automobile. Tried to drive 'em out, but they all protested they would sure die if they did not get to see the big telescope, so to get rid of them I showed it. They thought it was pretty and expressed admiration when told how heavy it was, and were justly edified when told its cost. They said we must have had a great time with it while Halley's Comet was around, and were much surprised when told that we scarcely looked at Halley through it when it was bright, and concluded that it must have been a waste of money to make it (the telescope, not the comet). One asked if Halley's Comet was inhabited, and how could we prove it wasn't. Another thought it must be pretty cold in the dome in the winter, " but of course you get used to it." " Yes! " I replied, "I expect to get used to being dead after I have been dead awhile."

IMPORTANT CORRECTION

With reference to the paragraph which appeared in yesterday's issue regarding Halley's Comet being visible at Nuwara Eliya, Mr. C. J. Lallyett telegraphs to say that the Comet's tail is 300–400 feet and not yards long as stated yesterday.

PHOTOELECTRIC CELLS

The potassium photo-electric cell is extremely sensitive to light, and is being used by Professor Rosing for experiments in television in place of selenium, its resistance when it is illuminated falling owing to the emission of electrons by the potassium cathode. It has recently been used for measuring the luminosity of terrestrial bodies, there being a direct relation between the resistance and the intensity of light falling upon it. Some important work has been done recently by Geitel and Elster in increasing its sensitiveness, this having been accomplished by passing an electric current for a short time through the tube. The tube containing the potassium is filled with hydrogen, and the effect of the current is to cover the potassium with a greenish film, which parts with electrons very readily under the influence of light. This film, and the extra sensitiveness due to it, are not permanent, but permanence has now been obtained by replacing the hydrogen, after the formation of the film, with argon.

AN UNUSUAL ASTEROID

An asteroid, to which the provisional designation MT has been assigned, has just been discovered by Dr. Palisa at the Vienna Observatory. It is of the 12th magnitude, and the remarkable feature is that, although in opposition, it is advancing in Right Ascension half a degree per day. There is only one known planet that can advance when in opposition; that is the planet Eros, discovered in 1898, which has been used to improve our knowledge of the sun's distance. Even it can advance only at a very slow pace; and it would appear that MT is still nearer to the earth than Eros.

It must also have an orbit of great eccentricity and be near its perihelion, since otherwise its velocity at a distance from the sun exceeding that of the earth would fall short of hers, instead of greatly exceeding it. The new body will therefore be followed with great interest till its exact orbit is determined in the expectation that, like Eros, it will prove useful in finding the sun's distance. There is just the possibility that the object is a comet without any nebulosity; some comets have presented an almost stellar appearance; Dr. Palisa, however, is a veteran asteroid discoverer, and as he uses the visual method, comparing the telescopic field with large star charts, with the aid of the great Vienna refractor, he is likely to have made a most careful examination before pronouncing the new body to be an asteroid.

THEORETICAL AERONAUTICS

a mathematician took upon himself to attack the engineers for daring to fly before mathematicians had fully worked out the conditions for stability. The practical man is certainly disrespectful to theoretical considerations at times; one of them wrote a few months back to give the following formula for the machine which he has been flying :—

$$\nearrow \quad \longrightarrow \quad ? ? \, OH! \quad \downarrow \quad **$$

fitted with an engine to which he has given the symbol:

$$ZIZ + POP_2 + . + DN_2 + B_3.$$

He says it is necessary not to omit the full stop, which always occurs in the middle.

UPWARDS OR DOWNWARDS ?

"Downwards" is the oposite of "upwards," and unless there is a centre inbetween these terms can have no existence. . . . .

When, however, a planet such as the Earth is taken as a centre, all that is above its North Pole is "upwards," all below its South Pole "downwards."

It is impossible to walk East from the North Pole because there is no line of longitude there.

THE IMPACT OF WAVES

It is a well-known fact that the impact of a wave on the face or top of a breakwater formed of blocks of masonry or concrete may cause such a block to start from its seat in the opposite direction to that of impact, and may actually cause its withdrawal from the face or top of the breakwater. Cases in which blocks weighing several tons and carrying the weight of several superposed courses have been withdrawn in this manner are of comparatively frequent occurrence. The explanations of this phenomenon which have been current in the past attribute it either to the fact that the wave pressure on the face of the breakwater at the instant of impact is transmitted hydrostatically, as in an hydraulic press, to every portion of the interior, thus giving rise to an excess outward pressure at those portions of the face over which the wave is not at the instant breaking, or to the compression of the air which is entrapped in the open joints. Such a compression gives rise to an internal pressure, which is maintained for some small but finite interval of time after the recession of the wave has relieved the pressure on the face. An experimental investigation recently carried out by members of the Engineering Department of University College, Dundee, and published in the Proceedings of the Institution of Civil Engineers, indicates that, assuming a *maximum* velocity of wave impact of 80ft. per second, the *maximum* internal pressure to be anticipated on the first of these assumptions is about 2·9 tons per sq. ft., while, assuming air compression, the pressure may amount to about six tons per sq. ft. The experiments, however, point to the fact that in favourable circumstances the impact of the wave on an open joint gives rise to a "water hammer" action, analogous to that following the sudden closure of a valve at the open end of a long pipe line conveying water. Where this action occurs, pressures vastly in excess of those possible either by simple hydrostatic transmission of pressure or by air compression, are produced, and, with the assumed velocity of impact, they may amount to some 40 tons per sq. ft., or even more in specially favourable cases. The results suggest the advisability of providing a free outlet, by means of a series of drains or weep-holes opening on its sheltered face, for such water as may percolate to the interior of such a structure. Such drains, preventing the accumulation of internal water, would be an effective guard against the production of internal pressures of sufficient magnitude to affect the stability of the structure, whether due to air compression or to water hammer.

THE EARTH'S RADIATION

Subsequently Professor J. C. M'Lennan read an interesting paper on " The Intensity of the Earth's Penetrating Radiation over Land and Large Bodies of Water." He remarked that the radiation possessed similar properties to Gamma or X-rays and was measured by the ionization of the air. His opinion was that it was mostly, if not entirely, contributed by soils and rocks. Many observers had recorded diurnal and other variations, but he had found no appreciable variation and believed the radiation to be constant. In support of this contention he gave a number of figures which also showed that there was a considerable decrease in radiation over water.

Professor E. Rutherford, in the course of the discussion, said he was inclined to support Professor M'Lennan's no-variation theory, but the Hon. R. J. Strutt, on the other hand, could not dismiss the idea of diurnal variations.

In reply to Professor Bragg, the Author said that he had not made his vessel of two different parts in order to ascertain the direction from which the radiation proceeded.

REMINDERS OF FAMOUS SCIENTISTS

A log of wood, supposed to be part of the famous apple-tree in Sir Isaac Newton's garden, has been presented to the Royal Astronomical Society. According to tradition, the scientist's attention was directed to the subject of gravity by the fall of an apple from the tree.

To this *Punch* replies as follows :—

We understand that a jet of steam, if not the same as, at least similar to, one which issued from the spout of the kettle of James Watt's mother, has been secured by the Amalgamated Society of Railway Servants, in whose museum it now rests.

MATTER AND LIFE

PROFESSOR SCHÄFER'S statement that the possibility of the production of life—that is, of living material—is not so remote as has been generally supposed continues to provoke discussion. It is publicly discussed by men of science, most of whom, no doubt, hope that the professor is right ; but it is also privately discussed by the general public, most of whom hope that he is wrong. They suppose that, if it became possible for human beings to produce living material, the last mystery would be gone from the universe, and that whatever was still unknown about it would only be outside the knowledge of man, not beyond his comprehension. But, if man were proved to be capable of understanding all that is to be understood in the universe, he would be revealed to himself as the highest form of being in it ; for whatever is understood is of a lower order than that which understands it ; and a universe in which man was the highest form of being would disappoint his expectations. This belief that the production of life would mean the complete understanding of life and of its purpose, if it has one, is a strange delusion, very characteristic of our time and of our half-ignorant worship of physical science. We see that science entirely occupied with reality outside us ; we are rightly impressed by its triumphant advances in the investigation of this reality ; and we suppose that what it investigates is the whole of reality. Now physical science very properly imposes certain limitations upon itself. PROFESSOR SCHÄFER, for instance, said in his address that the problems of life are essentially problems of matter ; that we cannot conceive of life, in the scientific sense, as existing apart from matter ; and that the phenomena of life are investigated, and can be investigated, only by the same methods as all other phenomena of matter. There he spoke as a physical investigator. He was concerned with life outside us, with that part of reality about which we learn by observation, and which consists for us of matter and material phenomena. But that is not the whole of reality as we experience it ; for there is also the reality inside us, the reality of our own consciousness, which exists for us just as surely as the other, but does not consist for us of matter or material phenomena.

TELEPHOTOGRAPHIC LENSES

The eclipse of to-morrow should, if it is visible at all, provide an opportunity for the use of telephotographic lenses, with which images of the sun, of satisfactory dimensions, can readily be obtained with quite ordinary cameras. In this case the extraordinary brilliance of the object renders instantaneous telephotography, even at high magnifications, an easy matter, and images over an inch in diameter will be obtainable by amateurs in possession of apparatus not at all costly or difficult to handle. But in any special work of this kind the adoption of one or two simple precautions may greatly help the business of exposure and tend to an improved result.

At very high magnifications it is necessary that both positive and negative elements should be fully corrected for colour. Most modern high-class lenses are so free from chromatic aberration that little or no trouble is likely to be experienced with them in this respect. But with lenses of doubtful quality high magnifications should not be attempted unless the difference between the chemical and visual focus has been ascertained and allowance made—a tedious and unsatisfactory process. If the magnification is very high it is hardly necessary to use smoked glasses for focussing, as the transmitted glare is proportionately diminished, but the precaution may well be adopted in any case.

WHAT IS WRONG WITH THIS POSTER ?

THE Editors have kindly consented to reproduce, though without its lovely colours, a poster discovered by Mr. Saunder, which he exhibited to our delighted eyes a few weeks ago. Please imagine a crimson glow in the sky, and say, from the lie of the coast-line, whether it is sunrise or sunset. Then interpret the appearance of the Moon, and say where the Sun is. Finally, calculate the chances of those of us who are crossing to France to see the eclipse of April 17 finding something wrong with the geography or the tables.

VICISSITUDES OF A TELESCOPE

Although only remotely connected with "discovery," and we trust entirely removed from "invention," an anecdote told at the recent dedication of Allegheny Observatory of Professor Pickering, of Harvard, seems worthy of preservation. The first Director of the Observatory became insane, and, before his removal, held the theory that the great telescope of the Observatory should not be used, a theory which he defended with a shot-gun. After his removal he wrote a poem, in which he predicted that the telescope's object-glass would be stolen. It was—although it was proved that the author of the prediction had no complicity with its fulfilment. Professor S. P. Langley, whose bolometer is only less well known than his classic experiments on the possibility of flight, was the next Director, and for a very long time carried out negotiations by post with the thief in order to get the object-glass back. The thief demanded a ransom. Professor Langley and the Trustees of the Observatory resolutely refused to give it, for otherwise every object-glass in the country would have been at the mercy of thieves. At length Professor Langley and the thief met by arrangement in the woods, and held a consultation, which ended by the thief saying to the Professor, "You and I are gentlemen, and we must trust each other." The glass was returned without any money being taken. The late Admiral Bob Evans said of curio-hunters that they would take anything except a cellar full of water, but one would have thought that the object-glass of a large telescope was almost equally outside predatory action.

MORE MARTIAN CONTROVERSY

A dispassionate article on Mr. E. W. Maunder's reprehensible book entitled 'Are the Planets Inhabited?' appears to-day in the *New York Times Book Review*. We call the book reprehensible, although the reviewer treats it with respect, because one of the author's avowed objects is to discredit the astronomical verity of Prof. Lowell, the American discoverer of Mars, deride his visual power, and poke fun at his engaging theories. Mr. Maunder, who gets his living by superintending the solar department of the Royal Observatory at Greenwich, does not believe that Mars is inhabited. Dr. Johnson, an earlier countryman of Mr. Maunder, did not believe that man could be carried over the Earth's surface at as high a rate of speed as twenty miles an hour, arguing convincingly that nobody could breathe while travelling at that speed. Mr. Maunder's argument is as convincing as Dr. Johnson's. Because Mars is ice-bound every night and its temperature is never much above the freezing-point in the daytime, this unimaginative Englishman refuses to believe that life can exist on that planet. He is entitled to his opinion, but when he proceeds to play havoc with the observations of Prof. Lowell, he must excite the ire of every true American. *　　*　　*　　*　　*

If Mr. Maunder continues on this course he may yet disturb the friendly relations of England and America.

PHOTOGRAPHY BY X-RAYS

At the present day X-ray photography has become so much a matter of ordinary routine at all big hospitals that it is no longer regarded with the wonder it originally called forth. The remarkable properties of the X-rays when they first became known were looked upon as inevitably leading to still fresh discoveries, but except from the standpoint of physical science little of general interest has developed, and the main progress made has been in their application to medical purposes.

To-day a radiograph can be taken through the thickest parts of the human body instantaneously—in the 200th part of a second. A few years ago an exposure of 20 minutes for a radiograph of a knee-joint or elbow was not remarkable. The methods of locating foreign bodies, of treating skin diseases, of isolating the rays and directing only where required have all attained a high point, hardly of perfection, but of technical soundness. The field of X-ray apparatus has grown to a surprising extent, and an industry has sprung up of the extent of which the general public has little or no idea.

One of the most interesting advances of the last year is perhaps the single flash apparatus, with which instantaneous radiographs can be obtained, giving very sharp definition in the photographs and, of course, a minimum of discomfort to the patient. This result—the high-water mark of modern X-ray science—is brought about only by a combination of all the most recent advances in radiographic apparatus, although some time ago Dr. Rosenthal and others obtained practically instantaneous results by the employment of very large apparatus and heavy currents. Apart from a specially wound induction coil taking 60 amperes in current and provided with a very heavy secondary winding of low resistance, a particular form of interrupter has been employed which produces a sudden discharge of current from the secondary winding of the coil to the X-ray tube of a duration of between 1-100th and 1-1,000th second, a larger amount of energy being discharged owing to the heavy iron core used in the construction of the coil.

The X-ray tubes used for these modern heavy discharges represent a distinct scientific achievement in themselves. The stream of rays from the cathode impinge upon the anticathode and are thence reflected down upon the object under treatment. A second or two will suffice to make even a heavy anticathode red hot, and various means of cooling have been adopted, such as forced air draught, water cooling, and irradiation. In certain recent tubes the cathode itself is cooled by special devices, so great is the heat produced.

The metal reflector used in the anticathode to deflect the rays from the tube has been a subject of considerable research, and a new material for the reflecting plate has been found in tungsten, the melting point of which is about 2,600deg. C. Tubes of this metal will withstand the heaviest currents, and its application is likely to prove extensive.

LOW TEMPERATURES

Low temperatures, as Sir James Dewar explained, were now of commercial significance, and not confined to laboratory experiments; very large quantities of liquid air were produced annually at the present time for industrial work dependent on intense cold.

The lowest temperature ever experienced on the earth, he said, was that of boiling carbon dioxide, —80deg. on the absolute scale, which had been met with on a high plateau in Siberia; this was considerably lower than any temperature met with in the upper atmosphere. A solid block of frozen carbonic acid gas, immersed in water, was projected on the screen, when the intense ebullition was seen to advantage, showing the interesting phenomenon of a boiling solid.

The wonderful changes in physical characteristics of bodies subjected to very low temperature was well demonstrated by showing the power of an egg and a candle to become luminous after having been subjected to a powerful light. The egg was placed in liquid air, and the rays from an arc lamp concentrated upon it; it was then removed, and when held in the darkened lecture theatre it was seen to glow brightly. A tallow candle treated the same way also glowed in the dark, the luminosity gradually dying away as the candle regained warmth.

The part played in the upper atmosphere by electrical charges was beautifully shown by moistening a highly cooled rod which had been electrified by rubbing it with a dry handkerchief. Snow crystals rapidly formed on the surface of the rod, and were seen in course of growth by projecting the image of the rod upon the screen. The delicate crystals, themselves charged with electricity, would occasionally fly off briskly through their breaking away from the rod and then falling victims to electrical repulsion.

Sir James Dewar concluded by showing some properties of liquid hydrogen contained in an outer vessel of liquid air; pieces of cork dropped into the liquid immediately sank owing to the lightness of the fluid hydrogen, while pith itself only floated if it absorbed sufficient fluid in its pores to make it buoyant.

THE SUMMER METEOR SHOWER

Every year an extensive and rich stream of meteors encounters the earth and distributes its particles into the atmosphere between the middle of July and end of the third week in August.

Anyone watching the heavens on clear, moonless nights during the period named may expect to observe several meteors in a brief period. The members of the chief shower may be known by the streaks they leave behind on their paths and by their directions from the constellation Perseus, which is situated in the north-east sky in the evening hours.

Meteors are nightly increasing in numbers and will continue to do so until about August 11 and 12, when the maximum is usually attained. Those who watch the firmament on the dates mentioned may count on perceiving at least 30 or 40 meteors per hour. These will not all be comprised of the class known as shooting stars, but will probably include several fireballs of rather brilliant type and magnificent aspect. They are well worth witnessing, and their lustrous flights produce an agreeable impression not soon forgotten.

An eminent mathematician once computed that in round numbers there must be about 300,000,000,000,000 meteors in the stream we are now encountering. We cannot adequately comprehend so vast a number, but can well believe that such an immense system can suffer little appreciable diminution, if any, by its annual expenditure of material upon the earth. Every year there are probably more than 50 millions of these meteors destroyed in our atmosphere, but this is little relatively to the total number of which the stream consists.

CATALOGUING THE STARS

Observations with the altazimuth have been made with the travelling wire micrometer and printing micrometer for declinations. The result of the introduction of this method has been practically to eliminate the "personal equation" of the different observers. It was found that the observer who had previously been selected as the standard observer by the tapping method had a "personal equation" of a quarter second. This correction has been introduced into the Greenwich time signals, which brings them into almost perfect accord with those received by wireless telegraphy from Paris and Norddeich, Germany.

Special attention has been paid to the determination of the photographic magnitudes of the stars on the series of plates of the entire sky taken by the late Mr. Franklin Adams at Mervil Hill, Surrey, and the Cape of Good Hope. This work has been carried out by Messrs. Chapman and Melotte, and published in the memoirs of the Royal Astronomical Society. They find that there are 689 stars in the whole sky of photographic magnitude 5 or brighter, 272 thousand of magnitude 10 or brighter, and 55 million of magnitude 17 or brighter; the faint stars are relatively much more dense in the Milky Way, as was already known. An attempt has been made to continue the series by extrapolation, and the authors conjecture that the total number of stars in our universe is about 1,100 millions, and that half of them are brighter than the 23rd magnitude.

The feature of the sunspot observations has been the abnormal quiescence of the sun. The mean area of spots appears to have been smaller than in any year of the past century. There are now signs of a recommencement of activity, spots of considerable size having been seen in March and April of the present year. The recent spots have been in high solar latitude, showing that they belong to the new cycle, not to the expiring one.

OBSERVING THE SUN WITH BALLOONS

SUPPOSE one sent up five valuable instruments by sounding-balloons, how many might one hope to recover, in a general way? It is good news from Mr. C. G. Abbot that he recovered all five of the pyrheliometers sent up in August last by sounding-balloons; and that all five of them have readable records, three of them very good. The highest altitude attained by pyrheliometer records was 50,000 feet, but the balloons went, of course, much higher. "We expect to repair the instruments," he writes, "attach constant temperature regulating devices, and send them up again next May. We hope then to get much better records, to at least 100,000 feet altitude, perhaps 125,000."

PROBLEMS OF LECTURING

Sir Robert Ball used to tell with relish an amusing story against himself. Visiting Stratford-on-Avon to give a lecture, he said to his landlady at dinner: "I will give you a lesson in Astronomy, Madam. Have you heard of the great Platonic year, when everything must return to its first condition? In 26,000 years we shall be here again, eating a dinner precisely like this. Will you give me credit till then?"

"Yes," was the prompt reply. "You were here 26,000 years ago and left without paying. Settle the old bill and I'll trust you with the new!"

Sir Robert also told a story of how, arriving in a remote town in Ireland to give a lecture, he could not find the promised conveyance. Presently, all the other passengers having cleared off, he was approached by an Irish servant, who rather timidly inquired whether he was Sir Robert Ball. Receiving the affirmative answer, he burst forth, "Sure, I am sorry to have kept you waiting, but I was told to look for an *intellectual* gentleman."

SAVED BY METEOR

Fiery Bolt from Sky Kills Fish for Starving Castaways.

(*From Our Own Correspondent.*)

LONDONDERRY, Saturday.

The four Irish fishermen marooned for 18 days on the uninhabited island of Roaninish—whose rescue was reported in "The Star" yesterday—described how they were providentially saved from starvation by a meteor.

Carried away in their small boat by the gale early last month, they were cast upon the islet, where there was neither food, fuel, nor shelter.

Interviewed after his rescue yesterday, one of the party said: "We had four days' provisions, so that we did not suffer from want of food till the end of the first week; but for the next seven days we were in great straits.

"We had sufficient fresh water, but nothing to eat, so we cut off pieces of overcoat, put them in our pipes and smoked them. This abated the pangs of hunger.

"On the tenth night of our stay on Roaninish, during the height of a furious gale, a great blazing meteor fell into the sea a mile to the westward.

"The next morning quantities of dead fish came from that direction."

COLOUR PHOTOGRAPHS

Some examples of "Kodachrome," a new process of photography in colours, were shown at the Royal Photographic Society yesterday. The process is the discovery of Mr. Kenneth Mees, an Englishman living in the United States.

The new method consists in exposing two plates, sensitized for all colours, either simultaneously or in quick succession, one through a red filter, the other through a green filter. The plates are developed and fixed in the usual manner, after which the black silver forming the image is removed by simple chemical means. The plates are then dyed—that taken through the red filter in a green dye, and that taken through the green filter in a red dye. The plates take up the dyes in inverse proportion to the original silver deposit. They are then brought face to face and bound up in register, the combination of the two-colour images by transmitted light giving a reproduction of the original.

In the portraits shown the flesh tints were especially well rendered, and the absence of the screen rulings and dots of the usual three-colour process undoubtedly adds to the effect. From the commercial point of view the most important factor is that any number of copies can be reproduced from the original pair of negatives. In that case the original negatives are preserved as negatives, master glass positives being made from them by contact in the usual way, and from these the required number of negative duplicates.

It is also claimed that the process only adds two very simple operations to the ordinary method of dealing with a photographic plate—namely, the bleaching and dyeing, and also that it has the fundamental advantage that the plates can be retouched, etched, and air-brushed, thus giving the artistic operator full scope for his abilities.

LIGHT FROM ELECTRICITY

Dealing with the production of light by electricity, he showed that a platinum wire through which a current is passed glows more and more brightly as its temperature rises, until it melts, and he explained that the defect of that metal as a producer of light is that it melts before its gets hot enough to give a good amount of light. The advantage of carbon as a material for electric lamps is that it can be heated to a much higher temperature than platinum before it melts, or rather volatilizes, but the vessel in which it is enclosed must be exhausted of air else it burns away. Tungsten, the metal now commonly used for incandescent lamps, does not melt at a temperature that would volatilize carbon, and it has lent itself to a still greater increase of efficiency in the half-watt lamp, the filament of which is enclosed, not in a vacuous space but in an inert gas like nitrogen. Other methods of producing light from electricity referred to were the electric arc, discovered by Sir Humphry Davy, also in the Institution, and the mercury vapour lamp.

Finally, the lecturer turned to the employment of electricity for heating and cooking, and after exhibiting a variety of appliances for use on the breakfast table and in the kitchen, concluded with a reference to the relative cost of gas and electricity for such purposes.

THE FLASH OF THE DIAMOND

He began with some experiments illustrating the optical laws of reflexion and refraction, and then gave examples of their practical application. The first was the "reflex light" sometimes used by bicyclists to show a light behind them, although they carry no rear lamp, and he explained that this is composed of a lens and a curved mirror, the effect of the latter being to send the reflexion back faithfully to the source of light, say, the lamps of a following motor-car. The mirror must be curved because a plane one would not reflect the light back to the source.

In order to illustrate the brilliance and flashing of precious stones, which depend on reflexion and refraction, Professor Boys used a diamond ornament lent by Lady Dewar. In a brilliant-cut diamond, he said, the facets are so arranged that practically all the light comes out in front of the stone by total reflexion, and when the back of the stone is looked at it is seen to be dark with the exception of a small window in the centre—the cullet. The colour effects are due to the refraction of some of the light.

After a reference to methods of comparing the intensity of different sources of light and a brief explanation of the principles of the grease spot and shadow photometers, the lecturer referred to the gradual improvements that have been made in the efficiency of gas lighting. He explained that with the introduction of the Welsbach incandescent mantle the light obtainable from a given amount of gas was increased sixfold as compared with the old flat-flame burner, and a well-adjusted inverted burner, which has several advantages over the vertical form, would give even better results. But the highest development of gas lighting is the high-pressure system, in which the gas is supplied at a much higher pressure than in ordinary cases, and this system represents the most perfect street lighting known, though possibly the future in this sphere may lie with the half-watt electric lamp.

Dealing with the modern science of illumination, Professor Boys explained that people do not want a source of light to shine directly in their eyes, and he reviewed the methods employed to avoid this result—shades of translucent material, reflecting globes cut on the principle of the total reflexion prism, and indirect lighting, where the source of the light is not seen at all and the illumination comes by reflexion from the ceiling and walls of the room.

After showing the different effects obtained by mixing coloured light and mixing pigments, the lecturer explained that some of the most brilliant colours in nature, such as those of the peacock's tail or the butterfly's wing, are not produced by pigments at all, but are the result of the structure of the object.

SCIENTIFIC GERMAN GUNNERY

He said that less than six months ago the artillery officer would have said there was no such thing as mathematics in artillery science. But that outlook was now ancient history. This was a mathematical war. Drawing upon his experience as a professor of artillery theory for instances where science could prove itself useful on service, he explained how particulars of the enemy's guns could be deduced from fragments of the wall of a shell and photographic pictures. From a fragment they could determine whether a shell came from one of the 42-centimetre howitzers, the very existence of which still appeared in doubt. Dealing with the calculations for ascertaining how far men should stand from a gun to avoid the danger of permanent deafness, he said they need not fear to stand 12 yards behind the 42-centimetre howitzer; and so the story was discounted of the firing party taking cover 100 to 200 metres away when this howitzer was fired. An application of the theory of the conduction of heat would have reassured our men that life in the trenches would not be too cold, or would at least be warmer than in the frost above, provided only the floor could be drained dry under foot. It had also to be borne in mind that the trench gave better cover than a tent.

Five years ago he had an invitation to Berlin to visit the Military Technical Academy there. It was a magnificent institution such as we could not afford, so our rulers assured us. Professor Cranz showed how in his department no money was spared in recent equipment, including a bomb-proof range available for artillery fire and yet in the heart of a big city. There were plenty of outdoor artillery ranges also to visit, where instructive work was in progress. The Perry system of education was adopted in Berlin. After a lecture on wireless telegraphy the class was set to work, as he saw, in making the antenna which had played such an important part in the war. Sixty officers were under instruction at a time for a course of three years, and he was assured their zeal was admirable. It was considered such bad form not to give the best in return for the honour and glory of the Fatherland. But our Regular was apathetic by comparison. We must put our trust in the junior ranks to push old Apathy from his stool and carry us through this war.

It was a mournful contrast to revert to Woolwich, shabby and undisciplined. There they had been evicted from their proper home and were told to found a new artillery college with the choice of a cellar under some stables or a kitchen and scullery and bare walls in a deserted hospital, there to organize victory and at no expense. (Laughter.) With the courage of an Austrian general compelled to maintain his muzzle-loading musket a match for the Prussian needle-gun, the Military Director assured them that there was nothing superior to be found at Greenwich, in the Naval College there lodged in the old Palace. Such dismal, penurious surroundings had a disastrous effect on the *genius loci*, and they never really recovered from a downhearted spirit not calculated for victory. Our military science was under the rule of Thumb, the official genius. His fumbling method was considered a match for disciplined theory.

We saw already how the cost had been well laid out in this war of the Berlin Military Technical Academy, the German jumping off with a lead he was able to keep so far. The finished article of the Academy was employed in the dissemination of true theory and in the scientific direction of warlike preparation as at Krupps. Assuming everything for the best for the Allies, and if we lived to go in again at Antwerp, an interesting match would be watched between our artillery science and the German, to see how long it would take us to get the other side out, compared with our own innings and the time we kept our wicket up. No long-range fire, he had been assured, was ever going to be of any use again, involving theoretical calculation. The word was, " Gallop up close, to 400 yards, and let them have it."

URANIA PUTS HER HOUSE IN ORDER

[The "new planet" discovered by the Barcelona Observatory turns out to be an old one which had been lost for 36 years.]

WHERE *have* you put that planet, Jane?
　　Lord, what a girl you are!
Don't tell me it's the cat again
　　That's gone and cracked a star!
Have you turned out the Milky Way?
　　It might have rolled in there,
Or on our last spring-cleaning day
　　Got muddled with the Bear.
Just run the carpet-sweeper round
　　The dark side of the Moon—
That planet simply *must* be found,
　　Or you must pay the tune.
What's that? It's just turned up in Spain?
　　Well, keep a better eye on
The planets in the future, Jane—
　　Now go and dust Orion.

SELF-MADE PICTURES OF ORGANS

A discovery which is, to say the least of it, of surpassing interest to doctors, and indeed to the world at large, is now on its trial in France. No public statement has been made about it, yet, according to the *British Medical Journal*, the work, "though very quietly performed, has given rise to a host of rumours, surmises, and conflicting views."

This is not surprising when it is understood that the discovery is quite as important as was the finding of X-rays, and much more remarkable. With the X-ray, in a darkened chamber, it is possible to obtain a picture of the bones and more solid structures, but by means of this discovery it is possible in broad daylight to obtain detailed pictures of any organ of the body, brain, liver, kidneys, spleen, and to see at a glance by what gross lesions they are affected. Thus it is possible to see the blood vessels in the brain, to observe a blood-clot in that organ, to detect abscesses in the liver and wounds or cuts in any organ. In one case a "concretion" in the appendix was seen clearly when the picture was complete.

The miracle is achieved by utilizing the currents of electricity which the body is believed to generate. This conception is not new, and we owe to it instruments like the electro-cardiograph. The difficulty was to make the current coming from any organ active and dynamic, i.e., to cause it to reveal its message, if that popular figure may be used. It is impossible here to enter into details, but it may be said that two electrodes, each ending in a perforated zinc plate or zinc wire screen, are employed. These are connected with batteries in the ordinary way, but are not placed in contact with the patient's body. One screen stands on a pedestal in a vertical position near the patient, the other is hung horizontally at right angles to the first one. In this way any electrical field coming from the first screen is always at right angles to that of the second.

The patient is placed with the part to be examined quite close to the first screen, and directly under but at a distance from the second. The current is turned on. Then the third electrical element, the current from the patient's own organs, is released. "The inventor," says the *British Medical Journal*, "believes that the results are primarily due to the fact that the process interposes between two alternating electric fields of equal strength—and at the precise point where they meet—a third electric field, whose facultative potential force is thus released, and can be converted into dynamic power." The third field is made to operate a sensitive needle, which works upon a revolving cylinder carrying waxed paper. The so-called "hammer needle" moves across the cylinder, tapping out little holes in the wax, and when later the wax is held up to the light a diagram is seen which "precisely resembles the outline of the living tissues lying vertically below the second screen." This diagram can be converted into an ordinary photograph by processes familiar in applied arts.

The actual finished photographs are remarkable, and show the blood vessels in the brain in detail, also wounds, but not the actual brain substance. In other words, differences, such as diseased areas, are shown.

The process of producing these pictures is not impressive. "There is no darkening of rooms, no flashing of lights, and no crackling of spark-gaps. In fact, the whole proceeding is so brief and seemingly so simple that when the results are observed the first sensation is one of bewilderment. A patient is laid on a plain deal table (insulated by standing it on glass), a little clicking is heard in a cupboard hard by, and after 60 seconds or so the bearers are directed to remove him. Nothing has been felt by the patient, little or nothing has been seen by the bystanders beyond what has been noted, yet a visible record of the outline of a living organ has been conveyed to a wax sheet."

The apparatus has to be specially timed for each organ. Thus the electrical force residing in blood is very small, so when blood vessels are to be delineated the alternations must be very rapid. Contrariwise, since the electric force residing in heart muscle is great, the alternations must be slow when a heart is to be delineated. The advantage of this is that when blood or pus occurs in areas where it should not be it at once shows on the picture as a blur, it is not delineated, and so its presence is made evident. The apparatus works so long as the body is alive and even till molecular death has occurred. But after that it does not work.

PLANCK ON 'KULTUR'

The well-known proclamation "To the World of Kultur," which, with the signatures of 93 German scholars and artists attached, was published in August 1914, has by its language, as I have repeatedly found to my sorrow, given occasion for unjust representations of the opinions of its signatories. According to my personal conceptions, which are also to my knowledge shared in all essentials by many of my colleagues, such as MM. Adolf von Harnack, Walter Nernst, Wilhelm Waldeyer, Ulrich von Wilamowitz-Möllendorff, that proclamation, which in its composition reflects the patriotic excitement of the first weeks of the War, must and could only have the significance of an act of defence: above all things, it was a defence of the German army in face of the bitter outcries raised against it, and a definite avowal that German scholars and artists would not separate their interests from the interests of the German army. For the German army is simply the German people under arms, and like the men of all other callings, so are also the scholars and artists inseparably bound to it.

That we cannot, of course, be responsible for every single act of every German, whether in War or Peace, I will gladly also specially emphasize, although that appears to me to go without saying, just as that we are not yet in possession of a final verdict, in a scientific sense, on the great questions of present history. Where the primary responsibility will be permanently assigned for the wreck of endeavours for Peace, and, for all the consequent trouble to mankind, can only be determined later from a comprehensive review of the evidence, the result of which we await with quiet conscience.

At present there is for us Germans, so long as this War shall still last, but one task set—how to serve the Fatherland with all our powers. But what I want further to emphasize, especially to yourself, is my firm conviction, not even at any time to be shaken by the events of the present war, that there are realms of the spiritual and moral world which lie beyond the strife of peoples, and that an honourable co-operation for the care of these international benefits of Kultur, no less than personal regard for the subjects of an enemy country, is in no way inconsistent with warm devotion and strenuous work for one's own Fatherland.   DR. MAX PLANCK.

NEGLECT OF SCIENCE AND FAILURE IN WAR

With a full sense of responsibility we submit the following memorandum on a subject which, we are convinced, requires immediate attention and drastic action. It concerns the public interest, and the public alone can deal with it.

It is admitted on all sides that we have suffered checks since the war began, due directly as well as indirectly to a lack of knowledge on the part of our legislators and administrative officials of what is called "science," or "physical science." By these terms we mean the ascertained facts and principles of mechanics, chemistry, physics, biology, geography, and geology. Not only are our highest Ministers of State ignorant of science, but the same defect runs through almost all the public departments of the Civil Service. It is nearly universal in the House of Commons, and is shared by the general public, including a large proportion of those engaged in industrial and commercial enterprise. An important exception to this rule is furnished by the Navy, and also by the medical service of the Army; in both these services success has been achieved by men who, while in no way inferior in courage, devotion, and self-sacrifice to their brethren elsewhere, have received a scientific training.

This grave defect in our national organization is no new thing. It has been constantly pressed upon public attention during the last 50 years as a cause of danger and weakness. In the whole history of British Governments there has been only one Cabinet Minister who was a trained professional man of science—the late Lord Playfair.

It is not our intention here to enumerate the catalogue of specific instances in which a want of understanding of "physical science" has led the Ministry and Executive into error. This has been done elsewhere, but as an example of the ignorance which we deplore we may instance the public statement of a member of the Government, unchallenged when made, that his colleagues should be excused for not having prevented the exportation of lard to Germany, since it had only recently been discovered that glycerine (used in the manufacture of explosives) could be obtained from lard. The fact is, on the contrary, that the chemistry of soap-making and the accompanying production of glycerine is very ancient history. In order that such serious blunders may be avoided, it is essential that we should have a proportion of men in the Government who, if not actual experts, yet have such a knowledge of science that will give them an intelligent respect for it, and an understanding of what it can do, how to make use of it, and to whom to apply when special knowledge is required.

Our success now, and in the difficult time of re-organization after the war, depends largely on the possession by our leaders and administrators of scientific method and the scientific habit of mind. They must have knowledge, and the habit of promptly applying known means to known ends. To trust to luck is a mark of the dangerous complacency bred of ignorance. The evidence of those back from the front makes it clear that, as of old, our "people are destroyed for lack of knowledge."

How can such a revolution as we desire in the higher and lower grades of the public service be brought about? Obviously it can only be effected by a great change in the education which is administered to the class from which these officials are drawn. The education of the democracy, which gives its consent to the present state of things, would follow the change in the education of the wealthier classes. For more than 50 years efforts have been made by those who are convinced of the value of training in experimental science to obtain its introduction into the schools and colleges of the country as an essential part of the education given therein. At first it seemed as though the effort had been successful, but it is clear that the old methods and old vested interests have retained their dominance, at least as far as our ancient universities and great schools are concerned. At Cambridge but four colleges are presided over by men of scientific training; at Oxford not one. Of the 35 largest and best known public schools 34 have classical men as headmasters. Science holds no place in the list.

The examinations for entrance into Oxford and Cambridge and for appointments into the Civil Service and the Army are among the greatest determining factors in settling the kind of education given at our public schools. Natural science has been introduced as an optional subject for the Civil Service examinations, but matters are so arranged that only one-fourth of the candidates offer themselves for examination in science. It does not pay them to do so; for in Latin and Greek alone (including ancient history) they can obtain 3,200 marks, while for science the maximum is 2,400, and to obtain this total a candidate must take four distinct branches of science. For entrance into Woolwich, science has within the last few years been made compulsory, but for Sandhurst it still remains optional. This college is probably the only military institution in Europe where science is not included in the curriculum. The result of this system of examination, not merely upon the successful candidates, but upon all the great schools and the old universities which necessarily (as things are at present arranged) work with them in aim and interest, is a neglect of the study of the natural sciences, and to some extent an indifferent, not to say contemptuous, attitude towards them. The one and effective way of changing this attitude and of giving us both better educated Civil servants and a true and reasonable appreciation of science in all classes is in the hands of the Legislature, and of it alone.

If a Bill were passed directing the Civil Service Commissioners and Army Examination Board to give a preponderating—or at least an equal—share of marks in the competitive examination to natural science subjects, with safeguards so as to make them tests of genuine scientific education and not an incentive to mere "cram," the object we have in view would be obtained. Science would rise in our schools to a proper position and gain the respect necessary for the national welfare. A popular appreciation and understanding of science would begin to develop; and our officials of all kinds, no less than members of Parliament, would come to be as much ashamed of ignorance of the commonplaces of science as they would now be if found guilty of bad spelling and arithmetic. Not at once, but little by little, the professional workers in science would increase in number and gain in public esteem. Eventually the Board of Trade would be replaced by a Ministry of Science, Commerce, and Industry, in full touch with the scientific knowledge of the moment. Public opinion would compel the inclusion of great scientific discoverers and inventors as a matter of course in the Privy Council, and their occupation in the service of the State.

With this object in view we urge the electorates to insist that candidates for their suffrages should pledge themselves to aid by legislation in bringing about a drastic reform in the scheme of examinations for all the public services in the sense we have indicated.

Our desire is to draw attention to this matter not in the interests of existing professional men of science, but as a reform which is vital to the continued existence of this country as a Great Power.

1917

THE ORBIT OF MERCURY

Speaking at the summer meeting at the Hampstead Garden Suburb last night, SIR OLIVER LODGE said he had been working on the variations in the orbit of the planet Mercury. This orbit revolved slowly, and there was a discrepancy known to all the great astronomers. Le Verrier found that it amounted to 43sec. It had remained a puzzle, and some had believed that it was due to the proximity of another planet, which could not be seen by us and which had been named Vulcan.

Sir Oliver Lodge said he believed the discrepancy was due to the electrical theory of matter. The inertia of a body moving at a high speed was greater than if it were at a standstill. Mercury, the fastest planet in the solar system, travelled 30 miles a second. One pound of matter moving at 30 miles a second, the speed of Mercury, equalled one pound *plus* the hundred-millionth part of a pound, and that fraction was not subject to gravity. This was, he thought, the probable cause of the variation of the orbit of Mercury.

ANOTHER SATELLITE OF JUPITER

A tiny orb, the ninth satellite of Jupiter, was discovered just before the outbreak of war by Mr. Seth Nicholson at the Lick Observatory.

Details of its orbit have lately been received. It goes round Jupiter, in a retrograde direction, in 745 days, or a week longer than the period of the eighth satellite, the two orbits being interlocked. The distance from Jupiter varies between 11 millions and 20 millions of miles. The magnitude of the satellite is 18½; it is the faintest known body in the solar system. Its diameter is estimated to be about 15 miles. Even as seen from Jupiter it would be a faint telescopic star of the 12th magnitude.

Its orbit is inclined about 24deg. to the ecliptic and about 10deg. to that of Satellite VIII. It is rather singular that Satellites VI. and VII. form a similar pair with interlocked orbits and periods almost equal. Their motion round Jupiter is direct, not retrograde.

THE ARABIAN NIGHTS BROUGHT UP TO DATE

During these moonlight nights our Astronomer Royal may thank his stars that we are not a Western Persia with a former Shah for ruler. Were we so we should simply petition such Shah to order an abiding eclipse of the moon, and then, alas for the Astronomer Royal! It is all in the book of records how the Shah, visiting Greenwich Observatory in company with the then Prince of Wales, bade the head of the establishment produce an instant eclipse. Sir George Airey, the official in question, was a little embarrassed.

* * * *

There was nothing doing in the way of eclipses, so he mumbled an apology for the ineptitude of the heavens. "Dog of an astronomer, produce me an eclipse," repeated the Shah. Upon a second gentle refusal, he turned to the future King Edward in regal wrath, saying, "You hear that? Cut off his head!" Sir George survived, but he was aged, and resigned that year. The Shah cannot have suspected that the veteran took his head upon his shoulders with him into retirement.

* * * *

The sequel to the visit of the Shah of Persia to Greenwich Observatory, referred to in this column, is worth telling. As we said, the Astronomer Royal, Sir George Airey, retired shortly after Nassr-ed-Din's return to Persia, and his successor, Sir W. H. Christie, was somewhat surprised to receive shortly after his appointment was announced a letter from the Persian monarch, the terms of which suggested that the writer was under the impression that Sir George Airey had been dealt with as he had proposed to the Prince of Wales.

SUMMER TIME PROPOSAL

In connection with "Summer Time" a remarkable suggestion has been made to the Paris Academy by M. Lecornu (*C. R.*, 6 Feb.). Not liking the sudden jumps of an hour, he proposes to use well-nigh imperceptible changes—say, of 30 seconds a day,—letting the clock slowly gain in one half of the year and lose in the other half. He would not touch the hands of the clock at all—only the pendulum. It is easy enough to see that this would produce the kind of effect required, and that there are several disposable constants: viz., the two dates when the pendulum is altered and the weights added or removed; making four constants connected by the condition that the total length of the year must not be changed. Now, I have not the hardihood, at present at any rate, to support this proposal; but I will state emphatically my opinion that, for our own sakes, we astronomers ought not to refuse to look at it. It is very difficult to distinguish the essential from the merely customary; and personally I have found it a most useful mental exercise to think out fully and completely the consequences of this proposal, both from the point of view of the world at large and from that of Science. More than once I found myself surprised at the turns of thought which followed in sequence.

CHRISTIAN SCIENCE

At a public meeting held in Cambridge Town Hall on Monday, Feb. 5th, a missionary from Warsaw, speaking about "Jews in Russia," gave the following ingenious proof, *à propos* of the statue erected to Copernicus in Warsaw, that Christ was aware of the rotundity of our Earth :—

"Christ said that at His Second Coming two women should be grinding at a mill, of whom one should be taken and the other left; while two men should be sleeping in a bed, of whom also one should be taken and the other left. Now we do not sleep in the daytime. Hence it must have been night where the two men were lying, and daylight where the women were at work. Hence we see that our Lord was aware that half our Earth was in darkness of night, while the other half enjoyed full daylight—*i. e.*, that He knew the Earth to be a globe and not a flat plain."

THE CAUSE OF THE COLD WEATHER

The severity of the present winter (one would have preferred to speak of it in the past tense) appears to have been due, primarily, to the state of barometric pressure over the North Atlantic, and more particularly in the neighbourhood of Iceland.

In a normal winter these regions lie in the pathway of large storm systems, travelling usually from south-west to north-east; and as the central areas of these disturbances pass in numerous instances directly across Iceland, the mean height of the barometer in that island is much lower than in any part of Western Europe. So far as the United Kingdom is concerned, the result of such conditions is seen in a strong drift of air from the south-westward; and, as a further result, this country experiences a succession of winters which are far more often mild than severe.

Records made in London show that of the past 45 winters there were six with a mean temperature differing but very little from the average. These may be classed as normal or ordinary seasons. Of the remaining 39 there were 23 with a mean temperature appreciably above the average, and of these five were unusually mild. The latter comprised the winters of 1876-7, 1897-8, 1898-9, 1912-13 and 1915-16. On the other hand, only 16 winters of the 45 had a mean temperature appreciably below the normal, and of these only six could be regarded as severe. The severe winters were those of the three successive years 1878-9, 1879-80, and 1880-1, together with those of 1890-1, 1894-5, and 1916-17.

Under anything like normal conditions the winters in this country have therefore a tendency to be mild rather than cold. Bearing this in mind, it is not surprising to learn that during the past four months the state of barometric pressure over the North Atlantic has been a long way from the normal. In place of cyclonic systems moving in a more or less regular north-easterly course the upper part of the ocean has been covered, far more often than not, by a large anticyclone, and in Iceland the barometer has been almost persistently above its average level. As a result of these exceptional conditions the normally mild south-westerly winds have been replaced by a current of air sweeping down from some point between north-west and north-east, and the temperature in this country has been, as we all know, greatly below the average. It is worthy of note that in the rare intervals during which a more normal state of things existed over the ocean the United Kingdom enjoyed some respite from the prevailing harshness. In the last 10 days of December, for example, the barometer in Iceland sank to about its normal level, and at the end of last year and the beginning of this the United Kingdom experienced about a week of unusually mild weather. The very brief run of mild days which prevailed around the middle of last month was also accompanied by a low or normal barometric pressure over the North Atlantic.

There can be no doubt that when the meteorologist is in a position to predict for three or four months, even roughly (and the thing seems within the range of probability) the mean height of the barometer in Iceland, he will also be enabled to predict the general character of the weather over the United Kingdom for a whole summer or winter, as the case may be.

FOCH'S TIME LIMIT

The time-limit fixed by Marshal Foch for a decision on the armistice terms—"eleven o'clock Monday morning, French time"—led more than one newspaper astray. The *Daily Express* interpreted it as equivalent to 10.51 A.M. English time, while the *Daily Chronicle* gave it as 11.10 A.M. English time. As a matter of fact, both were wrong, the *Chronicle* particularly so. Theoretically there is a difference of 9 min. 21 sec. between the time of the Paris and the Greenwich meridians, Greenwich being that interval slow compared with Paris; so that 10.51 Greenwich time would be the equivalent of 11 Paris time to the nearest minute. But the French no longer use Paris time. By the law of March 10, 1911, the "heure légale" throughout France was defined as the "temps moyen de Paris retardé de neuf minutes vingt-et-une secondes," which is, of course, the time of the Greenwich meridian. Eleven o'clock French time this morning was, therefore, exactly the same instant as eleven o'clock English time.

WHAT THE METEORS TEACH US

IN No. 11 of *The Chaldæan*, which has just appeared, there is an interesting article on "What the Meteors teach us," opening with the remark that they generally seem to disappear at a height of 80 km., where there is apparently a surface of separation, which may enter into twilight effects and the reflection of sound. Now when the "luminous night-clouds" (leuchtenden Nachtwolken) appeared after Krakatoa, they were found to be steadily at the height 82 km. during the years 1885–1891 (see *Ast. Nach.* No. 3347). If this confirmation of the surmises from meteor observation has already been noticed, it is curious that no mention of it is made in the article referred to. Possibly, therefore, it has been hitherto overlooked. Taking the proportion of free hydrogen present in the atmosphere at the Earth's surface as 1 in 100,000, it is found, by more than one investigator, that outside a radius of about 80–100 km. our atmosphere would be practically all hydrogen. It seems possible that the closely accordant measures of the height of the "luminous night-clouds" give us a definite inner radius for this envelope of hydrogen.

A LONG - RANGE GUN

The long-range gun which shells Paris is not so colossal as fancy painted it, says the German military writer von Ompteda in the illustrated weekly paper *Die Woche*, quoted by the Paris *Daily Mail*.

It is not a single gun, but a series of guns, of which some are held in reserve. To make the shell travel through as little air as possible, the gun is inclined at the greatest possible angle, and the low temperature which prevails at the height reached by the shell prevents it becoming red-hot from the friction of air. Calculations made on meteors by astronomers have given indications of atmospheric conditions at great heights, and these served as a basis of calculations for the "Paris gun."

NOVA AQUILAE – HOW TO IDENTIFY THE NEW STAR

The new star, which has made its appearance in the constellation of the Eagle, and hence has been named Nova Aquilæ, is at the present moment the brightest star in that region, but it will probably soon become fainter than Altair. It is likely to be seen clearly with the naked eye for several weeks, but will become much less conspicuous. This is the second new star to be observed this year. The first was noted in February, but, being much fainter, did not attract so much attention. The new star appears in the southern stream of the Milky Way, almost on the Celestial Equator.

It is interesting to note that it is in the Milky Way that practically all new stars make their appearance. The Milky Way is the most distant

(2550)

portion of our system. Although the star has made its appearance now for the first time, it does not follow that it is a new one in the sense of recent formation. On the contrary, it is quite probable that it is the product of a collision between two other stars thousands of years ago, certainly centuries ago. Another point about it is that, like all others of its kind, it is purely temporary. In a few weeks, in all probability, it will fade away and be seen no more.

Although its light will soon begin to fade, said Mr. Denning, the star would remain perceptible through telescopes as a small star varying in its light and possibly involved in a certain amount of nebulosity. These stars, which suddenly blaze up and thereafter dwindle after a few weeks' brilliant display to almost imperceptible points of light, convey the impression of a great disruption or disturbance. The distance of the stars is enormously beyond our powers of conception, and the outburst we behold in the sky must represent a conflagration or luminous upheaval on a stupendous scale. The astronomical world has been furnished with an object which will amply provide it with work, both of an observation and of a theoretical kind, for some time.

A number of theories have been formed to explain the phenomena of new or temporary stars. Seeliger thought they were due to the movement of a dark body through nebulous matter. Vogel considered that the appearance

of a new star could be accounted for on the supposition that an encounter occurred between a dark star and a decayed planetary system. In the case of the new star which made its apparition in Auriga in 1892, Huggins thought the outburst attributable to the near approach of two orbs having great velocities, the resulting tidal disturbances inducing great outbursts. Flammarion considered that a body involved in an atmosphere of hydrogen would, if it grazed a dark body surrounded by oxygen, at once occasion an explosion of sufficient magnitude to bring about all the appearances presented by a new star.

The present is certainly the most brilliant temporary star that has shone in the firmament since the celebrated one of 1604. There have been many new stars in that long interval but they have been fainter, and generally much fainter, except in the case of the one which made its appearance in Perseus on February 22, 1901. The new star will assuredly form one of the chief astronomical events of our time. It must represent an occurrence, possibly a catastrophe, of enormous magnitude in the inconceivably remote fields of sidereal space.

THE 100 INCH TELESCOPE AT MOUNT WILSON, CALIFORNIA

NEWTONIAN IDEAS OVERTHROWN

Yesterday afternoon in the rooms of the Royal Society, at a joint session of the Royal and Astronomical Societies, the results obtained by British observers of the total solar eclipse of May 29 were discussed.

The greatest possible interest had been aroused in scientific circles by the hope that rival theories of a fundamental physical problem would be put to the test, and there was a very large attendance of astronomers and physicists. It was generally accepted that the observations were decisive in the verifying of the prediction of the famous physicist, Einstein, stated by the President of the Royal Society as being the most remarkable scientific event since the discovery of the predicted existence of the planet Neptune. But there was difference of opinion as to whether science had to face merely a new and unexplained fact, or to reckon with a theory that would completely revolutionize the accepted fundamentals of physics.

Sir Frank Dyson, the Astronomer Royal, described the work of the expeditions sent respectively to Sobral in North Brazil and the island of Principe, off the West Coast of Africa. At each of these places, if the weather were propitious on the day of the eclipse, it would be possible to take during totality a set of photographs of the obscured sun and of a number of bright stars which happened to be in its immediate vicinity. The desired object was to ascertain whether the light from these stars, as it passed the sun, came as directly towards us as if the sun were not there, or if there was a deflection due to its presence, and if the latter proved to be the case, what the amount of the deflection was. If deflection did occur, the stars would appear on the photographic plates at a measurable distance from their theoretical positions. He explained in detail the apparatus that had been employed, the corrections that had to be made for various disturbing factors, and the methods by which comparison between the theoretical and the observed positions had been made. He convinced the meeting that the results were definite and conclusive. Deflection did take place, and the measurements showed that the extent of the deflection was, in close accord with the theoretical degree predicted by Einstein, as opposed to half that degree, the amount that would follow from the principles of Newton. It is interesting to recall that Sir Oliver Lodge, speaking at the Royal Institution last February, had also ventured on a prediction. He doubted if deflection would be observed, but was confident that if it did take place, it would follow the law of Newton and not that of Einstein.

Dr. Crommelin and Professor Eddington, two of the actual observers, followed the Astronomer-Royal, and gave interesting accounts of their work, in every way confirming the general conclusions that had been enunciated.

So far the matter was clear, but when the discussion began, it was plain that the scientific interest centred more in the theoretical bearings of the results than in the results themselves. Even the President of the Royal Society, in stating that they had just listened to "one of the most momentous, if not the most momentous, pronouncements of human thought," had to confess that no one had yet succeeded in stating in clear language what the theory of Einstein really was. It was accepted, however, that Einstein, on the basis of his theory, had made three predictions. The first, as to the motion of the planet Mercury, had been verified. The second, as to the existence and the degree of deflection of light as it passed the sphere of influence of the sun, had now been verified. As to the third, which depended on spectroscopic observations, there was still uncertainty. But he was confident that the Einstein theory must now be reckoned with, and that our conceptions of the fabric of the universe must be fundamentally altered.

At this stage Sir Oliver Lodge, whose contribution to the discussion had been eagerly expected, left the meeting.

Subsequent speakers joined in congratulating the observers, and agreed in accepting their results. More than one, however, including Professor Newall, of Cambridge, hesitated as to the full extent of the inferences that had been drawn and suggested that the phenomena might be due to an unknown solar atmosphere further in its extent than had been supposed and with unknown properties. No speaker succeeded in giving a clear non-mathematical statement of the theoretical question.

AETHER AND MATTER

Lecturing at the Royal Institution yesterday, Sir Oliver Lodge said that we were all moving through the aether of space at a rate which he roughly guessed as being 19 miles a second We were really ignorant of the speed and of the direction. The rate of motion might come to be measured, although the doctrine of relativity, which was at present complicating physics, seemed to allege that measurement was impossible, as the instruments themselves were stretched or shortened in accordance with their position along the direction of motion or across it. The exponents of relativity had had the courage to predict that a ray of light from a star, grazing the sun, would be deflected one and three-quarters of a second of an arc, and were going to test this by observations at the solar eclipse visible in Brazil next May. Sir Oliver ventured on a counter-prediction—that if there were any deflection at all, it would not be more than three-quarters of a second. If even this deflection occurred, he should think that it merely established that a ray of light had weight.

In conclusion, Sir Oliver Lodge said that he would pass from facts on which physicists were generally agreed to theories which were still in doubt. One of the most interesting of these was the conception developed by Professor Rutherford, and Mosely, a brilliant young physicist who had been killed in Gallipoli, as to the constitution of the atom. An atom consisted of a central nucleus with one or more electrons revolving in orbits around it. The nucleus consisted of positive electrons, the planetary electrons being negative. Each atom had the same number of positive and negative electrons, and these ranged from 1 in the case of hydrogen, to 92 in the case of uranium. For each of the intermediate numbers the existence of a distinct element might be inferred, and in actual fact there were at present only four gaps in the series, four elements yet to be discovered. The likeness between the constitution of an atom and that of our solar system was plain and it extended not only to general structure but to many other physical properties of the systems, so that we might speak of an astronomy of the stars and an astronomy of matter.

WIRELESS-STEERED BOMBER

an American correspondent describes a conversation he has had with M. A. G. H. Fokker, the Dutch inventor whose name is associated with the now famous Fokker aeroplanes.

" If the war had gone on for several more years, how far would the aeroplane have developed ? " I asked M. Fokker.

His answer revealed what he assured me had been a great military secret. " We would have put the artillery out of commission," he said, " We would have made the big guns as old fashioned as spears. It was all the fault of the Army ' red tape ' in Berlin that it was not begun sooner.

" It was like this : In 1916 the Army authorities asked me if I could make a very cheap aeroplane with a very cheap engine, capable of flying about four hours, which could be steered through the air by wireless waves. They intended to load each one of these aeroplanes with a huge bomb and send them into the air under the control of one flying man, who would herd them through the sky by wireless like a flock of sheep. He would be able to steer them as he pleased, and send them down to earth in just exactly the spot he selected.

" The German idea was that it was a tremendous waste to send shells through the air by means of explosives ; their idea was to put all their explosives into the shells, and then move the shells to their destination by gasoline power. They had really lost faith in the use of big guns. The ' Big Bertha,' which fired shells 75 miles to Paris, was probably partly intended to delude the Allies into believing that the Germans were developing their big guns instead of preparing to discard them, and if they had not got tangled up in their own ' red tape ' they would have rendered the big guns useless before the Armistice came.

" I prepared the plans they asked for. I found that we could make use of old engines that were not reliable for fighting planes. All we asked of an engine was that it should fly for about four hours at the most. Of course, each one of these aeroplanes with its engines would be blown up when the bomb exploded, but the whole thing was not much more expensive than firing long-range shells, and it would be far more sure and far more deadly.

" My plans were accepted by the authorities, and then the War Office made its great mistake—it decided to make the aeroplanes itself. The War Office bungled along with the manufacture of the planes for many months, and when they had finally turned out a few machines they found they could not be depended upon.

" In the summer of 1918, three months before the Armistice, they came to me and gave me a huge order for wireless-steered aeroplanes. I had just got ready to manufacture them in wholesale quantities when the end of the war came. Those aeroplanes would have worked havoc wherever they were used ; it would have been like shooting huge shells hundreds-of miles with a range that was absolutely accurate."

AIR SOUNDING OVER THE ATLANTIC

Having satisfied himself that his machine is in going order, the next matter of importance to the pilot is the weather. The Ministry has now decided to carry out an experiment in air-sounding over the Atlantic, and in testing the appliance with which the soundings will be made it will also prepare the way for cross-Atlantic flight by aeroplane, for, if the experiment is successful, a weather news service will at once be inaugurated, and pilots will be able to know without delay what lies ahead of them.

The s.s. Montcalm, of the Canadian Pacific Line, is the vessel selected to help in carrying out the experimental soundings, which will be made by Lieutenant Guy Harris, R.A.F., of the Meteorological Department of the Air Ministry.

The Montcalm is a 10,000 ton steamer, developing 12½ knots, and is now lying in the Surrey Docks, whence she will sail with cargo for America on Saturday.

When the soundings are taken two kites will be attached to the cable-end, a pilot of light make and, 400ft. behind it, one of the bigger box-kites carrying the meteorgraph. The air currents and varieties of atmospheric pressure vary in layers upwards, and soundings will be taken up to a height of four or five miles. The meteorgraph, a comparatively small instrument combining three devices in one, records at the same time, by pen points marking a chart on a revolving drum, the humidity of the air, atmospheric pressure, and the speed of the wind—all factors of the utmost importance in the consideration of flight.

If the sounding experiments made from the Montcalm are successful a number of ships will be fitted forthwith with similar gear to that which she is to carry. Soundings in great numbers will be taken, and the information obtained will be distributed by wireless from ship to ship and to stations at London, Lisbon, the Azores, and Newfoundland. There is also to be an intermediate station on a battleship cruising in a definite area between Newfoundland and the Azores. These are the points at which it has already been decided by the Ministry to establish ports of call for cross-Atlantic air traffic. At each there will be repair shops, spare parts, stores, and petrol for refuelling the aeroplanes. The type of machine to be used will be " a flying boat," and thus when the proposed method of gathering news about weather conditions is in full working order, and when in a month's time official cross-Atlantic flight by aeroplane begins, as it is hoped it will, pilots will be able to have their machines overhauled, to " refill," and to learn the latest tidings of weather ahead at five halts on the journey.

Ministry is reducing the risks of flight to a minimum.

It was at first proposed that these atmospheric readings should be made using balloons of the type employed by the Meteorological Office in carrying out daily readings over land, but there are many objections to flying balloons from ships. The kite, which is made of fine linen stretched between bamboos, is strong and serviceable, and can be folded away for storage in a very small space. It is also cheap to make and quickly turned out.

It may be remarked that the officer stationed at Lisbon is to be allowed full use of the observatory there for gathering weather news. The official view is that the prospects for cross-Atlantic flights by aeroplane are entirely favourable, and it is hoped to make the journey without touching the Azores. Early news of conditions ahead will have the greatest influence on the success of the venture.

The
1920s

The central theme of science reporting in the 1920s was undoubtedly Einstein's theory of general relativity. The observations of the 1919 eclipse were widely regarded as clinching evidence for his theory. During the 1920s, the popular image of Einstein, which is still the norm today, was created. He was presented as an archetypal professor—perpetually sunk in deep thought, difficult to understand, absent-minded. Stories that fitted this abstruse image began to circulate: for example, of Einstein failing to carry out everyday arithmetical calculations correctly. At the same time, there remained a group of scientists, mainly of an older generation, who found it difficult to accept a thorough-going commitment to relativity. A major stumbling block was their continued belief in an aether. Einstein's work had effectively destroyed the traditional concept of an aether, but they still saw a need for it. How otherwise, they asked, could forces, and especially wave motion, be transmitted through 'empty space'?

Einstein with President Harding in the United States

For a while, opponents of Einstein could point to the existence of specific difficulties. One of Einstein's major predictions was that light from the Sun would be shifted slightly to the red end of the spectrum as compared with light in terrestrial laboratories. Solar physicists immediately began to search for this 'red shift'. The work proved difficult, because of other factors which affected the solar spectrum simultaneously, and initial results were contradictory. However, by the mid-1920s, the leading experts were prepared to declare themselves satisfied that the Einstein red shift was present in the solar spectrum.

The mid-1920s also saw the first results of experiments by D C Miller, an American physicist. Miller repeated an experiment first devised by Michelson and Morley in the nineteenth century. This had tried to measure the motion of the Earth relative to the supposed aether. The negative result of the Michelson–Morley experiment became an important underpinning for relativity theory. Now Miller announced that he had obtained a positive result from his rerun: the Earth did move through an aether. It is a measure of how few sceptics remained that hardly any scientists were prepared to accept Miller's results at face value.

Another topic extensively described and discussed in the 1920s alongside relativity was the burgeoning new knowledge of atomic and sub-atomic physics. The 1920s were the heyday of quantum theory, on the one hand, and of the experimental study of nuclei, on the other. The matrix mechanics version of quantum theory was, hardly surprisingly, as difficult for the public to digest as were Einstein's tensors. But Schrödinger's wave model proved a good deal easier to expound. As with relativity, the interest of non-scientists was particularly engaged by the philosophical implications. The main points of debate were Heisenberg's uncertainty principle (enunciated in 1927) and the question of wave–particle duality.

In nuclear physics, work on isotopes generated special interest. Thus discussion was resumed on the old question of whether hydrogen might be the basic substance from which all other nuclei had been generated. The final blanks in the periodic table were being filled. In 1925, the discovery of the new elements, rhenium and masurium, was announced (the latter subsequently proved to be an incorrect identification). But it was the experimental work on transmuting elements in the laboratory that generated most comment. In part, this was concerned with potential future applications to energy generation in general, but one immediate application was to astrophysics.

During the 1920s, ideas on the internal structure of stars began to be put on a coherent basis. The big problem remained of the stars' source of energy. Nuclear transmutation offered a most attractive route, but it was not at all clear what specific transmutations might be involved. However, one point was clear. The lifetimes assignable to stars had always been uncomfortably short, whatever energy source was postulated. Deriving energy from the annihilation of matter à la Einstein could provide energy for almost unimaginably long periods of time. At the same time, observations in the USA were demonstrating that material in the Universe was much more spread out than had previously been supposed. More especially, stars were clumped together in galaxies which spread as far as the telescope could see. Scales for both time and space were therefore expanding rapidly during the 1920s.

Ideas concerning the origin of the solar system also underwent modification. It was suggested that the planets might have resulted from a near miss between our own Sun and another star. Even with the new expanded timescale for the Universe, such a near collision would be unlikely to have occurred more than once. The conclusion of particular interest to the general public was that life on Earth was probably unique in the Universe. Attention was already concentrating again on Mars as an abode of life, since Mars was scheduled to make one of its closest approaches to the Earth during the 1920s. On this occasion, the planet was observed not only optically, but also with radio transmitters and receivers. Despite various false alarms, no progress was achieved in settling the question of whether or not there was intelligent life on Mars.

Radio waves were increasingly dominating the world of communication. Radio broadcasting became widely established in the 1920s and experiments with the transmission of both still and moving pictures moved ahead rapidly.

Photographic techniques, stimulated by the boom in amateur photography and in films, continued to attract attention, and it was during this period that the basis of the 'talking picture' was laid.

Radio waves were also involved in another rapidly expanding area of investigation in the 1920s—into the nature of the upper atmosphere. Studying the upper atmosphere via the reflection of radio waves obviously had practical implications for long-range radio communication. Less obviously applicable was the beginning of systematic measurements of the ozone layer. Several areas of physics—upper atmosphere physics, astrophysics and nuclear physics—began to come together in the 1920s with the discovery that penetrating radiation, later called 'cosmic rays', was striking the upper atmosphere of the Earth.

1920s CHRONOLOGY

1921
- Bjerknes discusses large-scale motions
- Saha relates ionisation to temperature in stellar atmospheres

1922
- Richardson describes the use of numerical calculation for weather prediction

1923
- The Compton effect is discovered
- de Broglie introduces the idea of wave–particle duality

1924
- Bose–Einstein statistics are introduced
- Hubble shows that galaxies exist independently of the Milky Way
- The Stern–Gerlach experiment shows that neutral atoms can be affected by magnetic fields

1925
- Appleton uses radio waves to detect the ionosphere
- Heisenberg develops matrix mechanics
- Pauli enunciates the exclusion principle

1926
- Fermi–Dirac statistics are introduced
- Schrödinger develops wave mechanics

1927
- Davisson and G P Thompson show that electrons can be diffracted by crystals
- Heisenberg formulates the uncertainty principle
- Oort demonstrates that the Milky Way is rotating

1928
- Fleming discovers penicillin
- Raman scattering discovered
- Sommerfeld shows that electrons in a conductor act like a degenerate gas

1929
- Dirac predicts the existence of anti-matter
- Hubble establishes a distance scale for galaxies and demonstrates that the Universe is expanding
- Quartz-crystal clock introduced

1920s

1920

METEORITE HITS LAKE MICHIGAN

What both eye-witnesses and astronomers agree must have been a gigantic meteor fell in Lake Michigan last night. When it struck the water it sent up a pillar of flames 100 feet in height, and created at the same time a fog of steam. At South Bend, Indiana, Battle Creek, Grand Haven, and other western Michigan cities the earth trembled and buildings were badly shaken, the inhabitants rushing into the streets in panic. The country was illuminated as by the rays of a bright sun, after which came the rumblings of a terrific explosion.

The meteor, which was seen by thousands of persons at Fort Wayne, Warsaw, and other points of northern Michigan, and north-western Illinois, appeared as a great ball of white fire tinged with orange, descending towards the lake south of Grand Haven. The keeper of the lighthouse near the latter city describes how the meteor "rushed down with terrific speed." "I could," he said, "clearly hear the roar of its passage through the air." Telegraph lines and electric power stations were thrown out of order.

Sudden rainstorms were simultaneously reported from Pawpaw, Michigan. At South Haven, Michigan, a terrific wind, which lasted for a few minutes, raised the waters of the lake as in a storm. Captains of lake steamers report all kinds of strange phenomena.

CAMPAIGN AGAINST EINSTEIN

Professor Einstein is so much disgusted by attacks made upon him by his scientific colleagues that he proposes, says the *Tageblatt*, to leave Berlin altogether. The newspaper makes a strong protest against the annoyance to which Professor Einstein has been subjected, which it describes as disgraceful.

It is the duty of the Berlin University (it says) to do all in its power to keep Professor Einstein. Everyone who desires to maintain the honour of German science in the future must now stand by this man.

Einstein himself makes a reply in the *Tageblatt* to his assailants. He ends by saying that it will make a singularly bad impression on his confrères to see how the theory of relativity and its originator are being traduced in Germany.

EINSTEIN.

Twinkle, twinkle, little star,
How I wonder where you are!
'Cording to the new complaint,
Where you seem to be you ain't.

If your light-waves have a kink,
What, on earth, are we to think?
Are you here, or are you there?
You might be 'most *any*where.

Viewed from our terrestrial ball,
Some things are not there at all.
What, for instance, *is* Orion?
And the Bear? Perhaps a lion!

Twinkle, twinkle, little star,
How I wonder *where* you are!
You are less than ever fixed
I am more than ever mixed.

NO MESSAGE FROM MARS

Not a sound was heard from Mars by the scientists who at Omaha last night joined up the greatest wireless station in the world to a wave length of 300,000 metres in the hope of catching a message from that planet.

All night Dr. Frederick H. Milliner and Mr. Harvey Gainer, his electrical expert, listened for news from the Martians. Last night was chosen for the experiment because Mars was then closer to the earth than it will be for several years. Irreverent reporters describe the inhabitants of Mars as being like the subscribers to the New York telephones, insensible to all calls, but Dr. Milliner gives a highly dramatic account of his experiences as an interplanetary eavesdropper.

At first (he said) we used wave lengths of from 15,000 to 18,000 metres, and for several hours it seemed as if we heard everything that was going on in the world. We got Berlin, Mexico, and all the large stations. We got in on a thunderstorm somewhere, and the crackling lightning was like hailstones on a tin roof all around us. About 2 a.m. it cleared up and everything grew quiet.

Then we hitched up a long wave length, which took us into space—beyond anything that might be taking place on earth. There was a most deathly silence. We concentrated our faculties to catch the faintest sound, but there was nothing, nor was the silence broken during the entire time we had the long wave hooked on.

EINSTEIN'S THIRD TEST

In a discussion on the Einstein theory of relativity at Bad Nauheim on September 23 Professor Grebe, of Bonn, declared that the third test had been passed.

According to Einstein, there should be a "shift" towards the red of the lines in the solar spectrum of from 0·62 to 0·63. The absorption bands of nitrogen had been selected and compared with a spectrum of a carbon arc. More than 20 measurements of each line had been made, and the results were in close agreement with observations made in America. There were differences in the "shift" of individual lines, but when allowance had been made for disturbing factors the "shift" was found to be about 0·66—a sufficiently close agreement with Einstein's prediction.

A ROCKET TO THE MOON

The highest altitude ever reached with recording instruments is stated to be 19 miles, but Prof. Goddard believes he can send his rocket through the band of atmosphere surrounding the Earth, and even beyond the radius of the Earth's attraction. How he has attained to this is described in a statement issued by the Smithsonian Institution.

*   *   *   *   *

The great value of Prof. Goddard's experiment, says the Institution's statement, is the possibility of sending recording instruments to those higher levels of the air, the chemical composition, temperature, and electrical nature of which have for a long time been the subjects of much speculation, and a knowledge of which would be of the greatest assistance to meteorologists. The most interesting speculation in connexion with the rocket, it adds, is the possibility of sending a sufficient quantity of brilliant flash powder to the surface of the Moon to make its ignition on impact visible through a powerful telescope. This would be the only way of proving that the rocket had left the region of the Earth's attraction, as once it had escaped that attraction it would never come back.

The rocket will travel at enormous speed, it being calculated that it will rise to a height of no less than 230 miles in six-and-a-half minutes. Prof. Goddard is conducting experiments with the assistance of a grant from the Smithsonian Institution. He hopes to carry out a trial of his invention shortly.

WHAT ABOUT THE AETHER ?

Sir,—The all-too-short but brilliant article on Einstein's theory in your issue of the 29th inst. would have been additionally helpful if your scientific correspondent had given us some hint as to the solution of a difficulty felt by many in following Einstein's teaching. If this theory has made the hypothesis of the aether unnecessary, then we ask: how does it explain the transfer of energy through what we call empty space ? We know that the sun sends to us enormous quantities of energy in the form of heat and light, and that this takes some eight minutes to reach the earth. We ask, in what form is the energy stored up *after* it has left the sun and *before* it reaches the earth ? The old aether theories at any rate gave a plausible answer to this question, and the late Lord Kelvin, in a well-known paper, calculated from certain data the energy in a cubic mile of sunlight just reaching the earth. Wireless telegraphists whose daily business it is, so to speak, to handle the aether and dispatch this radiant energy will not readily give up their belief in its existence until some equivalent mechanism is suggested to take its place. Huxley once told us that the great and ever-recurring tragedy of science was that of a beautiful hypothesis killed by an ugly fact. It is just possible that the Einstein theory, by paying too exclusive attention to the ideas of space, time, and mass, and too little to the equally important conception of energy, will find some ugly fact waiting for it round the corner to inflict a serious wound on this attractive hypothesis which has no need for an aether.

I am yours, &c.,

J. A. FLEMING.
University College, London, March 29.

SUNSPOTS AND MAGNETIC STORMS

It is not always that a spot on the sun, however large it may be, is accompanied by a magnetic storm, but it is fairly safe to infer when the needles of the instruments that record the changes of earth-magnetism are unusually disturbed that there is a spot of some size on the solar disc, though it may not necessarily be of the largest. The group of spots now visible—it was quite near the western edge yesterday—was seen on Monday, May 9, and had probably been brought into view by the solar rotation a day or two earlier, but though this was quite apparent to those who watch the solar surface, it was only at the end of the week that the spots created general interest, for it was not until then that any coincident magnetic phenomena occurred.

On Friday, May 13, about a quarter past 9 by Greenwich time, there was a disturbance of the recording magnets, not of great amount nor of long duration, for they were fairly quiescent during the early hours of Saturday. The disturbance, however, was renewed on Saturday evening, and was excessive about midnight, remaining so for some hours, though at present it is not possible to contrast this as to its magnitude with previous storms. At this time there was a brilliant display of the Aurora Borealis, and, judging from news from various quarters, the magnitude of the disturbance appears to have been sufficient to disorganize telegraph services, which happens only in extreme cases.

The exact connexion between sunspots and the magnetism of the earth has not been decisively proved, but a suggested cause of magnetic storms, known as the hose-pipe theory, is in some accord with the circumstances of last week. This theory postulates the ejection of confined streams of electric corpuscles from disturbed regions of the sun with a speed comparable with that of light, and approximately in straight lines, which meet the earth and ionize the outer layers of the earth's atmosphere, or, in other words, convert it into a gaseous conductor of electricity. When the ionization is particularly intense, it is accompanied by luminosity which is seen as an aurora, and this change of state of the atmosphere, combined with other magnetic qualities of the earth, causes the storm.

LIFE ON MARS AND THE MOON

Explaining the broad curved streaks which he finds on Mars, PROFESSOR PICKERING said his theory was that in the Martian summer the moisture from the melting polar snow-cap is evaporated and finds its way down through the planet's atmosphere to the other pole, then in darkness, where it is condensed and frozen again. In the northern hemisphere there are, he says, three lake-like markings which appear to be centres of evaporation. From these three lakes the water vapour is carried southwards in broad bands, or currents, condensing by night and being partly precipitated as rain. The dark streaks visible to us are the strips of vegetation thus watered, and the reason why they are curved is that the vapour currents in the atmosphere are curved as the effect of the planet's rotation. Professor Pickering has not been able to explain satisfactorily the straight narrow "canals," but he showed as a possible suggestion a lantern slide of a long straight line of trees growing along a volcanic fissure in a desert tract of the Hawaiian islands.

In the case of the Moon Professor Pickering has detected changes occurring in many parts of the visible hemisphere which, in his view, strongly suggest that both snow and vegetation are to be found on that body. He showed some remarkable drawings of the changes that take place during the long lunar day.

SUNSPOTS AND THE WEATHER

In the middle of May a large spot on the earthward face of the sun was contemporaneous with a violent disturbance of terrestrial magnetism and many local outbursts of bad weather. The revolving sun is again showing us the same spot, smaller, but still the visible sign of a vast solar convulsion. Again magnetic needles have plunged and quivered ; again the swift havoc of unexpected storms has fallen on many places. The most serious calamity appears to have visited Pueblo, in California, where over 200 people have been drowned, and a considerable industrial city has been wrecked. The immediate cause has been wrongly described as a water-spout, a phenomenon due to the descent of a conical vortex of air over a lake or the sea, and not in itself giving rise to floods. More probably it was simply the violent and local rainstorm often called a " cloud- " burst." Is there a link of causation binding sunspots, magnetic and atmospheric storms ? We have rounded Astrology Point but have not yet weathered Cape Coincidence. Astronomers are unwilling to ascribe any terrestrial occurrence to the influence of the sun unless they can interpret the action in terms of known " laws " of physics. But they, and still more general opinion, are very easily impressed by coincidences.

COOKERY BY SUNLIGHT

The astronomical station of the Smithsonian Institute on Mount Wilson, in California, has been putting the sun to a new use. Dr. C. G. Abbott has devised a solar " cooker," by which, using only the sun's rays, he was able to cook meat, vegetables, fruit, and bread for his party. A description of the apparatus has not yet reached this country, but as it was invented by an astronomer, doubtless its main feature is the concentration of the rays by focal lenses or mirrors. It may be presumed, also, that the cooker, although convenient in the exceptional circumstances of a station on a lofty mountain peak, is too expensive for general use. But the world is living so largely on its capital of energy that any effort to use directly some of the daily income from the sun is of more than theoretical interest.

'ISOTOPES'

The address of a President of the British Association often introduces to general notice terms and conceptions formerly only within the cognisance of specialists.

Sir Edward Thorpe, speaking at Edinburgh, performed this service for the word "isotope," and for the chemical conceptions involved in it. Plainly he thought them of fundamental importance in theoretical physics and chemistry, and many who never heard the word before are now asking what isotopes may be. They are of introduction so recent that they have not yet found a place in the most comprehensive books of reference. They are so difficult that the Royal Society last spring devoted a session to discussing them, at which it was found that the highest authorities still differed not only as to their significance, but almost as to their existence. But at least there was agreement in using the word to express certain observations on the nature of the elements which must be explored until they have been fitted into the fabric of science.

Take the case of chlorine, referred to by the President of the British Association. Its weight has been determined repeatedly by methods in which great accuracy is possible. The weight is 35.46, accurate to the second place of decimals. All attempts to prove it a whole number, 35 or 36, have failed. By a method of analysis devised by Sir J. J. Thomson, Aston has shown that it consists of a mixture of two elements with the respective weights 35 and 37. The atomic weight as hitherto understood would appear, therefore, to be merely a statistical number, the expression of the proportions of the isotopes of different atomic weights contained in it. All the chlorine that has yet been examined and determined comes originally from the salt of the sea water. The separate isotopes in it may therefore have been mingled for aeons so that an invariable proportion has been reached. But chlorine examined from another source might show a different statistical result.

The determined atomic weights of the greater number of the elements are not whole numbers. Sir Edward Thorpe seemed to take the view that whilst those were probably mixtures of isotopes, the others were simple. Bolder chemists go farther and have already hazarded the opinion, for which there is some theoretical justification, that all the elements are mixtures of two or more isotopes.

GENERATING RADIO WAVES

It is still convenient to think of the æther as a kind of gas filling all space, but so much finer that it passes continuously through solids, liquids, and gases, through even molecules and atoms. When a stone is dropped in a pool of water, circular waves spread out from it. When a rocket explodes in the air, sound waves spread from it in every direction, travelling very much faster than the ripples in the pool, but still taking an appreciable time.

PRODUCING ÆTHER WAVES.

Electric waves can be set up in the æther. What corresponds in their case to the stone thrown into a pool or the explosion of a rocket in the air is the production of a current of electricity "oscillating" or vibrating at a high frequency in the wires stretched in the air at a transmitting station. The waves so sent out into space travel at the rate of 186,000 miles a second, a pace which allows them to pass from London to Australia in less than the twinkling of an eye. But to produce waves which will travel these great distances without fading out, very powerful oscillating currents are needed.

In the last five years this has become possible by three great inventions. First, there is the high-frequency alternator. The rotating dynamos used in towns produce what is called a low-frequency current, reversing or oscillating only from fifty to two hundred times a second. The high-frequency current required for wireless has to reverse from twenty thousand to one hundred thousand times a second. A grindstone turned too quickly will fly in pieces from centrifugal force, and it seemed impossible to imagine machinery which could rotate at the speeds required for wireless, until Alexanderson, an engineer of the American General Electric Company, and Béthenod-Latour, in France, solved the problem.

Next, there is the improved Duddell-Poulsen arc generator, used both in Germany and this country, by which a powerful high-frequency continuous current is generated by means of an electric arc in a chamber filled with coal-gas, kerosene, or alcohol vapour. Third, and most important, is the battery of "thermionic valves," a series of structures looking like rather complicated incandescent electric lights, each globe perhaps as large as a foot in diameter. By one of these three sets of devices, powerful transmission stations are able to produce æther waves which travel round the world.

IS THE UNIVERSE RUNNING DOWN

The conservation of energy is known in physical science as the first law of thermo-dynamics, because it became apparent as an exact statement chiefly from experiments on the energy of heat. It has been used in popular philosophy as an argument in favour of the permanence of the universe. The universe reveals itself as transformations of energy : as in none of these is there any gain or loss, the universe can have no beginning and no end. It is eternal, uncreated and indestructible. But there is a second law of thermo-dynamics, much more difficult to understand, still only in course of exploration and just beginning to creep into popular philosophy. Perhaps the simplest way of conceiving it is to state that energy may differ in quality as well as in quantity, and that whilst the total quantity may not change, the quality does change, and apparently in an irreversible direction.

The famous German physicist and chemist, W. Ostwald, in the comprehensive German fashion, elevated this second law of thermodynamics into a principle of conduct. He spoke of it as the "imperative of energetics," and in a volume of essays published in 1912 applied it to almost every department of human affairs. He used it as an argument against the waste of war, but this application of a categorical imperative, like many others, does not seem to have influenced the conduct of his country. More generally the doctrine of increasing entropy has been used to support a kind of cosmogonal pessimism, a theory that the universe, even although its total sum of energy may be indestructible, is moving inevitably towards quiescence. But at most the argument is only one derived from experience, and just as there are swiftly recurring phases in the life cycles of animals and plants, so it may be that the existing tendency towards the fixation of free energy is only a long phase in the infinite cycles of the universe.

MATHEMATICS AND REALITY

The present was a particularly happy moment for a pure mathematician, since it had been marked by one of the greatest recorded triumphs of pure mathematics. This triumph was the work of Einstein, a man who would probably not describe himself as a mathematician, although he had done more than any mathematician to vindicate the dignity of mathematics and to put that obscure and perplexing construction commonly described as "physical reality" in its place.

The mathematician, he continued, was in much more direct contact with reality than the physicist. This might seem a paradox, since it was the physicist who dealt with the subject-matter to which the epithet "real" was commonly applied. But a very little reflection would show that the "reality" of the physicist, whatever it might be, had few or none of the attributes to which the epithet "real" was commonly applied. A chair might be a collection of whirling atoms or an idea in the mind of God. It was not his business to suggest that one account of it was obviously more plausible than the other. Whatever the merits of either might be, neither drew its inspiration from the suggestions of common-sense.

We could not be said, either as physicists or as philosophers, to know what the subject-matter of physics was, although their task plainly was to correlate the incoherent body of facts confronting them with some definite and orderly scheme of abstract relations, such as could be borrowed only from mathematics.

A mathematician, on the other hand, was not concerned with physical reality at all. It was impossible to prove by mathematical reasoning any proposition whatsoever concerning the physical world. It was the business of mathematics to supply physicists with a collection of abstract schemes which it was for them to select from and to adopt or discard at their pleasure. A large number of different schemes of geometry had been constructed, Euclidean or non-Euclidean, of one, two, three, or any number of dimensions. All these were of complete and equal validity. They embodied the results of the mathematicians' observations of their reality, a reality far more intense and far more rigid than the dubious and elusive reality of physics. The old-fashioned geometry of Euclid, the entertaining seven-point geometry of Veblen, the space-times of Minkowski and Einstein, were all absolutely and equally real. When a mathematician had observed them his professional interest in the matter ended. It might be the seven-point geometry that fitted the facts best. There might be three dimensions in that room and five next door. The function of the mathematician was simply to observe the facts about his own hard and intricate system of reality, that astonishingly beautiful complex of logical relations which formed the subject-matter of his science.

25,000 PHOTOGRAPHS A SECOND

A remarkable step forward in the history of cinematography was announced to the Academy of Science yesterday. M. Bull, Assistant Director of the Marey Institute, has succeeded in photographing electric sparks whose duration was only 1-50,000th of a second. M. Bull employs a prism revolving 120 times a second. Images of sparks refracted by the prism are projected on a fixed film every hundredth of a second. Thus there are about 250 photographs of sparks per hundredth of a second.

THE 'C.G.S.' SYSTEM

A correspondent calls attention to an order issued at Baghdad at the end of 1918 and including the following recommendations:—

Adoptions of Continental Units Advocated.

Under peace conditions meteorological work will be much more in co-operation with Egypt, Turkey, and Caucasin than with India, so that the adoption of Continental units is practically essential. Indeed, if Egypt has adopted, or is likely to adopt, the "Chief of the General Staff" system of units with the millibar as the unit of pressure, it would be advisable to go one better than the Continental system and adopt the "Chief of the General Staff" system straight away from the beginning.

Instruments.

Instruments graduated for Continental or "Chief of the General Staff" units can be obtained in England.

Readers who are occasionally baffled by the use of unexplained initials in military intelligence will sympathise.

EINSTEIN CRITICIZED....

Sir,—It appears to be assumed that the confirmation by the recent American eclipse expedition of the value for the gravitational deflection of light obtained by the British astronomers nearly four years ago is the definite establishment of the Einstein theories relating to gravitation. I write because I have been responsible for publishing an argument to maintain that, if the subject is consistently reasoned out on Einstein's own point of view, one ought to arrive at a value equal to half the result asserted by him, notwithstanding that this result is confirmed substantially by observation. Indications have reached me that this contention has been interpreted as a reflection on the work of the British eclipse expeditions. On the contrary, the inference should be that if their determinations stood firm, as there was every reason to expect that they would, the Einstein theory could not stand—at any rate, without serious modification. Soon afterwards French critics, led by Professor Le Roux, of Rennes, were guided by similar reasons to the more drastic view that the Einstein ideas were incapable of leading to any definite result at all ; and considerable recent discussion in Paris does not seem to have shaken their position. The conflict of opinion in Central Europe has long been in evidence.

This minute gravitational deflection of light is no new idea. In the eighteenth century Cavendish and Michell deduced, on the basis of the Newtonian corpuscular hypothesis, a definite value for the deflection, which was exactly half of the value now found by observation. From a different point of view, Einstein has arrived at twice Cavendish's formula ; but his argument, though fascinating, has been widely challenged as unsound on its own premises. Other points of view may in future lead to the same or to other multiples of the Newtonian value. The critics would doubtless be glad to be convinced ; but so far as I am aware no adequate answer to their entirely independent questionings has yet been forthcoming.

Faithfully yours,
JOSEPH LARMOR.
Cambridge, April 14.

.... AND DEFENDED

EINSTEIN AND GRAVITATION.

The gravitational deflection of light as calculated by Einstein in advance has now been three times verified. Sir Joseph Larmor's individual criticism is that " if the subject is consistently reasoned out on Einstein's own point of view, one ought to arrive at a value equal to half the result." It is clear, however, that the French critics, to whom he subsequently alludes as being " led by Professor Le Roux," do not agree with Sir Joseph's own calculations, since " similar reasons " have led them to the conclusion that no definite result from Einstein's ideas is possible. I merely point out that there is this divergence of opinion among the critics of Einstein's work.—Mr. R. LEETHAM JONES, London, S.W.5.

EINSTEIN'S THEORY CONFIRMED

It will be remembered that Einstein suggested three crucial tests of his theory, which experience could make. The first concerned the movement of the planet Mercury, and had already been satisfactorily made. The second could be made at a total eclipse of the sun, and concerned the bending of light rays from a star ; at the eclipse of 1919 the English astronomers obtained a clear answer in favour of the theory, very satisfactorily confirmed by the American observers in 1922. The third test concerned the apparent length of the waves of light as affected by gravitation.

In this third case experiment gave at first very dubious results, some observers even declaring against the effect suggested by the theory. Moreover, some mathematicians challenged the correctness of the inference from the theory, though Einstein never wavered in his declaration that it was a necessary inference. These clouds which have hung about the third test have now been dissipated. Mr. C. E. St. John, of Mount Wilson, who had thrown the gravest doubts on the experimental facts, has now come round definitely in favour of the Einstein result. He makes his own announcement in *Science* for September 28. Mr. Evershed (who has just retired from a long and able Directorship of the Kodaikanal Observatory in Southern India) had already given very strong evidence in favour of Einstein, but the conversion of Mr. St. John is of obvious importance, and the joint testimony of these former opponents leaves the matter now in no reasonable doubt.

THE EARTH WOBBLING

The Earth's spin is wobbling.

Lieut.-Colonel P. Jensen, a Danish scientist, who returned recently from a degree-measuring expedition in Greenland, reports that Greenland is moving westward at the rate of twenty yards a year.

This confirms the recent reports of surprising climatic changes at the North Pole.

It is now established that there is a periodic shifting of the latitude of the North Pole. The movement is difficult to detect because of the small area of the pole—about the size of a tennis court.

RISING FROM THE SEA.

Potsdam Observatory reports a slow march northward. The observatory of Ukiah, California, reports a similar change. The coast of Labrador is rising from the sea.

Many authorities say that the poles are gradually changing their positions, and that this alteration to the world's axis will in time mean that regions which are at present ice-bound will become warm, habitable countries.

SUMMER TIME

There is apparently some divergence of opinion even in the medical profession on the question of the advantages of "Summer Time." Dr. J. W. Bone, on behalf of Bedfordshire, put forward a proposal regretting that "Summer Time" had been curtailed this year, and expressing the opinion that "Summer Time" was most beneficial to the health of the nation.

There were cries of "Agreed," but several doctors rose to oppose it. They were mostly from agricultural districts, and were not treated very seriously by their colleagues.

TRANSMUTATION OF THE ELEMENTS

Next month, at Liverpool, we shall expect to hear from Sir Ernest Rutherford, president-elect of the British Association, some account of his recent progress in transmuting elements and liberating sub-atomic force. What he is known to have accomplished already may be stated shortly. Of the eighty-seven known chemical elements, hitherto in practice, if not in theory, ultimate pieces of matter, distinctive moulds into which the stuff of the universe is cast, he has succeeded in transforming six—boron, nitrogen, fluorine, sodium, aluminium, and phosphorus.

His method has been to knock chips from the atoms of these elements; the fragments removed are not scraps of unknown or indeterminate matter but are definitely known substances; the mutilated atoms also are not unknown, or indeterminate matter, but are the atoms of other known elements, lower in the scale than those from which they have been dislodged. In the process energy is liberated, which Rutherford did not put into his apparatus, and which can have no other source than the sub-atomic energy until now securely locked up against human effort.

But let it not be thought that his discovery is yet ready to form the basis of economic processes. He has had to employ radium, the rarest and most valuable of all known substances, and he has worked on a scale infinitely minute. The triumph, so far, has been purely scientific, but it is among the three or four greatest and most startling of the conquests of natural knowledge.

1924

SHORT WAVE RADIO

Marconi's recent work is a return to his earliest experiments. In 1895 and 1896, using waves of a few inches in length, he was able to receive and transmit intelligible signals over distances of one and three-quarter miles, although with antennæ and long waves he found it difficult to reach a mile. In these early experiments, moreover, he sent the short waves in a beam, so that the receiver was "called up" only when it was directed towards the source of the message. But at that time there was difficulty in producing short waves of sufficient strength and the spectacular success of waves with lengths up to thousands of feet was so great as to attract all research work.

The new investigations were begun during the war partly with the objects of minimizing interference and the chance of tapping. They have been continued since 1916, with the help of the Italian Navy and the cooperation of Mr. C. S. Franklin. The reflectors of solid metal have now been replaced by screens of wires placed parallel to the antennæ and arranged in a parabolic curve of which the receiving and transmitting antennæ form the axis, as was shown in a photograph recently published in *The Times*. With a reflector of that type, using a wave length of three metres, a range of 20 miles was obtained in 1917, and in 1919, using a 15-metre wave, clear and strong speech was got over a distance of 70 miles. With reflectors used at both ends the energy received is now 200 times that received without reflectors. On May 30 of last year Australia was asked to "listen in." Messages were sent out from Poldhu on a 100-metre wave and for the first time in history intelligible speech was transmitted from London to Sydney.

THE ZODIACAL LIGHT

I have to-day (March 13) seen a most marvellous example of the zodiacal light. Unfortunately I had not a clear view of the sunset at first, when it began, but at about 6.20 came from a valley to the sea shore, and saw at once this wonderful column of light, standing straight up from the point where the sun had set, and glowing like a huge torch. Calculating roughly, it seemed to extend one-fifth of the distance to the zenith, but perhaps this is too large an estimate. The glow lasted quite clearly till 6.30, and then gradually faded.— A. D. Burgess, The Ladies' College, Jersey.

DEATH RAYS

So much has been written of what is dramatically called "The Death Ray" by those who have no special claim to scientific knowledge that perhaps another layman may, without absurdity or presumption, contribute a word.

When one reads of all the powers attributed to this remarkable Ray, one becomes lost in admiration of the diverse genius of its inventor. Not only is it said to destroy life, set fire to anything inflammable, wreck aircraft, stop motor-cars at a distance of 58 miles, and explode ammunition or other explosives, it is even claimed that it will cure cancer and detect submarines. It is sublime to think of this death-dealing weapon turning aside from its special lethiferous task to cure the sick ; to realize the delicate adjustment of means to ends which enables it to set fire to anything inflammable and yet arrest the magneto in a car without igniting the petrol. It is also a poignant reflection that all these incredible benefits are denied to an unworthy country because its purblind authorities refuse to put a considerable sum at the inventor's disposal. This gentleman must indeed deplore the publicity, which even his reticence has been unable entirely to avoid, now alas! surrounding a discovery too confidential apparently to be set forth even in a secret patent.

Ray is, of course, a loose term, applied sometimes to a stream of particles : at others to a stream of energy. The latter type has been explored and its characteristics have been determined from one end of the scale to the other. Even beyond the region directly investigated, enough is known even to laymen to enable them to exclude anything resembling the alleged Death Ray. Streams of particles are, I believe, equally familiar to science ; and while it is rash for a layman to dogmatize on such a subject, I question with great scepticism whether any competent physicist would seriously consider that they could, in any circumstances conceivable to science, possess the properties claimed for the "Ray" in the daily Press.

Physics, after all, is, I believe, an exact and also an exacting science. That an unknown amateur should stumble upon an epoch-making discovery is about as likely as that a child of five should defeat either a champion chess-player or Mlle. Lenglen.

THE MARTIAN LANGUAGE

New York.—Two radio operators at Newark, New Jersey, declare that they heard a weird succession of sounds on a wave-length of 2500 metres which might have been from Mars. The sounds started abruptly, stopped suddenly, and then resumed. They were like a bass note on a piano. anoLTishrdlu aoinshrdlu altled dawadel.—*Daily Paper*.

THE LATEST PLANET

Mars is the only planet which has ever appealed with any force to the average man. One never hears Sirius mentioned on an omnibus.—*Evening Paper*.

EINSTEIN'S ARITHMETIC

Prof. Einstein, famous as the discoverer of the theory of relativity and one of the world's foremost mathematicians, had an argument the other day in a Berlin tramway-car with the conductor regarding the amount of change that had been given him.

The professor said it was a halfpenny short, but the conductor proved that he was wrong, and concluded the argument with the remark, "I see you are weak at arithmetic."—*Reuter*.

MONT BLANC OBSERVATORY

THE recent history of the Observatory put up by M. Janssen on Mont Blanc is curious. We remember how the veteran astronomer was carried up the mountain by an army of guides, so that he arrived at the summit with his mind unclouded by physical exertion, and was thus able to give a unique account of the effects to the French Académie des Sciences; how later he had the materials for a small wooden observatory carried up, and, finding it difficult to get through the snow to solid rock, ultimately placed it on the snow itself. But the foundation was not sufficiently solid: the hut sank into the snow, and disappeared completely. But Dr. Tutton, F.R.S., tells me that after an interval it emerged some half-mile lower down the mountain, and has been dragged by the guides about another mile to the Col just below the Vallot huts, which were put up before the War for meteorological purposes. Dr. Tutton saw it there on August 9 last.

THE PLACE OF HYDROGEN

It was soon found that whole number atomic weights, worked out with great precision on the new conception of isotopes, formed an entirely satisfactory series if the atomic weight of oxygen were taken as being exactly 16. But 16 what? The theory to which all the most recent work pointed was that hydrogen should be the unit, the primitive element. But hydrogen, investigated by the most refined methods, insisted on giving an atomic weight of 1.0077, a small deviation, but seemingly ruling out the conception to which all other work pointed. Sir Oliver Lodge has recently put the difficulty and its explanation clearly.

The element next to hydrogen is helium, which has an atomic weight of 4 on the oxygen scale, and on the scale to which all the other atomic numbers except that of hydrogen have been simply adjusted. If helium is built up of four atoms of hydrogen it should have an atomic weight of four times 1.0077, that is to say 4.03, instead of its ascertained weight of exactly 4. What has happened to the missing fraction? The explanation brings in some of the most difficult theories in modern physics.

We have to suppose that hydrogen in combination is different from hydrogen when it is free, that something disappears in the process of combination. We have learned now that what we call matter is electrically constituted and that its inertia or mass on which the atomic weight depends is an electro-magnetic phenomenon. Every electric charge has a definite mass associated with it, and in an aggregate of electric charges their masses are added together. Each atom consists of a central nucleus with a positive charge of electricity, the nucleus containing practically the whole mass of the atom. The nucleus of helium contains four hydrogen nuclei. But these are tightly packed together in an area almost incredibly minute compared with the area of the whole atom, the relative dimensions being comparable with those of the sun and the whole solar system. When electric charges are brought into very close contiguity they interfere with one another to some small extent. Theory points indeed to the possibility that if brought into actual approximation they would disappear altogether. So far it is not known what is meant by this disappearance of inertia, mass, or electric charge which would seem to violate the old axiom of the indestructibility of matter and energy. Sir Oliver Lodge probably would invoke the ether, and those who think the ether to be no longer a necessary hypothesis can summon the quantum and the mathematics of relativity to their aid.

But whatsoever be the explanation, the fact seems to remain that four hydrogen nuclei, combining to form one helium nucleus, lose the odd decimals in the process and that on this basis hydrogen and helium can serve as the bricks of the other 90 elements and their isotopes with whole number atomic weights.

RELATIVITY TESTS: CHICAGO PROFESSOR'S EXPERIMENTS

Professor Albert Michelson, Professor of Physics in the Chicago University, and a noted scientist, has made a provisional announcement of the result of his test of Einstein's relativity theory. Professor Michelson hitherto was one of the most cautious in admitting the theory of relativity, hence the importance of his declaration at the conclusion of his lecture in Chicago yesterday that "There is no question that these tests furnished another striking confirmation of Einstein's brilliant work."

The Professor desiring to determine whether the earth's rotation influenced the velocity of light as the relativity theory predicted chose an experiment which consisted of the simultaneous transmission of light beams in opposite directions around a rectangular course. According to the theory of relativity the earth's rotation should cause a slight retardation of the return of one of the beams, but it was only possible to measure this tiny difference by observing the so-called interference fringes. For observing these fringes Professor Michelson had been many years constructing an interferometer which measures light waves of 1-50,000th of inch.

The apparatus for the recent experiments, which was completed last autumn in the fields outside Chicago, consisted of a vacuum tube, 5,200ft. long, made up of tile pipes, with a diameter of 1ft., laid as level as possible and hermetically sealed. The tube formed a rectangle running 1,200ft. north and south, and 1,400ft. east and west. Iron boxes containing mirrors were placed at three corners, whilst at the north-east corner were two shanties, one containing a huge air pump and the other containing the control instruments. Within the control room an arc lamp threw by means of mirrors two light beams in opposite directions through the tube.

Professor Henry Gale, the Dean of the Science School, sat underneath watching the returning beams. The desired fringes were visible many times, the best experimenting usually occurring at night. The apparatus needed the most careful construction and adjustment, in which Professors Fred Pearson and H. Lemon assisted. Professor Michelson intends to continue his experiments for a few months.

NEW ELEMENTS DISCOVERED

An announcement made by Professor Walter Nernst at the last meeting of the Prussian Academy of Science dealt, according to the *Börsen Courier*, with the reported discovery of two new elements by the physicist, Dr. Walter Nonnack, and Fräulein Eva Tacke, assisted by Dr. Otto Berg.

The elements, which are said to have been detected by chemical analysis, and also by the Röntgen spectroscopic process, had been sought unsuccessfully by several earlier investigators. In minute quantities they may be obtained from a number of mineral substances, of which the principal is platinum ore. Their atomic numbers are 43 and 75, and they have been named masurium and rhenium, after the East Prussian borderland and the Rhine. As the result of this discovery only three elements still remain to be found.

IS THERE AN AETHER DRIFT?

On the theory of the aether, light waves travel through that medium with a uniform velocity of about 300,000 kilometres a second. But the earth is rotating on its axis, is revolving round the sun, and travelling with the whole solar system in some unknown direction through space. The aether itself is supposed to be motionless, solid bodies passing through it as if it did not exist. With regard to the ocean of aether, the movement of any point on the Earth must be a complicated path depending on rotation on the axis, rotation round the sun, and movement of the solar system through space. If the aether does exist, the Earth must be rushing through it, or to put it in another way, the aether must be drifting past it, the rate of the drift at any point varying according to the cooperation of the three movements, but always existing unless the almost incredible chance happened that at any one moment the three motions exactly neutralized one another. Michelson and Morley came to the conclusion that if there were an aether drift due to the Earth's velocity through space it must be less than $7\frac{1}{2}$ kilometres a second.

Professor Dayton C. Miller has been repeating the experiments with a number of improvements in the apparatus, and using a light path of 224ft. With the collaboration of Professor Morley, he has constructed and used an inferometer four times as sensitive as that employed in the Michelson-Morley experiments. In a recent communication to the National Academy of Science at Washington he has given the results of his investigations, which have led him to the direct conclusion that a measurable aether drift does exist.

He points out that although the results of the Michelson-Morley experiments have generally been taken as negative, the authors themselves suggested that relative motion imperceptible near sea-level might be detected by their instruments at even moderate distances above it. In the autumn of 1905, Michelson and Morley removed their plant to a site 800ft. above Lake Erie, where five sets of observations gave a positive effect of about one-tenth of the "expected" drift. Since then Miller has made approximately 5,000 observations with his new apparatus at the Mount Wilson Station near Pasadena, California, at an elevation of over 6,000ft. The observations were in groups at four seasons of the year. The first sets showed a drift of about ten kilometres a second. Since then they have been reinforced by a further series, with special precautions against error due to temperature. Miller has reached the definite conclusion that the displacement shown by his inferometer is such as would correspond to a definite aether drift of ten kilometres a second, and that there is a partial drag of the aether by the Earth, so that the original Michelson-Morley results made at Cleveland were not really negative. The implications of this new work are far-reaching; there will be much re-examination and criticism before the existence of an aether drift will be accepted; but if it be confirmed, Dayton Miller's achievement will rank among the most memorable and important contributions to knowledge of the universe.

VISIT BY KING GEORGE V TO GREENWICH TO CELEBRATE THE 250th ANNIVERSARY OF THE FOUNDING OF THE ROYAL OBSERVATORY

1926

RAIN FROM ELECTRICITY

So long ago as 1884 Sir Oliver Lodge discovered that even the infinitesimal globules of mist or of visible steam, if electrified, would unite and fall as fine rain. He was able to show the phenomenon on a lecture scale at a meeting of the British Association in Canada. Since then, by himself, his son, and workers in this country, America, and Germany, the process of electrical precipitation has been applied on a commercial scale to the removal of dust from blast-furnace gases. The advantage is sometimes to produce better combustion, sometimes to recover products, such as potash, which are of commercial value, as much as 20 tons a day of dust being removed in one installation from a stream of gases rushing past the electrodes at the rate of 50,000 cubic feet a minute.

Could an electrical plant of this kind be applied to cleaning the air of cities from the suspended particles of dust which darken our skies, in themselves and act as nuclei for the precipitation of water? Sir Oliver Lodge is doubtful. Even if it were practicable, the rain of soot would not be welcomed, and the more obvious remedy is a form of fuel consumption that does not discharge soot into the air. But, although the practical difficulties are equally great, more is to be hoped from efforts to induce clouds to precipitate their moisture as rain. It seems a hard saying after one wet and another broken summer and a rainy winter, but rain is more often coveted, and in more parts of the world, than drought. Even we in the South of England, remembering the summer of 1923, from which deep springs are hardly yet recovered, knew something of the eagerness with which dwellers in a parched land watch the occasional futile drift of clouds across the hungry plains. And yet, could a slight difference in potential be communicated to one part of the cloud, cohesion of the droplets would follow and the rain would fall. At present, rain-making experiments have been limited almost wholly to persons who have concealed their methods, and who, it may not uncharitably be supposed, have depended chiefly on luck for their results. But many diseases, not long ago combated only by invocation, by magic, or by quacks, have yielded to science, and it is at least within the range of possibility that experiments on the precipitation of the moisture of clouds, conducted on a large scale under proper direction, might yield results of untold value.

OPTICAL GLASS PRODUCTION

" Some recent developments in the art of production of glass for optical purposes " was the subject of a paper submitted by Mr. W. H. S. Chance and Mr. W. M. Hampton, of Chance Brothers and Co., Ltd., Smethwick and Glasgow. Mr. CHANCE, who presented the paper, stated that in the last few days his firm had received an inquiry from Germany for optical glass. This was the first inquiry of the kind that had been made for 40 or 50 years. The paper showed that in 1914 Chance Brothers were listing approximately 26 types of optical glass ; in 1918 they produced an output equal to four or five times the estimated pre-war world demand, and at present they were listing 112 varieties. With regard to the production of new types with optical properties essentially different from those at present available, the authors considered that such was not probable in the present state of knowledge, but they believed that further improvement could be effected in transmissibility both in the visible and ultra-violet portions of the spectrum, and they were now, with the help of the British Scientific Instrument Research Association, engaged in a programme of research with that end in view. Reference was made to the fact that the " distant " signals of British railways were being fitted with a special orange-yellow glass developed by the firm. This glass enabled the drivers to distinguish " home " from " distant " signals, and thus conduced to further safety in railway travel.

" VITAGLASS."

A new form of window glass, termed " Vitaglass," which had the property of transmitting the curative ultra-violet rays present in sunlight, was being manufactured and was being used in hospitals and private houses.

LORD RAYLEIGH, F.R.S., in a paper entitled " Notes on Silica Glass," discussed the question why silica glass made from sand was only translucent, while the silica glass made from rock crystal was transparent. The difference was shown to depend on the fact that sand grains when heated developed gas bubbles internally. This behaviour was traced to the geological conditions attending the formation of the material.

A 'CEILING' IN THE UPPER ATMOSPHERE

When " Big Bertha " bombarded Paris from a distance of 76 miles, the curved path of the shell at its highest point reached a height of from 20 to 25 miles above the surface of the earth. This trajectory depends on the construction of the gun and shell, the nature of the charge and the initial elevation, and not on any peculiar condition of the atmosphere at the highest level reached by the projectile. Evidence is accumulating from many sources that wireless waves sent out from a transmitting station in a given direction take huge leaps round the curved globe roughly comparable with what would be the path of a " Big Bertha " shell if, after hitting its mark, it were to rebound and take a second leap, a third leap, and so on. The length of each leap varies with the kind of wave, short waves apparently taking short leaps and long waves very long leaps, but the height reached by the different waves at the summit of each leap appears to be about the same, and, roughly speaking, to be from two to three times the height of the " Big Bertha " shell's leap. This height, moreover, seems to depend on a peculiar layer or stratum in the atmosphere, which may be called the wireless ceiling.

The evidence as to the wireless paths comes from many sources. An Admiralty vessel, sailing across the ocean, sent out pre-arranged signals at four-hour intervals along its course. The receiving station, watching for these, found that there were points of strength and long gaps of weakness regularly arranged as if the signals were received in full strength only when the distance of the vessel from the station was a multiple of the length of a single leap. In a paper published by the Royal Society last week, Mr. R. L. Smith and Mr. R. H. Barfield described some of the results of their observations on wireless waves arriving from the upper atmosphere, made under the Radio Research Board. Working with waves of short length and using a beautiful combination of observation and mathematical analysis, they were able to show that these waves behaved as if they were the descending limbs of curved leaps of approximately 90 kilometres high. Other work, although pointing in the same direction, suggests rather a greater elevation for the ceiling and the existence of a peculiar stratum rather than a limiting level.

Millikan and others have suggested that the peculiar conditions of the upper atmosphere may be due to the arrival of short-wave, extremely penetrating radiation from some extra-terrestrial source. But this remains a guess rather than a well-supported theory. The close relation between daylight and the vagaries of wireless suggests strongly that the effect of the sun's rays on the atmosphere directly influences the propagation of wireless waves. Every amateur knows the differences in reception that come about near sundown, and exact observations show variations in the strength and in the direction of waves beginning with darkness. Under the influence of the sun impurities of a nature to disturb the ordinary course of ionization may perhaps rise to an average level.

But what has to be reckoned with most of all is the effect of thunderstorms on the upper atmosphere. Mr. C. T. R. Wilson has recalled calculations showing that at any moment about 1,800 thunderstorms are in active operation on the whole earth, and that the number of flashes of lightning is about 100 per second. Potentials may rise to 1,000 million volts, and 100 million billion ergs are, so to say, wasted every second. Incidentally, this waste is of little concern to us, as it is only about one ten-thousandth of the energy we receive from the sun. But a very much smaller voltage would be sufficient to give electrons energy corresponding with the most powerful extra-terrestrial radiation that has been suggested. A discussion at the Royal Society last week showed that from a number of different points of view a concerted exploration is being made of the upper atmosphere, and that the practical problems of wireless are leading us to a new knowledge of this difficult and interesting physical region.

MARTIANS AS ASTRONOMERS

ONE well-known astronomer received a telegram from the Editor of a daily paper as follows :— "Would you state very briefly your idea of a Martian, presuming you suppose such a being to exist? Should be most glad to publish your news." Twelve words were prepaid: the necessary address absorbed five: and the selected seven remaining were "Mars sky so clear Martians all astronomers." But the editor did not print the reply.

THE END OF SCIENCE

A long lifetime later, it is curious to contrast men's present attitude to the newest triumphs of research or discovery. As one succeeds another, all are taken more or less for granted. The metaphysical stage is over, its battles have died down, and the practical era has set in with a vengeance. The arcana of power are no sooner mastered than they are harnessed to the most prosaic needs. Does a criminal want catching? His finger-print is instantly given to the police in another hemisphere. The most trivial music or conversation is broadcast. The most vapid of human faces can be photographed across oceans. We can do now, at the last moment, simply by catching up time and abolishing space, things that ought to have been done weeks ago. But of awe, wonder, or the pangs of doubt, not a word is heard. Literature seems unable to digest all these devices, at least it takes no decided colour from them, perhaps because there is no time for the impression to fix itself. Very useful, of course, the sceptic will say, are many of the latest dodges for doing necessary but undistinguished things ; but how dull! Science has not found out everything, but there is a strong inkling that the thrills and raptures of scientific adventure are over. No future NEWTON or DARWIN is expected whose discoveries will seriously affect the framework of men's thought on the things that really matter. The universe turns out to be a less exciting place than it once promised to be now that such commonplace transactions as writing a cheque can be projected into it.

OBSERVING THE ECLIPSE

All that had gone before seemed insignificant in the amazing moments that followed. The sinister yellowish twilight faded away as though some unseen hand were turning off a gas-jet. The shadow rushed at us and enveloped us. We turned and found the sun blotted out in a sky akin to that of a full moon. To the extent of another diameter the disc was surrounded by a glowing, iridiscent, irregular circle of fiery light from which shot red and yellow flames. The spectacle surpassed anything that imagination had shaped for the eclipse. Half consciously one heard Mr. Horrocks counting seconds while mind and vision were concentrated on the glory of the corona. As "Twenty-three" was called a dazzling flash of reddish-white light, brilliant as that of molten steel run from a crucible, burst out from the right upper rim of the darkened sun. Watched by the naked eye, the flash appeared to bulge into a blazing oval. Suddenly there came realization that the darkness was passing, that the corona had vanished, that the moon's black disc on the sun could no longer be seen. From somewhere behind the camp came a sound of cheering. The eclipse was over, or as much of it as anyone now wished to see. Looking at a white sickle through a plate of smoked glass was only mildly interesting. Birds recovered their desire of song, the gathered sheep began to separate, and from the hillsides the crowds moved slowly to the highways.

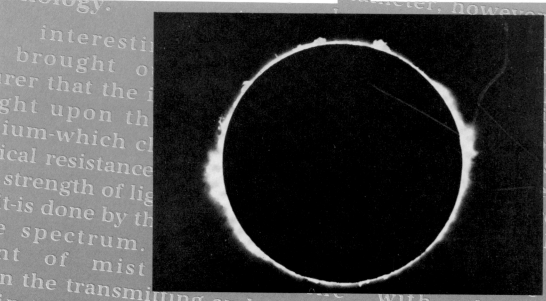

PHOTOGRAPH OF THE ECLIPSE OBSERVED FROM YORKSHIRE, ENGLAND

TAKING SOUNDINGS BY ECHOES

There are almost no more marvellous adaptations of science than the various types of apparatus for measuring the depth of water under a ship by measuring the time taken for a sound-wave to reach the bottom and be echoed back to the ship. Soundings are required for many purposes. From the earliest days of navigation approach to shoal water was ascertained by "heaving the lead," that is to say by throwing overboard a weighted line and noting the length that had been paid out when it ceased to sink. Until about 1921 sounding devices were little more than arrangements by which the primitive method could be carried out with more speed and accuracy and with more exactly calculated allowance for deviation from the vertical due to currents and the drifting or the actual motion of the vessel. For shallow depths great progress had been made. Down to about 40 fathoms, for example, the Schäfer gear enabled sounding to be carried out at great speed, even when the ship was under way.

About 1921 the British Admiralty and the United States Navy began to use methods on a new principle, that of the acoustic echo. A shock or impulse was produced on a ship, directed towards the bottom, and its echo, reflected back to the ship, was timed on a receiver. The velocity of sound through water is known, and therefore the time taken between the emission and the reception of the sound-wave gives twice the distance to the bottom.

Special Publication number 14 of the International Hydrographic Bureau (Imprimerie Robaudy, Cannes) gives a fully illustrated technical description of the newest French apparatus for echo-sounding. The whole apparatus is arranged to give one sounding per second. As the velocity of sound through water is approximately one-hundredth of a second for 24½ft., the operation is practically instantaneous in all shallow depths, so that the officer of the watch knows the depth of the water under his vessel at every instant. He is "feeling" the bottom at every moment, and thus in unknown waters or in fog the dangers of navigation are reduced, as running aground accidentally becomes practically impossible. The soundings are made between the depths of 4 and of 360 metres, although a special plant can be supplied for much greater depths, as for cable-laying or pure exploration. No corrections are required for rolling or for speed, but there is a possible error of one per cent. with an absolute maximum of one metre, due to changes of temperature and salinity of the sea altering the rate of propagation of the sound-wave through the water. The sound-wave is sent in a beam of such a nature that it returns unaltered from a flat surface, but is indented for unevenness such as may be caused by reefs, submarine cliffs or even submerged wrecks.

WIRELESS AND THE ECLIPSE

The results of recent research in wireless communication have shown that the action of sunlight on the upper layers of the earth's atmosphere has a most marked influence on the transmission of wireless waves. The effects of the transition from daylight to darkness are very noticeable under certain conditions in causing the fading of wireless signals, variations in the readings on direction finders, and so forth. It was therefore natural that the opportunity should have been taken of studying the effect of the eclipse of the sun upon wireless signalling in general. The results of the observations obtained should provide a valuable check of existing theories of the propagation of wireless waves and also give new information as to the effect of completely cutting off the sun's rays from the earth for only a few seconds instead of for several hours as normally occurs during every night.

The results obtained through all phases of the eclipse showed, as was expected, that the after-effects of sunrise had not ceased when the eclipse began to have an appreciable influence. At a station in Yorkshire practically on the centre line of the totality shadow, a direction-finder was used to study the signals received from the London broadcasting station at a distance of 200 miles. Readings taken at sunrise showed that the apparent bearings of the transmitting station were subject to the variable errors usually occurring at night for such ranges of transmission. As time progressed these errors decreased to an appreciable extent, and then began to increase considerably, as the phase of the eclipse approached totality. At the same time the signals received from London showed a very marked increase in strength. At 6 o'clock the time signals and the announcer's voice were only just audible, whereas about 20 minutes later the strength had increased so much that speech was audible a distance of several feet from the telephone.

A SPLIT STAR

Recently the Union Astronomer, Dr. Wood, received a message from Dr. Bernard Dawson, of La Plata Observatory, Argentina, reporting that the star *Nova Pictoris* was looking so strange and nebulous that he was unable properly to study it with his rather small telescope, and asking the Union Observatory, Johannesburg, to make an examination with its 26½in. instrument. This was done, and the Union Observatory discovered that *Nova Pictoris* was split in two.

The star is now being studied closely, as it may afford important information in connexion with the composition of matter and the structure of the universe.

In a statement published in the *Cape Times* this morning Dr. Spencer Jones, Astronomer Royal, Cape Town, points out that it is incorrect to say that the star Nova Pictoris has split in two. "There are two stars now," he says, "and there were two stars before, although we did not know that." According to Dr. Spencer Jones what has happened is a collision or a "grazing impact" between Nova Pictoris and another star.

TELEVISION

A remarkable achievement in the practical applications of physical science seems to be well on the way. Belin in France and Alexanderson and Jenkins in America have been able to transmit somewhat crude shadowgraphs over the telephone line and in the course of the year Mr. John L. Baird, a former student of the University of Glasgow, has given demonstrations which seem to show that he is well on the way to transmitting recognizable images of moving objects. An image of the object to be transmitted is passed over a light-sensitive cell in a series of strips. The modulated current from the cell is transmitted over a telephone line, and, after amplification, is used to control the light of a glow discharge lamp. By a revolving slot-shutter a point of light from the lamp is caused to travel over the field of vision in exact synchronism with the passage of the image over the cell at the transmitting station. The amount of light and shade in the image is sufficient to secure the recognition of the features of the person serving as the image. Future developments will be awaited with the keenest interest, but at least the possibility of "television" has been demonstrated.

WIRELESS MESSAGES TO MARS

Dr. Mansfield Robinson, a London lawyer, believes that he has been in communication with Mars, and at his request wireless messages will be sent out from Rugby on an 18,500-metre wave-length on Wednesday morning in an effort to get official proof that "contact" has been established. "The Post Office has agreed to accept the message for Mars, not as a scientific experiment for which it is responsible but as a commercial proposition," an official explained. "Obviously we do not guarantee delivery. No doubt Dr. Robinson has chosen Rugby because it is the largest and most powerful wireless station in the world and therefore most likely to reach Mars. We shall send the message in the same way that we send to ships in all parts of the world every day—and the charge will be the usual 1s. 6d. a word, which is admittedly a low cost for such a long distance! But it will cost no more to transmit than an ordinary message. At Dr. Robinson's request a watch will be kept for half an hour at St. Albans wireless station on a wavelength of 30,000 metres, and for this service of listening for a reply a special charge will be made. The experiment will probably be about 2.15 a.m."

SNAPSHOTS IN COLOUR

I have just been shown the methods and some of the amazing results of a new process by which the amateur photographer, using the camera he now possesses, will be able to take photographs in colour with the same ease that he now takes black prints, and at a cost only 3½ times as great. He can buy a special roll film, insert it in an ordinary camera, give one exposure at a 75th or less of a second, in any reasonable light, producing simultaneously three negatives, approximately blue-sensitive to be printed in yellow, green-sensitive to be printed in red, and red-sensitive to be printed in blue. He can send the exposed rolls through the ordinary agencies to service stations arranged by the new organization, and in due course will receive as many prints as he wishes of his snapshots in colour. If he likes to do his own developing and printing, he can buy the necessary materials and produce his own results with no apparatus that he does not already possess.

The roll-film or the plates of the new system are composed of three layers, being, in fact, an adaptation of the well-known "tri-pack" device in which three emulsions are used, sensitive respectively to three wave-length bands within the range of the visible spectrum. It has usually been the practice to interleave the sensitive surfaces with dyed filters to act as compensators or correctors to the emulsions, but in the new system the emulsions have been improved so that they themselves suffice. Hitherto, moreover, the blue-sensitive (yellow-printer) film has been coated with a chloro-bromo emulsion which gives an extremely close approximation to the theoretically required cut-out of rays. But chloro-bromide emulsions are very slow, and it has, therefore, been necessary to place it in front of the pack. But this film is relatively unimportant with regard to definition, the blue and red printers being the elements which go to make up the outline and so to form the ultimate picture. In the new system a bromide emulsion of extremely high sensitiveness has been selected, and this yellow-printer is placed at the back of the pack, where its unimportance in definition is of no moment, and its value in giving colour tone has full scope. Either the red or the blue printer may be placed in front, as they are practically equal in definition. The speed sensitivity of the tri-pack is determined by the sensitivity of the front emulsion, and with this ingenious arrangement, the overall speed is as high as in ordinary roll films.

STELLAR ENERGY

Every star is continually radiating energy into space, and we have no knowledge of any appreciable part of the energy coming back or of the [stars replenishing themselves. Every square inch of the sun's ten thousand million million million square inches of surface is radiating energy at a rate comparable with the output of a 50 horse-power engine. If the sun got this energy from a power station, coal would have to be burned at the rate of about a million million tons a minute. All the recently current theories of how the sun got its store of energy cannot account for reserves sufficient to supply this output, even for the relatively short past of the earth. Even radio-activity, the solution of many difficulties, will not meet this case. Sir Ernest Rutherford has calculated that if the sun had originally consisted of pure uranium it could not have given off its known output for more than, say, five million years.

It is an inference from the theory of relativity that a body losing energy by radiation must also be losing mass. The possible, and so far only sufficient, theory of this production of energy and concomitant disappearance of mass is to suppose that the positive and negative charges of the electrons and the proton in the atom rush together as when the positive and negative knobs on a Leyden jar are put in contact. The result is radiation, but in the case of the atom there is no Leyden jar left behind, as all matter, so far as is known now, consists of charges of electricity in opposition. This would mean that the daily loss of mass of the sun, and of every star, implies an absolute annihilation of matter and a stage in the disappearance of the universe.

TREES FELLED BY THE TUGUSKA METEORITE IN SIBERIA

VIEWS ON MATTER

Physical science was never in a more exciting state. Its recent discoveries have outrun the power of composing them into a consistent scheme of the universe. In his presidential address to the British Association at Glasgow, last year, Sir William Bragg gave a striking example of the present confusion. In the 19th century observation and theory combined to support the belief that light was a series of waves in an all-pervading ether. Great advances in knowledge were based on that belief, and many inferences from it were verified. There was no question of its truth in the ordinary sense. But in the 20th century new observations and new theory combined to support the belief that light was particulate, a stream of small particles. This belief also passed the experimental test, inferences from it being justified by new observations. But could radiation be at once a continuous wave and a stream of discontinuous particles? It seemed incredible, and yet physicists had to accept both, working on one hypothesis, say, on Mondays, Wednesdays, and Fridays, on the other on alternate days, in each case being justified by their results.

Last month Dr. Irving Langmuir, of the American General Electric Company, one of the most distinguished of living physical chemists, discussed the changing concepts of physics in a striking essay contributed to the American journal *Science*. He recalled that in 1913 Bohr had developed a new theory of the atom by combining the "quantum" theory of Planck (on whom the Copley Medal, the highest honour of the Royal Society, has just been conferred) with Einstein's theory of relativity. The result was a mechanical model of the atom as consisting of electrons revolving in orbits about the nucleus, according to laws which were partly those of the classical mechanics and partly inconsistent with these laws. The model enabled him to form certain mathematical equations from which he could calculate the frequencies corresponding with the different lines in the spectra of hydrogen and other elements. The agreement between the theory and experiment was practically perfect, and this success made most physicists and chemists believe that Bohr's model, at least for the hydrogen atom, was substantially correct, that in fact in the normal hydrogen atom the electron really described a circular orbit round a nucleus with a diameter and a frequency given in the model. But almost insuperable difficulties arose, and Heisenberg and Born found it necessary to discard the mechanical model and to develop a mathematical theory of the atom which was carried farther by Schroedinger, de Broglie, and others, until it led to a conception of the whole atom with its electrons as a wave phenomenon. The long-standing conflict between the wave theory of light and the corpuscular theory, Langmuir thinks, has now dis-

appeared. To ask whether an electron is a particle or a wave or whether light consists of waves or of particles are meaningless questions. Both are particles or waves according to the kind of operations that we may perform in observing them. Sir William Bragg's jest about the views to be taken on alternate days is a simple statement of fact.

In a part of his essay Langmuir insists that the conception of the relation of cause and effect hitherto universally accepted in science has lost its meaning. The argument is not easy to follow. It seems to depend partly on the changed conception of time. If time is no longer to be accepted as an arrow with a point and a tail, a past and a present, but as a fourth dimension, varying for any event with the other dimensions according to what may be called crudely the position of the observer, then cause and effect are merely two points in a system. But there is a more difficult conception. Langmuir argues that a question has no scientific meaning unless an operation is conceivable by which the answer could be given. Atomic processes, he says, seem to be governed fundamentally by the "law of probability." In the past men of science chose for their operations phenomena in which such enormous numbers of individual quantum phenomena were grouped together that the result could be determined only by their averages. For example, when the variation of the pressure of a gas with its volume was studied the forces that were measured resulted from the impacts of great numbers of molecules, the average force remaining steady and definite. But if there were only one molecule in a small volume the pressure exerted on the walls would be zero, except for those instants at which the molecule struck the wall. It would be impossible to predict in advance what the pressure would be at any particular time. By varying the conditions the probability could be altered that there would be a certain pressure at a certain instant, but in no absolute sense could the chain of cause and effect be predicted.

He admits a possible escape. If phenomena are to be described in terms of ordinary space and time causality has to be abandoned. A helium atom, for instance, can be described in terms of the motion of two electrons in ordinary three-dimensional space, if we are content to know only the probabilities that the electrons may be found at any point in this space. But the motions of the electrons, or at least of the waves corresponding to them, can be completely described in the case of the helium atom by a quantity which has a definite value at each point in six-dimensional space.

The sequence in the hunt after reality is interesting. The relativity theory, with its transformation of time into a dimension, seemed to abolish any old-fashioned conception of reality, to the great joy of metaphysicians. Then the stubborn "quantum," an obstinately recurring fact of observation, came on the scene and led to wave mechanics, and wave mechanics are now recovering cause and effect as steady and calculable relations in a six-dimensional reality.

MOVING PICTURES OF THE MOON

Cinematograph pictures of the moon are now an accomplished fact according to an announcement made yesterday by Dr. John Q. Stewart, Associate Professor of Astronomical Physics at Princeton University. A wonderful panorama of the early light of dawn creeping across the lunar landscape at the rate of only nine miles an hour, and showing the valleys still in shadow when the mountain peaks are already illumined, there being no glow of twilight as on the earth, has been secured at Princeton University Observatory. The giant lunar crater of Copernicus looms clear in the centre of the visible area, with its walls rising two miles from the level of the moon's floor, and forming an almost perfect crater circle.

The mechanism for taking these moving pictures, which was designed by Mr Robert Fleming Arnott, a consulting engineer, of Montclair, New Jersey, and his assistants, consists of a Victor motion picture camera operated by a special electric motor. Both camera and motor are fastened to the eye of a 23in. telescope, the lens of which is nearly 2ft. in diameter.

The pictures were taken one at a time every six seconds, which is one hundred times slower than the normal rate, and 2,000 pictures are contained in the 50ft. long film of Copernicus Peak. The device discloses in a few minutes that which the astronomer sees only during many hours of careful watching through the telescope.

METALLIC ALLOYS

Dr. Guillaume is known to the world for the invention of three metallic alloys of great importance—invar, elinvar, and platinite. The latter is a nickel-iron alloy having approximately the same temperature co-efficient of expansion as glass, so that it can be fused into glass and used as the wire for introducing the current into electric incandescent lamps. The wire is frequently covered with a thin coating of copper to which the glass adheres, and in this form the wire is known as "red platinum." As about one thousand million lamps are made annually the saving between the cost of platinum wire, which was the only suitable material formerly available, and the alloy approximates to £1,000,000 per annum.

Invar, a nickel-steel discovered in 1896, has practically no temperature co-efficient of expansion; so that if a surveying tape is made of it the length is almost unaffected by temperature. This quality is of the greatest service in accurate surveying work and has reduced the time taken in such a survey to 1-50th of that taken in a survey of a few years ago. The pendulum rods of all modern first-class clocks are made of the same material.

Elinvar was invented by Dr. Guillaume for the manufacture of the balance springs of watches. The co-efficient of elasticity of elinvar does not change with temperature, so that the control exerted by the spring does not vary with temperature, and this has immensely improved the behaviour of the watch to which it is fitted. It is estimated that about 5,000,000 watches are made annually in which the balance springs are of this material.

Dr. Guillaume was appointed Director of the Bureau des Poids et Mesures in 1915 and received the Nobel Prize for Physics in 1920.

What will Yale say ?
UNIVERSE FOCUS THOUGHT FOUND.
**Middle of Sidereal Galaxy
Is Located at Harvard.**

Newspapers, like physicists, saw nuclear physics as the key area of development in physics during the 1930s. The flood of new results in nuclear physics stemmed, in part, from the appearance of new instrumentation for accelerating particles to high energies (high that is by comparison with what had gone before). Collisions between these particles and the nuclei in a target could lead to a variety of transmutations. Of these devices, the one most referred to by the newspapers was the 'cyclotron', introduced by E O Lawrence and his colleagues in California. However, Cockroft and Walton's linear accelerator in England also gained publicity, mainly from its use in the discovery of the neutron in 1932.

Antenna developed for new short-wave radio transmissions in Germany

Despite this growth in laboratory-based methods of disintegrating nuclei, recording the effects of cosmic-ray impacts continued to be an important alternative source of information. The most important advance made in this way during the 1930s was the discovery of the positron by C D Anderson, whilst studying cosmic-ray tracks in a cloud chamber. Though newspapers were impressed by the new accelerator—'atom smashers', as they sometimes called them—the mysterious nature of cosmic rays intrigued them still more. Varying and usually conflicting views of cosmic rays and their origin continued to appear during the 1930s. It was known that the particles observed at the Earth's surface were not the same as the primary rays in space. As the latter ploughed through the Earth's atmosphere, they interacted with the nuclei of the air to produce different particles at the Earth's surface. One reason for the continuing interest in high-altitude balloon flights during the 1930s was that they could be used to gain a better knowledge of the influx of primary cosmic rays.

A number of researchers continued to use naturally occurring radioactive materials as their source of particles for nuclear bombardment. The Joliot-Curies in France, using a polonium source, detected the occurrence of artificial (or induced) radioactivity in 1933–34. Their work was rapidly followed up by E Fermi in Italy, who showed that most stable elements could be transmuted to radioactive forms by neutron bombardment. The developments in this area occurred so quickly that physicists, and consequently newspaper reports, are sometimes hazy as to the precise interpretation of the observations.

One constant newspaper query was whether advances in nuclear physics would lead to methods of producing large quantities of energy. Most leading physicists doubted this until near the end of the 1930s, when radioactive fission was discovered (but not widely reported). At the same time, it had been accepted since the 1920s that stars must be obtain their energy from burning some kind of sub-atomic fuel. The problem was pinning down the exact process at work. In 1938, H Bethe, a German physicist who had recently emigrated to the USA, proposed the first feasible (and still accepted) mechanism. This involved the transmutation of hydrogen to helium in stellar interiors.

If atomic nuclei and fundamental particles dominated much physics reporting of the 1930s, Einstein had not been forgotten. Newspapers noted his attempts to find a unified theory that would bring together quantum and gravitational forces. He was also involved in a change of view regarding the observable Universe. The idea of a finite universe, which Einstein had supported in the 1920s, became less compelling in the 1930s. Work by the American astronomer, E Hubble, showed that the Universe as a whole was expanding—an observation that could be reconciled with infinite dimensions.

So far as astronomy was concerned, much newspaper attention was concentrated closer to home. The major scientific event of 1930 was the discovery of a new planet beyond the orbit of Neptune. (This planet, subsequently called 'Pluto', is the last to have been discovered to date.) It was found by a young assistant in Arizona. Lowell, himself, had initiated the search for the planet: though he died in 1916, the search was continued. Pluto quickly became the subject of controversy. Lowell had predicted its position from apparent gravitational effects on the orbits of neighbouring planets. His prediction had assumed that the new planet would be massive. In fact, Pluto proved to be much fainter, and therefore presumably much less massive, than had been expected. This led to the immediate question (still unanswered)—was Pluto really the planet that Lowell had predicted?

Even closer to home, interest in the physics of the terrestrial atmosphere remained strong during the 1930s. Studies of the upper atmosphere using radio waves continued to expand. Such work provided the background and skills for the development of radar (though this aspect was not noted by the pre-war press).

Another atmospheric concern related to the study of the past climate on Earth. Attempts to explain climatic variations continued to differ. One of much importance later, proposed by the Yugoslav, M Miklankovitch, was ignored at the time—an indication of the geographical limitations of scientific communication and, hence, of its reflection in the press.

1930s CHRONOLOGY

1930
- Astbury begins x-ray diffraction studies of proteins
- Pauli postulates the existence of the neutrino
- Superfluid helium is detected
- Whittle applies for jet engine patent

1931
- First electron microscope developed
- Gödel's theorem propounded
- Lawrence works on first cyclotron

1932
- Anderson discovers the positron (predicted by Dirac)
- Chadwick discovers the neutron
- Jansky detects radio noise from the Milky Way

1933
- Meissner effect discovered in superconductors

1934
- Cherenkov radiation detected
- The Joliot-Curies produce the first artificial radioactive isotope

1935
- The Turing machine is proposed
- Watson-Watt and colleagues develop the first workable radar
- Yukawa predicts the existence of mesons

1936
- DNA is isolated
- The prototype of the field-emission microscope is developed

1937
- Anderson discovers the mu meson
- First rocket tests are carried out at Peenmünde

1938
- Fission of uranium achieved
- Hydrogen-burning process in stars identified

1939
- Second World War starts
- It is confirmed that uranium fission reactions can be made self-sustaining

1930s

GENERAL RELATIVITY RIOTS

A mob of 4,500 persons rioted to hear about the Einstein theory at the American Museum of Natural History at 8.10 o'clock last night.

The crowd, which had gathered in the main auditorium among the big meteorites, resented the fact that the uniformed attendants were trying to exclude those who did not have tickets. Fearful of being excluded from the lecture altogether, a group of young men suddenly charged the four or five attendants who were guarding the door which leads into the Hall of the North American Indians.

SOMETHING NEW UNDER POLICE SUN.

Shouting exultantly, the rioters swarmed between the canoes, totem poles and walrus tusk ornaments to the main auditorium, where they seized the remaining seats. After the attendants had once been butted aside, the men, women and children in the meteorite hall surged through.

The less agile were knocked down and stepped on. Women screamed. The man-handled attendants, as soon as they could find an opening, ran for help. The doorman telephoned for police, and in a few minutes uniformed men were rushing into the great scientific institution on a mission that was new to Police Department history—quelling a science riot.

The battle was not over with the victory in the first skirmish. Frantic knowledge-hunters begged to be admitted to hear about Einstein. Bruised and tearful women pleaded with the overflow mob which blocked the doorway not to bar them from a chance to learn if Einstein were right or wrong.

THE NEW PLANET

The discovery of a new planet beyond Neptune was the result of 25 years' systematic search by the staff of the Lowell Observatory at Flagstaff, Arizona, which began when Dr. Percival Lowell, the founder of the Observatory, calculated that irregularities in the movements of Neptune indicated the presence of a body beyond that planet. A search of the skies by photograph, was started then in accordance with Dr. Lowell's calculations, and has continued to this day.

Early last year, according to the announcement made by the Observatory yesterday, the Lawrence Lowell telescope, a highly efficient special instrument for the search, was put in operation. Some weeks ago (January 21) an object was detected by telescopic photography on a plate which has since been carefully followed. It has been observed photographically with the large Lowell reflector by Mr. C. O. Lampland, and it has been observed visually with the large refractor by various members of the staff. All the observations indicate that the object is the one which Dr. Lowell saw mathematically. Two years before his death Dr. Lowell published his " Memoir on a Trans-Neptunian Planet," in which he called the body yet to be discovered " planet X." Referring to one of his mathematical calculations, he wrote, " It indicates the existence of an unknown body with a mass between those of Neptune and of the Earth, with a visibility of 12-13 magnitude, according to the albedo, and a disk of more than one second in diameter."

The Harvard astronomers give the position of the new planet as " close to delta Geminorum, within 7 minutes time from Geminorum," or technically, as Right Ascension 7 hours 15 minutes, and Declination North 22.5deg. They believe that because of its tremendous distance from the sun it receives the light of sun with a brilliancy at most hardly exceeding that of moonlight, and they put the time it takes to go round the sun at not less than 330 years.

The distance of " planet X " from the sun has been estimated at Flagstaff as 45 times the distance of the earth from the sun, or 4,185,000,000 miles. It will be at least a month before the actual size and mass of the planet can be roughly determined, and probably several months before an accurate measurement of its dimensions may be had.

Yesterday—when the announcement was made of the discovery of the planet which Dr. Lowell had predicted would be found—was the birthday of the late astronomer.

CONTROVERSY OVER NEW PLANET

Dr. V. M. Slipher, director of the Lowell Observatory, Flagstaff, Arizona, to-day reaffirmed his belief that the new astral body photographed at Lowell Observatory at the end of January is actually a planet. He said :—

On March 13, when the announcement of the discovery was made, we believed it to be the ninth member of the planetary family and so stated. To-day we are more firm in that belief in that all our observations and computations confirm the first announcement . . . Our orbit computations and observations since February show definitely it is trans-Neptunian and afford strong evidence of planetary nature. It has no cometary appearances. Computations have shown that the planet's orbit is more eccentric and extensive than was first believed. The orbit is in the form of an ellipse. . . .

Computations have reached a point where it can be said that the distance from the earth is approximately 41.35 astronomical units, that is, 41 times as far as from the earth to the sun.

Professor Leuschner, director of the Students' Observatory, University of California, yesterday announced his agreement with Dr. Slipher as to the trans-Neptune nature of the object but declined to accept it as a planet. He based his view on calculations made at Swarthmore University Observatory and at his own observatory. He said : "Computations of the mass indicate that the object is too small to have caused the deviation in Neptune's orbit which originally led to the suspicion that a trans-Neptunian planet exists." He thinks it may be a large asteroid lately disturbed in its orbit by a close approach to major planets such as Jupiter, or, maybe, one of many long period planetary objects yet to be discovered, or a comet.

Both Professor Leuschner and Dr. John Stewart, Associate Professor of Astronomical Physics at Princeton, agreed yesterday that although the latest evidence failed to support the original planetary theory, the new discovery is of the highest significance because it extends the limit of the solar system, and brings to knowledge a celestial body of possibly unique characteristics.

MICHELSON AND INTERFEROMETERS

The interferometers invented by Professor Michelson, of which the first was used for carrying out in 1887 the famous Michelson and Morley experiment, have been applied by him, always with complete adequacy of design, to other important and difficult problems, most of them of audacious novelty. These problems included the measurement in 1892 and 1893 of the metre in wave-lengths of light ; the measurement of the diameters of stars ; the re-measurement of the earth tides ; and the testing of the effect of the earth's rotation on the velocity of light.

The measurements have had great consequences. For example, the difficulties of reconciling the result of the Michelson-Morley experiment with the then prevailing physical conception of the nature of the universe were the direct cause of the inquiry by Einstein, which resulted in the theory of relativity. The measurement of the metre in wave-lengths of light resulted in establishing a standard of length free from the uncertainty concerning possible variation which attaches to all material standards. The interferometer for the measurement of the diameter of stars, suggested by Michelson in 1890, and first applied by him to Betelgeuse, has not only confirmed the correctness of the previously almost incredible dimensions yielded by indirect means of calculation, but has detected fresh stellar phenomena in the variable diameter of Mira Ceti, and the separation of double stars too close for resolution by the unassisted telescope.

DISINTEGRATION OF A NITROGEN NUCLEUS BY AN ALPHA PARTICLE

" Cold weather was experienced in Edinburgh yesterday. The Sun shone, but its altitude was low for this time of the year."—*Local Paper.* Depressed by the cold, no doubt.—*Punch.*

CHURCHILL VISITS LICK LAB

THE visit of Mr. Winston Churchill to the Lick Observatory seems worthy of a Note, if he will permit us to transcribe a few of his words to the *Daily Telegraph* of Dec. 23. He had been to the top of the tallest building in San Francisco, and thence telephoned to his wife in England. " We hear each other as easily as if we were in the same room, or, not to exaggerate, say about half as well again as on an ordinary London telephone." And then he recounts his visit to Mount Hamilton :—

I sit upon a ladder. The planet Saturn is about to set ; but there is just time to observe him. Of course I know about the rings round Saturn. Pictures of them were shown in all the schools where I was educated. But I was sceptical. We all know how astronomers have mapped the heavens out in the shape of animals. We can most of us—by a stretch of imagination—recognise the " Great Bear," but still one quite sympathises with those who call it " The Plough." " Bear " or " Plough "—one is as like it as the other. So I expected to see, when I looked at Saturn, a bright star with some smudges round it, which astronomers had dignified by the name of rings.

SATURN'S RINGS.

In this mood I applied myself to the eyepiece. I received the impression that some powerful electric light had been switched on by mistake in the Observatory, and was in some way reflected in the telescope. I was about to turn and ask that it might be extinguished, when I realised that what I saw was indeed Saturn himself. A perfectly modelled globe, instinct with rotundity, with a clear-cut lifebuoy around its middle, all glowing with serene radiance. I gazed with awe and delight upon this sublime spectacle of a world 800 million miles away.

THE AGE OF THE UNIVERSE

If the expanding universe were accepted as an established fact, its most immediate reaction was on the time-scale of evolution. Three main time-scales had been favoured at one time or another which we might distinguish as " short," " intermediate," and " long." No one now had a good word for the short Kelvin time-scale ; and practically our choice lay between the intermediate scale giving the sun an age of the order 10^{10} years and the long scale giving an age of 5.10^{12} years. When there was no definite evidence one way or the other the longer time-scale naturally got the preference. The more time allowed, the more could happen ; so the policy of the evolutionist was to grab as much time as possible. That, rather than any striking success, accounted for the popularity of the long time-scale in recent years.

The hypothesis of the long time-scale came about through Professor Einstein's theory which gave the total amount of energy in a given mass of matter. We knew just how much energy there was in the sun and could calculate how long it would maintain the radiation if it could all be released. To release it all it was necessary that protons and electrons should annihilate one another, thereby undoing the lock which fastened the energy in. That idea of the source of a star's energy seemed to have been first mentioned by the lecturer in 1917. It was the only adequate source that could be suggested at the time ; but in 1920 a possible alternative was recognized in the energy released by the transmutation of hydrogen into higher elements. That alternative, however, sufficed only for the intermediate time-scale.

With the universe doubling its radius every 1,300,000,000 years, it was obvious that the long time-scale of billions of years was altogether incongruous. It was true that we could not set any definite limit to the time occupied by the first slow development of the expansion. But if there were billions of years to choose from it was strange that the evolution of our own solar system should coincide with the relatively short interval between the " bursting of the bubble " and the complete dispersal of the galaxies.

CURVATURE OF THE EARTH

What was described as the first photograph to show the curvature of the earth's surface was exhibited last night to members of the American Association for the Advancement of Science at their convention in Cleveland, Ohio, by Dr. C. E. K. Mees, Research Director of the Eastman Kodak Company.

The photograph was recently taken from an aeroplane by Captain A. W. Stevens, United States Army, over a small town in South America. By using supersensitive plates he succeeded in photographing what was invisible to him, a range of the Andes some 300 miles distant and the peak of the volcano Aconcagua, 320 miles distant, as well as the unbroken Pampas intervening. The mountain range appeared as a straight horizontal line, but the distant horizon of the Pampas was not straight, but bent slightly downward at one end.

When Jeans grows too didactic
Or Friedman makes too free
Among extra-galactic
Clusters of nebulæ—
When life is full of trouble
And mostly froth and bubble,
I turn to Dr. Hubble,
He is the man for me.

WHAT IS A QUANTUM ?

What is the quantum ? Is it possible for those who are not modern mathematicians to get any approximate notion of a conception which has caused a more revolutionary change in physical reasoning than Einstein's relativity, and that is taken, probably rashly, as a scientific reinforcement of our instinct to believe in the freedom of the will ? It is not easy, and any rough picture of the kind of idea the quantum is can only be temporary, because although the facts which led to the theory began to be known about 1900, the quantum is assuming novel, enlarged, and more abstractly mathematical interpretations or applications almost day by day.

Begin with a crude image. Imagine a large reservoir filled with fluid and provided with a tap. By manipulating the tap you can get the liquid to flow in a steady, unbroken stream. By alterations in the tap you could get similar steady streams whether the liquid were a very light substance like alcohol, or whether it were thick oil, or molten lead, or mercury. Such a steady stream is an example of continuous motion, and for long it was thought that the radiation of light, or of heat, and many other physical occurrences was continuous. The classical electro-magnetic theory of light assumed, and at the same time seemed to prove, the continuity of radiation.

It is possible to adjust a tap so that the continuous stream will be transformed into a set of drops. The size and weight of the drops, of course, will vary with the kind of liquid, the head of pressure, and many other factors. Now imagine a kind of tap, which doubtless is impossible, that lets nothing but drops through, whatsoever the liquid. Suppose next, what would be quite within the power of physicists, that you weighed the drops of all the different kinds of liquids, and set them down on paper—alcohol drop so much, water so much, mercury so much, molten lead so much. Naturally you would not expect to find any arithmetical relation between the different figures you had set down. But suppose you found, to your great surprise, that there was a numerical factor common to all the numbers which we will call " X," of such kind that alcohol, say, was one time " X " and the drops of all the other liquids twice or three times or fifty times " X," but never " X " multiplied by anything except a whole number ; you might then call " X " a " constant " because it was always turning up, faithfully and exactly, or a " quantum " because it appeared that a drop must be of that weight or some exact multiple of that weight to get through the tap. In this imaginary case you would probably attribute the quantum to some magical property of the tap, and in so doing you would be acting not very differently from some of the interpreters of the quantum in physics.

So far as I can ascertain, the first appearance of the quantum theory came, like most great discoveries in science, from an endeavour to measure accurately a well-known physical phenomenon, just as Ramsay discovered argon in the course of experiments whose object was to measure more accurately than had been done before the exact composition of the atmosphere, and in particular the proportion of nitrogen it contained. Experiments were made on the radiation of heat from a " black body "—that is to say, from a body which absorbs completely all the heat which falls on it and which thus reflects, transmits, and scatters none. A uniformly heated enclosure was constructed with blackened walls having a very small opening. The radiation emitted through the aperture was as nearly as possible a typical " black radiation." It was then found, partly by observation and partly by calculation, that bodies did not grow so cold as they ought to do if they parted with their heat continuously and not by jumps.

Professor Max Planck, of the University of Berlin, calculated the dimensions of the jump that would be necessary to account for the facts, and afterwards it was found that the size of this jump reappeared in a large number of quite different physical phenomena, either directly or multiplied by a whole number. The " jump " is Planck's constant and is the basis of the quantum theory. It appears that energy almost, or, perhaps, actually of every kind, does not act continuously, but in jumps, and the size of the jump is a multiple of the quantum.

The quantum is an almost incredibly small quantity. If we take the word billion in its proper sense of a million of millions, then a billion billion times Planck's quantum would be a quantity just appreciable in the everyday world outside a laboratory with its instruments of amazingly minute precision. But it comes to be of great importance in the theory of the electron. On the supposition that an atom is a kind of solar system of electrons revolving round a central proton, then Planck's quantum appears in the following form. If the electron is in its smallest possible orbit, then the quantum is exactly the circumference of the orbit multiplied by the velocity and again by the mass of the electron. Moreover, an electron cannot pass from an orbit of one circumference to an orbit of another circumference except by a jump which is the quantum multiplied by a whole number. The electron appears to have no existence in any intermediate condition. And at least for the time being there is the principle of "indeterminancy" taken as a basis for free will by certain persons; in the words of Eddington, " a particle may have position or it may have velocity, but it cannot in any exact sense have both," or, in other words, it is supposed not to be determined by any chain of causation.

So far contemporary physicists; but to those who are not physicists it would seem clear that this obstinate quantity points to some relation not yet understood, and the " indeterminancy " of electrons and of waves below the electrons may mean only that science has not yet understood the relations, rather than that the relations are lawless.

PROTONS AND NEUTRONS

The simplest of these systems was that of the hydrogen " atom," its nucleus consisting of one proton with a unit charge of positive electricity, with a single electron as its satellite, so that when the electron was detached the system had the single positive charge of the proton. Practically all the mass was assumed to be in the proton, which was calculated to have about 1,850 times the mass of the electron, but the dimensions of the proton and the electron were assumed to be roughly identical. The proton or hydrogen nucleus was therefore regarded as a particle of unit mass, carrying a unit charge of positive electricity. But with the arrival of the conception of protons and electrons as waves, as well as particles, the Solar system model is also well on the way to be little more than an honoured ghost.

The next most simple system was the helium atom, consisting of a nucleus of four protons and two electrons very intimately bound together, and two loosely held satellite electrons. The helium nucleus, known also as the " alpha " particle, which appears in many forms of radiation, had a positive charge of two units and a mass four times that of the hydrogen nucleus, the mass of the electrons being taken as relatively negligible. The helium nucleus appeared to be a system of extreme stability, and it was a useful supposition that all the elements in the periodic table had nuclei consisting of the greatest possible number of helium nuclei with hydrogen nuclei to make up the difference. Thus phosphorus, with the atomic weight, 31, could be built up of seven helium nuclei giving 28 of its mass, and three hydrogen nuclei to bring up 28 to 31.

But this attractive picture may be only another ghost. Already in 1920, in a Bakerian lecture to the Royal Society, Rutherford suggested the possibility of a proton or hydrogen nucleus having intimately associated with it, in the fashion of the electrons of other nuclei, a single electron. Such a combination would form a new kind of body having mass 1 and no charge. For such a body he proposed the name " neutron " and predicted that it would have properties of a novel kind. As its external field would be practically zero, it would be able to move freely through any kind of matter and it might be impossible to retain it in any sealed vessel. During last year, Dr. Chadwick, working in the Cavendish Laboratory, has shown that it is possible to excite from Beryllium a radiation of great penetrating power. The most probable interpretation of this is that it consists of particles of mass 1 and of zero charge, that is to say, of Rutherford's " neutrons." Thus a third arrangement of matter, the neutron, may have to be added to the proton and the electron in future models of atomic systems.

WHEN A COMET HIT THE EARTH

According to a statement made at the Convention of the American Association for the Advancement of Science, which is meeting in Atlantic City, between 100,000 and 1,000.000 years ago the earth was in collision with a giant comet, larger by one-third than Halley's. The comet was smashed to pieces after it had devastated an area of about 190,000 square miles of the earth. This, it was urged, was the most probable explanation of the elliptical depressions, averaging half a mile in length, which are to be found in numbers in the Carolinas.

Dr. F. A. Mellon and Dr. William Schriever, of the University of Oklahoma, who put forward this hypothesis, and photographed the depressions from the air and measured them on the ground, pointed out that the long axes of all the elliptical depressions were, as far as could be seen, parallel, running from north-east to south-west; that elevated rims completely encircled some of the depressions; and that there was a general increase of ellipticity with size. These facts and many others they cited seemed to permit the interpretation that the depressions were made by a shower of meteorites striking the earth at an angle.

COCKCROFT AND WALTON'S APPARATUS FOR NUCLEAR DISINTEGRATION IN THE CAVENDISH LABORATORY AT CAMBRIDGE, ENGLAND

THE SUN AND RADIO

The total eclipse of the sun which is to be observed in Canada in August is to be made the occasion of an experiment to decide between rival wireless theories.

It is now well known that the two ionized layers in the upper atmosphere, which reflect wireless waves, have their electrification replenished every day by means of solar influence, but the exact nature of the solar stream which is effective is by no means so certain. There is a good deal of evidence which suggests that the upper or Appleton layer is caused by ultra-violet light, but in the case of the lower Kennelly-Heaviside layer it is not yet settled whether ultra-violet light or swiftly moving atoms constitute the agency in question.

At a discussion which took place at the Royal Astronomical Society in January of this year Professor Appleton pointed out that the question of ultra-violet light versus moving particles could be settled by making wireless observations during an eclipse, for the effect of the cutting-off of the ultra-violet light would be immediate, whereas the eclipse of the hypothetical particles, because of their lower velocity, would not be so closely correlated to optical totality.

The question was further discussed by Professor S. Chapman, who is the chief exponent of the particle theory, and he has since shown that, in consequence of the motions of the moon and the earth, the stream of supposed particles will be interrupted by the moon more than an hour before the ultra-violet light is stopped, so that what may be termed the "particle eclipse" may, in fact, be over before the "optical eclipse" begins. The "particle eclipse" region, it has also been calculated, should lie to the east of the band of optical totality.

On August 31 the "optical eclipse" will be observable in Canada, but the "particle eclipse" effects, if they occur, are to be expected in Greenland and Western Europe. Special wireless observations at a site on the track of the ultra-violet light eclipse are to be made in Canada by Professor A. S. Eve and Dr. J. T. Henderson. If they experience a partial return to night-time conditions round about totality it will show that ultra-violet light is the principal ionizing agency; but if, on the other hand, the same kind of effect is observed in Europe about two hours before the totality occurs in Canada the experiments will decide in favour of the moving particle theory.

Although the atmospheric conditions which prevailed during the recent solar eclipse in America prevented successful astronomical observations, the reports which have now reached this country show that the special radio experiments carried out on that occasion were completely successful. Since the object of these investigations was to examine the effect of the eclipse on the electrical conditions in the upper reaches of the atmosphere their prosecution was not dependent in any way on the weather.

To test the ultra-violet light theory special observations were made in America under the auspices of the Canadian National Research Council and of the Bureau of Standards in Washington. In both cases these radio observations confirmed the British results obtained in 1927, for a partial return to night-time conditions was experienced at totality in the shape of a diminution of the electrification in both the Kennelly-Heaviside and Appleton regions of the upper atmosphere. The ultra-violet light theory of Professor Appleton can therefore be regarded as having received complete support.

The results of experiments carried out on more easterly sites to test the corpuscular hypothesis have not yet been completely worked out. Observations were made at the International Polar Year stations at Angmagssalik and Scoresby Sound, in Greenland, and at Tromsö, in Norway, as well as in Holland and South-East England. The results communicated so far from these stations have not shown any definite support for the corpuscular theory, but, on the other hand, because of the unfavourable situation in which they were carried out, cannot yet be regarded as providing a completely positive disproof of the theory.

THE ATMOSPHERE OF VENUS

The possibility that there is life on Venus has been opened up by the discovery of two astronomers of the Carnegie Institution at Mount Wilson Observatory that there is probably carbon dioxide in the planet's atmosphere, one of the essentials of life on this earth. Drs. Walter S. Adams and Theodore Dunham, who made the discovery, did it by focusing sunlight reflected from Venus through their 100in. telescope on to a slit in the spectroscope. The light as it passed through the atmosphere of Venus was partly absorbed, as was shown by a dark band in the spectrum. The position of the band indicated that the gas which absorbed the light was carbon dioxide.

THE COSMIC RAY MYSTERY

Cosmic rays are still something of a mystery, but the *Boston Herald* has made this very important contribution towards solving the problem : " What their effect upon mankind may be, they (scientists) can only guess, but for a starter they have attributed the operation of Neon lights to cosmic rays, alone." In case of any confusion we should state that these are not the same as the cosmetic rays referred to in another American paper.

BIRDS AND SOLAR ECLIPSES

THE tendency of animals and birds to behave during an eclipse as though it were nightfall is well known, but we think the following description, which refers to August 31 last, is worth quoting. It is by L. M. Terrill, President of the Quebec Society for the Preservation of Birds, and appeared in the Canadian *Farmer's Advocate*.

"Starlings are noted for their regular habits and the cloudiness of the sky (at Montreal) served to accentuate their reactions to the accompanying phenomena. At two o'clock at their ordinary roost at St. Lambert not a starling was in sight. At 2.20 thirty starlings appeared and alighted as usual on the wire along the railway, then started for the clump of dead trees where they habitually swarm before retiring for the night, but abruptly they swerved as the shadow spread swiftly and into the roosting grove they hurried. For a space I saw nothing, then the sky was filled with onrushing bands of starlings hurrying from every direction. There was no time for the usual manœuvres, for the wonderful mass evolutions and the night-cap from the pond. Their curfew had sounded, and they responded with all possible speed. At least 15,000 came hurrying home, but by the time the last arrived the early birds were growing restless, the darkness was ending and soon they were all off again, on their endless quest for food. I watched through their usual bedtime, and in spite of the afternoon nap they were all in their roost."

GAS-DRIVEN CARS

Very satisfactory results are said to have been obtained during the week-end from the installation in touring motor-cars of a gasogene machine producing gas generated from any kind of charcoal. The charcoal employed is not, it is said, necessarily that formed from wood, but may also be formed from vegetable refuse or other normal waste materials.

The gasogene apparatus in these experiments was fitted to a "baby" car, and the machine is said to have been driven from Rome to Ostia at rates varying between 40 and 48 miles an hour. The principal advantage claimed is that of great cheapness. It is stated that the cost of running the car with three persons on board from Rome to Ostia works out at 75 centesimi if wood charcoal is used, whereas with charcoal of vegetable refuse the cost is only 30 centesimi or one centesimo a kilometre. The cost of running a similar car with petrol over the same stretch of road would, it is declared, be six lire.

LIFE FROM SPACE

The presence of living bacteria—cocci identical with terrestrial cocci—deep inside aerolites is reported by Professor Charles B. Lipman, of the University of California. He made the discovery in specimens of meteorites obtained from the American Museum of Natural History, the Field Museum at Chicago, and the Colorado Museum of Natural History.

After scrubbing several aerolites with soap and water and treating them with mercuric chloride of superoxol, he soaked them in alcohol and finally subjected them to flame in order to destroy any micro-organisms which might be on the surface or in cracks of the aerolites. Then, under sterile conditions, he ground them to powder. In nine out of 24 cultures made with this powder he obtained globose bacteria, and in all the aerolites found organic nitrogen on which they might feed.

In explaining how these germs had survived their fiery journey through the sky Professor Lipman said that the heat generated by a meteorite passing through the Earth's atmosphere was not sufficient to kill all bacteria, as the passage is made so fast that the interior of the piece is relatively cool, although the exterior may be glowing.

THE AURORA

In the explored part of the spectrum of the aurora, from low in the infra-red to the limit of atmospheric transmission in the ultra-violet, 85 bands of lines have been detected. Apart from a strong green line and two red lines, probably due to oxygen, the auroral spectrum is dominated by nitrogen bands, and the type of nitrogen spectrum agrees with Birkland's theory that the luminescence is produced by electric rays from the sun. The spectrum gives no indication of any upper atmospheric layer dominated by hydrogen and helium. The spectral altitude effect, seen in relation to the height, extension, and luminescence of the streamers and to the low temperature observed, showed that nitrogen must be carried to high altitudes through the effect of an electric state set up by the action of solar radiation of short wavelength. The paper described the resulting state and distribution of the matter of the aurora, suggesting a resemblance to the sun's corona, and showed it to fit in with the results of radio-echo work and to give a simple explanation of the zodiacal light and the luminescence of the night sky.

SRATOSPHERIC EXPLORATION

Yesterday the Soviet balloon built for the exploration of the upper air, " Stratostat S.S.S.R.," successfully ascended from Moscow into the stratosphere, attaining, according to provisional official reports, a height of 19,000 metres (nearly 12 miles). It descended near Kolomna after being in the air about seven hours. The balloon, which carried three persons, maintained wireless communication with the earth. The minimum temperature reported was minus 67deg. centigrade on the outside of the apparatus, which was comfortably warm inside.

Preparations for the ascent began last December. The spherical gondola of the balloon was of duralumin, and had nine glazed windows, which were shut automatically with safety caps if the glass was broken. The external instruments and the ballast discharger were worked electrically from inside. The occupants had chairs and could stand upright and walk about a few paces.

The Soviet authorities state that the objects of the ascent were to study atmospheric conditions in the stratosphere, check existing theories, and prepare data bearing on the possibility of flying in the stratosphere. It was intended that the balloon should remain aloft for 18 hours. Why it descended earlier has not been explained.

The voyage of the balloon and its gondola in which Lieutenant-Commander Settle and Major Fordney rose yesterday to a height in the stratosphere measured as 59,000ft. by their altimeter ended at 5.50 in the afternoon in a salt hay marsh on Delaware Bay, near Bridgeton, New Jersey. The boggy spot on which they came down was completely surrounded by a natural moat 10ft. wide and 6ft. deep.

In landing some photographic plates were lost, but the scientific instruments suffered no harm and the balloon was not much damaged. Not knowing where they were, and judging from a survey by flashlight that it would be foolishly hazardous to attempt to reach a settlement, Settle and Fordney wrapped themselves in the folds of the balloon and went to sleep. This morning Fordney swam the " Moat End " and made his way through a bog three miles to a farmhouse where there was a telephone. Thence he reported to Washington the success of his voyage.

Major Fordney said to-day that Commander Settle and he had had a delightful trip, except that they came down so fast that " we had to throw things overboard as fast as we could to lighten the ship." The landing was " pretty rapid," but neither of them was hurt.

EINSTEIN ON THEORETICAL PHYSICS

If one wished to know what a theoretical physicist had done, he suggested, one should not hear him speak, but should consider his achievements. To a physicist, his own theories appeared so natural and obvious that he could not understand how it was possible not to accept them. It was impossible to avoid a certain amount of bias. In dealing with the development of theoretical physics, Dr. Einstein said that the Greeks had laid the foundations of Western science by the creation of a system in which each part followed logically on another, and could not be attacked. That had been an encouragement to scientists to continue. The Greeks, however, had not appreciated that actual knowledge about reality was entirely derived from experience, and that no conclusion reached through a theoretical system could add to it.

What, then, was the place of reason in modern science ? Reason enabled us to form concepts and laws for a theoretical system, and the consequences of these laws and concepts ought to correspond with the results of our experience. The basic concepts of a system were entirely fictitious and created in the mind of the theorist. In the eighteenth and nineteenth centuries that had not been understood. Newton was the first to offer a comprehensive theory of physics, but Newton believed that his concepts could be derived from an abstraction of the data given by experience. From the way in which Newton expressed his theories, however, it was clear that he was by no means comfortable about the concept of absolute space, because nothing in his experience seemed fully to correspond with it.

Physicists in the eighteenth and nineteenth centuries did not recognize their basic concepts as a free invention of the human mind, and believed they could be derived by a logical process from the data of experience. Was it possible for the physicist to create a correct theory that would be a model of reality, or did such a correct theory not exist at all except in the imagination ? He firmly believed that it was possible for the theorist to create such a perfect system. Our experience justified us in thinking that in nature could be seen the ideal of a mathematical simplicity. It was within the power of the theorist to discover the laws and concepts which would give us the key to the understanding of the phenomena of nature. Experience could not provide the key, although it could guide one in the choice of the mathematical processes to be used.

Dr. Einstein dealt with some of the difficulties of the Quantum Theory, which, he said, was only made tenable to-day by a daring interpretation that had been recently set out by Max Born. He believed that such a daring interpretation should not be necessary, but that there was a possibility of creating a theory that would be a true model of reality, and that would not merely say what was probable. In such a theory he did not think that there would be any mathematical localization of particles, but without such a localization a theory of this kind might still be satisfactory.

DR HUBBLE'S CALCULATIONS

New dimensions have been calculated for the physical Universe on the basis of astronomical observation by Dr. Edwin Hubble, of the Mount Wilson Observatory. Last night in an address to the National Academy of Sciences in Washington he said that the Universe was a finite sphere 6,000,000,000 light years in diameter and that it was composed of 500,000,000,000,000,000 nebulae, each unit being 80,000,000 times as bright as the sun and about 800,000,000 times as massive.

A year ago Dr. Hubble and his fellow observers came to the conclusion that the nebulae 150,000,000 light years distant were rushing away from the Earth and from each other at the rate of 15,000 miles a second; but now nebulae are being observed through the 100in. telescope at Mount Wilson at distances as great as 300,000,000 and 450,000,000 light years, and their speed of recession varies from 30,000 to 40,000 miles a second. It has been observed that the greater is the distance of the nebulae away the greater is the speed at which they are receding. Their speed, in fact, increases about 100 miles a second for each million light years of distance. As the latest calculations show a diameter for the Universe of 6,000,000,000 light years, the nebulae at its outer rim are rushing away at the speed of light, 186,000 miles a second.

NEW HEAVY ELEMENT CREATED

Signor Fermi claims to have discovered a new element of atomic number 93, and therefore the most complex element known. The new element was produced when uranium, of atomic number 92, was bombarded with neutrons. The new element is radioactive and is stated to be extremely unstable—half its atoms changing their nature within about 13 minutes.

Professor Corbino observes that the artificial radioactivity recently discovered by M. Joliot and M. Curie in Paris, and realized upon a relatively large scale by Signor Fermi, is the manifestation of a rejuvenation of old, stabilized matter brought about by a nuclear clash. He considers that the scientific possibilities opened up by this latest discovery are of incalculable importance. Men of science, he thinks, may well be on the road to proving a process of automatic rejuvenation of the material universe, since a constant creation of new atomic elements would imply a constant renewal of forms.

MILLIKEN ON COSMIC RAYS

Professor R. A. Millikan (Pasadena) presented on behalf of himself, Mr. I. S. Bowen, and Mr. H. Victor Neher the results of a survey at very high altitudes of the effect of latitude upon cosmic ray intensities, together with a general interpretation of cosmic ray phenomena. The survey extended from Northern Canada to Peru. It was carried out in aeroplanes at altitudes up to 22,000ft. and in three stratospheric flights over the United States, including that made by Major Fordney and Lieutenant-Commander Settle, who reached a height of 62,000ft.

The survey, whose results were yesterday communicated to the scientific world for the first time, led Professor Millikan to reaffirm emphatically his belief that the primary cosmic radiation is of a photon nature (like light), not of a particle nature. But as cosmic ray photons traversing space in all directions must in any case produce secondary electrons wherever they traversed matter (whether that were in inter-stellar space, in nebulae, or in the atmospheres of suns through which they must sometimes pass in their journey of billions of years), space must be traversed in all directions to some extent both by photons and by their secondary electrons. The only question for experiment was to determine their proportion, and that Professor Millikan and his co-workers had tried to do in many parts of the world and at different altitudes.

A NEWSPAPER placard of May 7 :—
TOBACCO MAGNET'S SUDDEN DEATH.
The permeability of tobacco is notoriously variable.

DEVELOPMENTS IN NUCLEAR PHYSICS

DR. H. R. CRANE and PROFESSOR C. C. LAURIT-SEN (California Institute of Technology) considered gamma-rays emitted during nuclear transformations. These rays had been observed when lithium and fluorine had been bombarded with protons, and when beryllium, boron, and carbon had been bombarded with deutons (diplons). Attempts had been made to determine the quantum energies by means of absorption measurements in lead and in copper, and also by means of cloud chamber photographs of the recoil electrons and electron pairs ejected from a lead plate by the radiation. On the origin of the gamma-rays the author held that one or more of the newly formed particles in atomic transformations would sometimes be formed in an excited state, and would subsequently fall back to its normal state with emission of a gamma-ray. A list of excitation levels which had been found was given.

PROFESSOR G. GAMOW considered general problems of the stability of atomic nuclei. According to modern views, he said, the nuclei of the different elements were built up of elementary heavy particles: neutrons, ordinary protons, and probably to some extent unobserved negative protons. From considerations of energy, he decided that, especially in the region of the heavy elements, reactions could occur which, although possible, possessed such small probabilities that they had no chance of being observed. Carbon nuclei, for example, could be emitted from radium, but the number would be ten-thousand million-million-million-million-million times smaller than the number of alpha-particles. He explained his hypothesis of a negative proton. He thought there was a certain periodicity in the nuclear structure, most probably connected with saturated neutron shells.

PROFESSOR MAX BORN, in a paper on "Quantum Electrodynamics," pointed out the difficulty in current theories that for every kind of particle there had to be invented a separate kind of electromagnetic field, and that these fields had to be connected by laws specially invented. The reason was the generally accepted dualistic idea according to which the material particles were something essentially different from the electromagnetic field. They were assumed to be charged mass points. But charge meant electromagnetic energy, and energy was equivalent to mass according to Einstein's law. Therefore all dualistic theories were in contradiction with Einstein's law, and that was the root of the difficulties. A way of avoiding these seemed to him the introduction of a unitary theory which assumed only one kind of field, a modified electromagnetic Maxwell field, the singularities of which played the rôle of particles. There had been several attempts to produce a unified field theory, and the failure of those attempts started by Professor Einstein, who tried to connect the cohesive forces in the electron with gravitation, did not seem to him astonishing owing to the factor 10^{40} (10 followed by 40 noughts), which expressed the relative magnitude of electric and gravitational effects. He believed that all the unified field theories which tended to combine electromagnetism and gravitation into a single formalism were on the wrong track.

EINSTEIN ON CONTEMPORARY PHYSICS

In an interview at Pittsburgh to-day Professor Albert Einstein said the universe might be infinite after all. There was a very good chance that it was. Asked if science might some day convert matter into energy for practical purposes, he expressed doubts. Bombarding atoms with sub-atomic particles in order to break them up was like " shooting birds in the dark in a country where there were very few birds."

Speaking of Heisenberg's Principle of Indeterminacy, he said nobody could know whether science would eventually state its laws in a statistical fashion such as that of Heisenberg or whether it would use a deterministic form based upon the law of cause and effect. Most physicists believed the final form of laws would be statistical. He did not. He believed it would be found simpler to state laws in a deterministic form. In the long run a statement in terms of cause and effect would be simpler, but to-day most physicists did not believe so.

COLLECTING SOLAR HEAT

A device for making the sun do a part of man's work was shown to-day to the American Association for the Advancement of Science at the beginning of its winter meeting in Pittsburgh.

It is a solar heat collector, which in 20 minutes develops temperatures in excess of 400deg. Fahrenheit. The inventor is the Secretary of the Smithsonian Institution, Dr. C. G. Abbot. The device can be used for ordinary cooking purposes. Dr. Abbot has been using it on Mount Wilson, California, for the generation of steam for power. It is hoped that ultimately it can be made to drive large steam engines at a practicable cost.

The apparatus employs a number of small mirrors fixed at right angles to the earth's axis and mounted on a frame which will turn on an axle parallel to the earth's axis. There are a number of tubes filled with black oil and protected from loss of heat by double vacuum jackets. These transmit the collected heat to a reservoir of oil, and into this reservoir ovens are inserted for cooking purposes, or a tubular boiler for generating steam.

Africa up to Date.—A correspondent sends us this True Story from the Sudan :—

English Official : " If you do not comply with my orders, I shall put Big Magic on you. I will bite a piece out of the Moon to-morrow."

Chief : " If you are referring to the eclipse, that doesn't happen until the day after to-morrow."

THE PLANETARIUM

It is in a sense an accident that the Munich planetarium happens to be in the Munich Museum. Elsewhere the planetarium may be on a water-tower, or on a special site of its own, or in a zoo, or on top of a newspaper office. The point I wish to make is that Munich has a planetarium: that a planetarium, properly handled by a skilled lecturer, makes astronomy come alive and intelligible in an immediate way, providing a rare combination of emotional and intellectual interest: that Germany possesses a dozen planetaria, America four, Italy two, even impoverished Austria and struggling Russia one each: that there are over a score now in existence, but that England, and indeed the whole British Empire, is without a single one.

A NEW SCIENTIFIC SOCIETY

A. I. U. R.—One of our highly reputable contemporaries publishes in it current issue a letter announcing the formation of this important new scientific society. We are told that " There has long been felt the need for an efficient governing body to organize the vast quantity of useless research that is being pursued day by day and hour by hour in the many institutions of higher learning. . . . It is the fervent hope of the founders that this worthy movement will spread its tentacles throughout the land and be an ever present aid to those endeavoring to unscrew the inscrutable ". It then goes on to mention some activities of the society, including a lecture by an authority on Banned Spectra on *Some Higher Harmonics in the Brass Bands,* and the sponsoring of certain publications including " *The Nastyphysical Journal,* and for those who are unable to read English, . . ., the *Makeshift für Physik* ".

The mystic letters, by the way, stand for " American Institute of Useless Research ".

Dr. Albert Einstein announced to-day that, in collaboration with Dr. N. Rosen at the Institute for Advanced Study at Princeton, he had opened a possibility —stupendous in its implications—of accounting for atomic phenomena by the method of General Relativity. Hitherto a great gulf has separated the theory of Relativity and the Quantum Theory, but now with a first successful step already taken there seems to the two collaborators to be a chance of developing an all-embracing physical and mathematical theory which would include macrocosm and microcosm, the universe as a whole as well as the atom.

The " theoretical method " which Dr. Einstein and Dr. Rosen are exploring is published in *The Physical Review* of the American Physical Society and the American Institute of Physics. The collaborators " justify," as they say, its publication now " because it provides a clear procedure, characterized by a minimum of assumptions, the carrying out of which has no other difficulties to overcome than those of a mathematical nature." They frankly admit that though the first step towards formulating a universal theory has been taken successfully they do not know yet whether the next ones will follow logically ; but it is plain that they are more than merely hopeful that they will.

In moving towards the formulation of his new theory Dr. Einstein has had the aid of some startling new " pictures of space," photographed with the mathematical " lenses " of a " cosmic camera " designed by himself and Dr. Rosen. These " photographs " indicate that space is not one thing but " two identical sheets joined by many new bridges." A new particle of matter " without gravitating mass " is found to be, by mathematical calculation, the most natural electrical particle. Electricity and mass are found to be not related but independent constants in nature. And the " bridges " linking the two " identical sheets of space " are atoms.

Prof. H. N. Russell.—Analysing a spectrum is exactly like doing a crossword puzzle, but when you get through with it, you call the answer research.

MODERN ALCHEMY

The most recent advances, which hold out promise of rapid and sensational extension, have depended on the artificial transmutation of elements. Lord Rutherford in 1919 was the first to effect such a change, which had previously been thought a dream of the alchemists, by bombarding nitrogen with alpha-particles (helium nuclei) from radium. Some of these became incorporated in nitrogen nuclei, which rendered them unstable, so that they disintegrated into new types of structure. This method has since been used to effect the transmutation of several light elements.

A later and more efficient method is to use as bombarding agents the charged particles (protons, deuterons, and alpha particles) in hydrogen or helium at low pressure and exposed to a powerful electric discharge. This demands very high voltages (up to 2,000,000 volts are now being employed in Pasadena), and the technical equipment needed is elaborate and costly. A third and still more recent method is the employment of neutrons as projectiles. These are uncharged particles, of mass 1. Being uncharged, they do not encounter so much disturbance on approaching a nucleus, and have a greater chance of entering it. This method is being successfully used even with the heavy elements.

A modification of this method is to slow down the neutron projectiles by passage through paraffin or other substances containing hydrogen. By this means their speed is reduced from about 10,000 miles a second to speeds of the same order as the ordinary thermal speeds of particles which determine the temperature of a body. At this low speed neutrons can be absorbed in relatively enormous quantity by certain elements such as boron and cadmium. In certain cases cooling of the neutrons to the extremely low temperature of liquid oxygen may actually enhance their effectiveness. In these ways a new field of investigation of very great promise has been opened up. Finally, there is the method, studied at Cambridge during the past year, of using high-energy gamma-rays (X-ray waves) to effect the disintegration of nuclei.

The most striking recent result of these various bombardment methods has been the production of artificial or induced radio-activity, first discovered about 18 months ago and now being intensively studied in many laboratories. Strictly speaking, the new substances thus produced are not new chemical elements but new isotopes of known elements—i.e., substances with the same charge on the nucleus but of different nuclear mass and therefore different nuclear construction. These are unstable in the same way as radium and other normally occurring radio-active elements are unstable, and disintegrate, with the emission of particles, to form stable substances.

They are of special interest in several ways. In the first place, although now no longer found on the earth, they must occur, it can confidently be stated, in the sun and other hot stars where a natural nuclear bombardment must be constantly proceeding. They must therefore have existed in the substance of the earth when it was first flung off from the sun, but have disappeared with the cooling of our planet. The only survivors of these unstable elements are radium and other heavy radio-active elements with long periods of disintegration (together, of course, with certain unstable products of their disintegration). The physicist, therefore, is now artificially re-creating substances which used to exist on the earth.

Lord Rutherford also made the interesting suggestion that natural neutron bombardment might prove to have importance in the problem of cosmic rays. Neutrons must be emitted in large quantities by the sun and other lost stars; but none had ever been detected in our atmosphere. It was reasonable to suppose that they were slowed down in their passage through the outer layers, and finally all absorbed before reaching the neighbourhood of the earth's surface. He suggested that future explorers of the stratosphere would do well to take or send up apparatus for the detection of neutrons; if these were really present, they might account for many of the phenomena of cosmic rays which still puzzled physicists.

Dr EINSTEIN'S NEW THEORY

THE NEW ATOM SMASHER

The physics laboratories of Columbia University have recently been equipped with the largest instrument for the "splitting of the atom" which has yet been produced. Hitherto only the lighter elements on the atomic table have been split, but it is hoped that the new apparatus will be able to disintegrate the atoms of the heavier elements such as gold, silver, and lead, a task which has been beyond the powers of any instrument yet used. Physicists expect that it may now be possible to create radioactive elements more powerful than radium and much less costly, which would be within the reach of all sufferers from cancer and similar diseases.

The heart of the new instrument is an electro-magnet, weighing 58 tons, which was in use until last year at the naval wireless station at Annapolis. A more modern transmitter has now been installed there and the magnet has been transferred to Columbia University by virtue of an Act of Congress which authorizes the use of obsolete naval apparatus for scientific purposes. The magnet can create a magnetic field 75,000 times greater than that of the earth. Whereas the strength of the earth's magnetic field is only one-fifth of one gauss, the great electro-magnet can produce a field of from 14,000 to 15,000 gauss. Protons and deuterons will be introduced into this magnetic field and directed under the "accelerating chamber" designed by Professor Ernest Lawrence, of the University of California. The instrument will be capable of emitting atomic projectiles with an energy of 15,000,000 volts and of producing as much as 20,000,000 volts.

THE GREATEST TELESCOPE

The first stage in the great project for the construction of a 200in. telescope has been successfully completed, though not without some vicissitudes, by the production of a thoroughly annealed 20-ton disk of pyrex glass. The first disk that was cast was injured by some defects in the mould. It would have been possible, if attempts to cast another disk had failed, to have utilized this disk, but, fortunately, at the second attempt a perfect disk was obtained. The imperfect disk was subjected to a process of rapid annealing, to serve as a guide to the time of annealing to be given to a second disk. When it was removed from the annealing oven after a couple of months it was found to be unbroken, though, naturally, after so short a period of annealing, it was not entirely free from strain.

This provided a striking illustration of the advantages of using pyrex glass, the glass that is widely used for cooking utensils because its low coefficient of expansion enables it to withstand rapid changes of temperature. A disk of ordinary plate glass of the same size would have split into pieces, even though the annealing had been prolonged for a period of several times as long.

The second disk was annealed for nearly a year. The temperature of the annealing oven was maintained at a uniform value by electric controls, the temperature being slowly decreased from day to day. It safely survived the extensive floods in New York State last autumn, but disaster was only narrowly averted, for the water rose almost to the level of the annealing oven. A heavy earthquake also occurred in New York State while the annealing was in progress, but fortunately did not damage the disk.

For the 3,300 mile railway journey from the glass works to California the disk was encased in a steel crate, weighing 10 tons, with the face of the disk protected by a 4in. blanket of cork and the rim protected by five layers of heavy felt. It was carried on edge in a specially constructed well-hole truck, the weight of the crated disk being supported by steel beams, covered with cushions of compressed cork. The bottom edge of the crate was only 6in. above the level of the rails, and at certain tunnels and bridges there was a clearance of only 3in. In order to minimize the risk of damage the speed of the train did not exceed 25 miles an hour.

The next stage in the construction will be the lengthy one of grinding and polishing the surface gradually changing from the spherical to a paraboloidal shape. The amount of material to be removed in this last process is extremely small; it will amount to a maximum of five-thousandths of an inch at the centre of the mirror. This figuring of the mirror, as it is called, must be done with very great precision, and may easily require three or four years. The process is one of trial and error, the degree of precision attained being frequently checked, as the work proceeds, by optical tests of extreme delicacy. In the finished mirror the actual curve of the glass will not differ from the theoretical curve by more than about two-millionths of an inch at any point of the 24 square yards of its surface.

The requirements can perhaps better be appreciated if expressed in this way. Suppose the mirror to be enlarged 200,000 times, so that its curvature becomes about equal to the curvature of the earth's surface. The diameter of the mirror would then be about 630 miles, so that it would cover the whole of Great Britain and Ireland. The depth of its concave surface would be about 12 miles; the greatest difference between the initial spherical surface and the final paraboloid would be about 90ft., while the divergence of the finished surface from its true shape would at no point be greater than $\frac{1}{4}$in.

HUBBLE'S VIEW OF THE UNIVERSE

This lecture described the distribution of nebulae as indicated by the two corrections. Counts of nebulae to various limits of apparent faintness indicate the numbers within various limits of distance, or various volumes of space. The uncorrected data for five independent surveys, to distances ranging from about 150,000,000 to 400,000,000 light-years, show that the apparent distribution of nebulae thins out with distance. The true distribution is derived by correcting the limits of apparent faintness of the surveys (in other words, the estimated volumes of space included in the surveys) for effects of red shifts. The simple "energy" correction, $m = 3d \, l/l$. required by the mere presence of red shifts, leads to uniform distribution; over the entire range of the surveys, the numbers of nebulae in the various counts are directly proportional to the volumes of space which they occupy. Thus, on the assumption that red shifts are not velocity shifts (the universe is not expanding at an appreciable rate), the observable region of space is thoroughly homogeneous, and no characteristics are observed which vary systematically with distance.

On the other hand, if we assume that red shifts measure the mutual recession of nebulae in an expanding universe, and the additional correction for recession is applied to the surveys, the true distribution departs from uniformity. The departures are in the sense that the density (number of nebulae per unit volume of space) increases outward in all directions, leaving our own stellar system in a unique position. This conclusion is simply unwelcome to our sense of proportion, and would be accepted only as a last resort in order to "save the phenomena." Therefore, the departures from uniformity, introduced by the assumption of a rapidly expanding universe, must be compensated by still another assumption. The only available factor in relativistic cosmology (in its present state) is spatial curvature, and the necessary curvature would be positive and very great—the present radius of curvature would be comparable with the range of existing telescopes (order of 500,000,000 light years).

The departures from uniformity, which indicate the numerical value of the radius of curvature, together with the departures from linearity in the relation between red shifts and distance, determine the " cosmological constant." These data identify the only possible model of a homogeneous expanding universe among the many that are offered by relativistic cosmology (in its present state) The model is of the " ever-expanding " type that will always be associated with the name of Lemaitre. It is closed (finite but unbounded) and small, and the rate of expansion has been slowing down. A large fraction of its present volume and a considerable fraction of its past history (time since the expansion began) can be explored with existing telescopes.

These remarkable features follow from the assumption that red shifts measure recession. The distance relation deviates from linearity by the amount of the postulated recession. The distribution departs from uniformity by the amount of the recession. The departures are compensated by curvature, which is the equivalent of the recession. Unless these coincidences are evidence of an underlying, necessary relation among the various factors, they suggest a forced interpretation of the data.

If red shifts are not primarily due to recession, but represent some hitherto unknown principle, the observable region loses much of its significance. The distance relation is linear ; the distribution is uniform : there is no evidence of expansion, no trace of curvature, no restriction of the time-scale. The sample, it seems, is too small to indicate the particular type of universe we inhabit

FROM NEWTON TO THE PRESENT DAY

◆

After paying a tribute to the memory of four noted members of the section who had died in the past 12 months—Sir John McLennan, Sir Richard Glazebrook, Sir Joseph Petavel, and Professor Karl Pearson—he reminded his hearers that these men had been trained in a physical and philosophical tradition wholly different from that in which the present generation lived and moved.

Their scientific world-picture was based on the simple notions of velocity, acceleration, momentum, and force which were first formed into an ordered scheme by the genius of Newton. Looking back on this scheme, we to-day found it philosophically naïve. There was also the striking, though temporary, success of the theory of the ether. Lord Kelvin went so far as to assert that " this thing we call the luminiferous ether . . . is the only substance we are confident of in dynamics." Such a conclusion bade them beware of some of the present-day confident certainties.

The closing years of the nineteenth century began to show cracks in the superstructure and weaknesses in the foundations of this system of thought. The most notable new discovery was that of Planck, with his quanta of radiation— a fact quite inexplicable on the older theories. The quantum theory, once introduced, proved valuable elsewhere. The new concept of the complex atom, necessitated by the facts of radio-activity, would not work on Newtonian principles ; the quantum principle, embodied in the Bohr model, saved it.

Quite early in twentieth-century physics, however, an alarming dualism reared its head. First of all it was suggested that the classical waves of radiation might for certain purposes be better represented as particles ; and later that the classical particles of matter might for certain purposes be better represented as waves. As one writer had put it, the universe was composed of particles on Mondays, Wednesdays, and Fridays, but of waves on Tuesdays, Thursdays, and Saturdays. Alice seemed to rule in physics' Wonderland. This dualism, however, was now disappearing under the analysis of the last few years—an analysis which was essentially mathematical and introduced the notion of probability.

After describing the recent development of our atomic models, Professor Ferguson went on to discuss that great twentieth-century achievement, an achievement comparable with the nineteenth-century identification of heat and energy, the identification of mass and energy. He next mentioned the remarkable discoveries of recent years concerning artificial radio-activity, and passed on to consider the uncertainty principle, which had had such a potent effect on transforming our world-picture. As was well known, it had been established that it was impossible to determine both the position and momentum of a particle with complete accuracy. Increased accuracy in the determination of one was accompanied by a calculable decrease of accuracy in the determination of the other. It was to be noted that this principle applied only to the determination of the two quantities. The word " indeterminism," Professor Ferguson was careful to point out, should be reserved for the facts thus subsumed, and not applied indiscriminately in all kinds of irrelevant philosophical senses. Conversely, when they spoke of the disappearance of causality, they should be quite certain of the limitations of their meaning.

The resolution of the difficulty and the bringing back of common-sense into scientific philosophy was, for Professor Ferguson, to be sought in Planck's distinction between the perceptual and the conceptual worlds. We lived among a mass of perceptions. In order to correlate and " explain " these perceptual facts of experience, the physicist was driven to devise a number of concepts, quite out of the reach of direct perception, such as molecules, atoms, and electrons. Purely as a matter of measurement we could not predict any physical event with entire accuracy ; but in our conceptual world we could make accurate predictions. " Planck thus retains the principle of causality . . . in the conceptual world, remarking that the relation between events in the perceptual and conceptual worlds is subject to a slight inaccuracy." This applied to classical physics ; it applied also to modern quantum physics under the sway of the uncertainty principle : although the resemblance between conceptual and perceptual worlds was less close, the philosophical principle was the same.

Professor Ferguson ended with a brief mention of some recent advances in large-scale and applied physics—the flotation process for separation of minerals, the recent extensions of the theory of Brownian movement, electron diffraction, and low-temperature research— and of the recognition that the social impact of science might be harmful just as much as beneficial. But the main impression left was of the theoretical study of pure physics resolving its contradictions, and once again permitting a philosophical view of reality which, however complex and difficult, is compatible with our reason and our common-sense.

SPOTS ON THE SUN

There is a large group of spots on the sun at the present time which to-day will be placed across the middle of the disk rather above its centre, having been brought into that position by the solar rotation.

An estimate on Tuesday gave the area as 1,000-millionths of the sun's visible hemisphere, so that, apart from any possible increase in size, the group may be expected to be visible to the naked eye or by use of a field-glass, shielded, it is hardly necessary to say, with a dark cap, unless the sun happens to be seen through hazy atmosphere. This follows a still larger group that came on the visible disk on July 22, was on the central meridian on July 28, and has now passed round the western edge. The area of that group, which was probably the largest that has been seen since 1926, was measured as 2,000-millionths on July 24 and 3,000 on July 26. As a millionth of the sun's hemisphere is equivalent to rather more than 1,000,000 square miles, the extent of the phenomenon on a terrestrial scale may be inferred.

The group of last week was a stream of usual type consisting of large spots preceding and following, both of which could be distinguished separately by naked eye, with smaller spots between. As answer to the question, "What is a sunspot?" it is generally said that it is a hole or depression in the solar atmosphere caused by vortex motion, but the further reason for this is obscure or unknown. It is known with a certainty hardly to be contested that they appear on the sun in great numbers at intervals of approximately 11 years, the period being slightly variable, and that this maximum epoch is now recurring or is about to recur. Such spots as these are therefore not unexpected, but in one respect they are abnormal.

At the beginning of the sunspot cycle (that is to say, just after the time when the sun has been spotless), the spots that appear are in high solar latitude—30 to 40 degs., perhaps, north or south of the solar equator. As the cycle progresses, and the spots become larger and more numerous, the regions in which they appear converge continuously to the equator, till at sunspot maximum the mean latitude of the spots in the two hemispheres is 14 or 15 degs. The group of last week was in heliographic latitude 31 deg. N.; that now on the sun has latitude 24 deg. N.

It may be said with little fear of contradiction that the hot weather now being, or to be, experienced is not to be attributed to these spots. Though the appearance of sunspots is spoken of as a sign of solar activity, that activity has no immediate effect on weather, which is mainly due to terrestrial and local causes, though it almost certainly has a connexion with earth magnetism, which is often shown a day or two after a large spot passes the central meridian, the aurora, and some other terrestrial happening. The spot now on the sun will probably be seen until August 12, when it will pass round the western edge.

NEW THEORY OF THE SUN

Photographs and films of the total eclipse of the Sun on June 8 made from an aeroplane flying at heights of from 25,000ft. to 30,000ft. over Peru will probably compel astronomers to revise their conception of the solar corona and of the Sun's composition, for they seem to prove conclusively that the Sun is surrounded by a globular envelope or atmosphere more than 1,000,000 miles deep. This discovery is announced jointly by Harvard University and the American Museum of Natural History.

Taken from such a height, the pictures were exceptionally clear, for there was no dust to contend with, and only a little atmospheric or sky haze. Without exception they differed widely from pictures of the eclipse taken from the ground, for the coronal streamers appearing as tongues of flame in the latter were completely dominated in the stratosphere pictures by a giant halo in which they became relatively insignificant.

The photographs were made by Major A. W. Stevens, of the United States Army, a veteran stratosphere flyer.

THE DISTANT NEBULAE

By using a photographic lens which is claimed by its makers to be the fastest in the world Dr. M. L. Humason, of Mount Wilson Observatory, California, has been able to photograph, through a great telescope which has a 100in. reflector, spectra of nebulae 30,000 times fainter than the faintest star visible to the unaided eye. These nebulae have been calculated to be 80,000,000 light years distant from the earth.

The lens was made by the Bausch and Lomb Optical Company, of Rochester, New York, and is called Rayton, after its designer, Dr. Wilbur Rayton. It was necessary to make an exposure of 60 hours with it to get a photograph 2-25in. long of a spectrum. From this spectrum Dr. Humason found that a faint cluster of nebulae in Ursa Major, probably one of the most remote clusters known at present, was running away from the earth at the rate of 26,000 miles a second.

Dr. Humason stated that he had made a photograph and calculations to verify the theory of the Dutch astronomer, the late Dr. William de Sitter, that the universe was expanding and that the recessive velocity of the nebulae was proportional to their distance.

USES OF THE PHOTOCELL

Mr. WHITELEY said that there had been an unfortunate tendency to hail the photocell as a near substitute for the human eye, but he foretold that its greatest successes would be where the human eye was ineffective. The photocell was instantaneous in action, and a relatively simple electric " relay " could be made to respond to a light impulse lasting only one ten-thousandth of a second.

The photocell could control the folding and cutting of printed wrappers and paper bags from a continuous web of printed material. It could distinguish between plain and printed paper or cellophane and operate the folding mechanism accordingly. The photocell could ensure also that the perforations on a sheet of stamps corresponded accurately with the white spaces between the stamps. So far this had only been applied abroad, but each control might be expected to reduce the number of rare oddities available for collectors.

In such applications the photocell might be required to observe paper, &c., travelling at 600ft. a minute, a speed at which a man would have to run to keep up with the moving stream. The limit was set by the mechanical task to be performed, and not by the sensitiveness of the photocell.

PROFESSOR G. I. FINCH, of the Imperial College of Science and Technology, London, described the results obtained with the electron diffraction camera. This uses an electron beam, which penetrates less than the first millionth of an inch of the material examined. Investigation had revealed, he said, that cylinder wear by aluminium pistons in aero engines was increased because the surface of aluminium had the same structure as sapphire. Consequently white polishing resulted in the momentary production of a glasslike layer. Recrystallization took place immediately. There would be a large number of small hard crystals, calculated to give an abrasive surface.

Professor Finch suggested that burnishing with a magnesium rod might have a remedial effect, by enabling sufficient magnesium to be absorbed to give a surface which would polish without crystalline structure forming.

PHOTOGRAPHS OF ATOM TRACKS

Most of the exhibits consist of photographs of the tracks of single atoms and electrons moving at high speeds in air or other gases. A large proportion of these represents work done in the Cavendish Laboratory, Cambridge, but others come from America, France, Germany, Russia, and elsewhere.

It is just 25 years since Professor Wilson first succeeded in making visible the tracks of individual atoms and electrons. The essence of his method was to force water-vapour to condense momentarily along the track of the moving atom or electron, thus forming a tiny streak of cloud which could be seen and photographed. The original apparatus with which he worked is the first exhibit in the exhibition.

PREVENTING RUST IN A METEORITE

An interesting experiment in preserving a perishable mineral in gas is being made at the Natural History Museum, South Kensington. The large piece of meteoric iron found at Cranbourne, Mornington County, Victoria, some 30 miles south-east of Melbourne, about 1850, and presented to the museum by James Bruce in 1862, has been placed in an air-tight glass case containing dry nitrogen, which is admitted from a cylinder concealed below.

The Cranbourne meteorite is the largest of a number of pieces of meteoric iron found in the same area. When discovered it was almost completely buried in sand. It is composed mainly of two nickel-iron alloys, kamacite and taenite, which are intergrown into a regular network. Among its other components, however, is a proportion of lawrencite (iron chloride), which is especially liable to rust, and as it rusts it expands and so causes splitting. The weight of this meteorite is approximately $3\frac{1}{2}$ tons, and since its arrival in the museum 70 years ago the rusting and splitting have reduced it by about 4cwt. Such wastage, it is expected, will be completely overcome by enclosure in nitrogen.

The meteorite, which is of a dark reddish-brown colour, is now flood-lit, and forms a handsome object at the eastern end of the Mineral Gallery. The gas is kept under a pressure which is said to be equivalent to a 45-mile-an-hour wind blowing outwards against the glass sides of the case.

An introductory section, consisting of a few photographs specially chosen for simplicity and clarity, and of explanatory labels, expounds the elements of the subject, and is subdivided into groups relating to alpha-rays, beta-rays, and cosmic-rays. This is followed by other sections more technical in character and appealing especially to the expert. A stereoscopic apparatus enables one to see certain of the tracks in three dimensions, as they actually occur.

It is believed that this is the most comprehensive collection of the kind yet shown in this country. Included in it is the well-known photograph taken in 1924 by Professor P. M. S. Blackett, the first which showed an actual atom in collision with the nucleus of another atom and disintegrating it. Another, taken by Professor W. Bothe, of Heidelberg, shows a pair of electrons created out of nothing by gamma-rays. Further especially striking photographs represent " cosmic-ray showers," in which a number of electrons diverge from a small region of space.

EARTH ESCAPES COLLISION

Johannesburg has been startled by the revelation that the Earth recently escaped collision with a minor planet by little more than five and a half hours. On the night of October 30, at the very small astronomical distance of 400,000 miles, a planet travelling at about 20 miles a second crossed the path of the Earth, which was speeding on its way at the rate of 18 miles a second.

"This was the narrowest escape from collision which the Earth has ever had within the period of astronomical observations," declared Dr. Wood, the Union Astronomer, in giving an account here of observations in which the Union Observatory cooperated after a discovery made on the night of October 28 by the German astronomer Dr. Reinmuth. In photographing the sky at Heidelberg Dr. Reinmuth found a streak on the plate revealing some body moving across the face of the heavens. From a single observation the path of the visitor, which from the appearance of the photographs was a minor planet, could not be determined, but a similar streak was found on plates exposed at the Union Observatory and also at another observatory in Germany. Investigation of the orbit, which from these three positions was then possible, showed that the planet was travelling almost straight towards the Earth, and that it shot past on the night of October 30. "From the point of view of the universe, where these bodies usually are such tremendous distances apart, this was a close shave for the Earth," said Dr. Wood.

THE SOURCE OF THE SUN'S ENERGY

Professor Hans Bethe, an exile from Munich University, now on the faculty of Cornell University, makes a notable contribution to astrophysical theory in announcing his conclusions from researches into the source of heat in the sun and the stars.

He suggests that just as carbon is burned by man on earth for heat and power, so it is carbon which is responsible for the heat in the sun and the stars. But in the sun carbon is endlessly regenerated by the utilization of energy within nuclei or cores of atoms of carbon and hydrogen. On earth only the outside carbon atom is burned; in the sun and the stars it is the nucleus which is burned.

Professor Bethe believes that in the constant collisions going on in the sun between carbon and hydrogen atoms both are annihilated, and nuclear reaction is set up, giving quantities of heat coinciding with known facts about the sun. The carbon nucleus is pierced by the hydrogen nucleus and a nitrogen nucleus is formed. This again is bombarded by hydrogen nuclei, and finally a helium nucleus is produced and the original carbon nucleus is regenerated.

VARIABLE STAR'S HUGE ATMOSPHERE

Experiments with infra-red light by Dr. Otto Struve in America throw new light on Epsilon Aurigae, the variable and nearest bright star to Capella. Its variability was suspected as long ago as 1821, but this was not confirmed, or in any case only a slight change was noticed. In 1902 Vogel, of Potsdam, found that the star is a spectroscopic binary, or that its spectrum is that of two stars superposed, the two spectra being of different types and their lines showing periodic relative displacement.

Ludendorff, also of Potsdam, made a complete investigation of the light changes from 1842 to 1903 and found that the variability is of the Algol character; that is to say, it is caused by the eclipsing or partial eclipsing of one component by the other rotating around it. He found that the period is 27.14 years, or perhaps 54.28 years, and that the middle of the last minimum occurred on March 31, 1902.

The magnitude of Epsilon Aurigae is now said to range from 3.3 to 4.1, and its spectrum classed as F5, peculiar. It has three faint companions of 12th to 14th magnitude, but these are not close.

These are the facts known about the star. It now appears that Dr. Otto Struve, whose skill and acumen are beyond question, has by means of infra-red light detected an atmosphere, or surrounding, to this star which would make the body, including its atmosphere, almost comparable in size with our solar system.

NOISY WATER PIPES

Experiments have recently been carried out at the National Physical Laboratory, Teddington, on reducing the noise transmitted along water pipes. The sound of a hissing tap, the hum of a circulating pump, and noises made in the boiler house by stoking are often carried to distant rooms via the pipes. Such annoyance is sometimes considerable in large blocks of residential flats.

A successful method of reducing these noises almost to vanishing point was found by inserting a length of rubber hose in the water system, i.e., replacing the metal pipe by rubber for a short distance. Precise measurements of sound intensity showed that the length of the rubber hose was important; a length of a few inches was sufficient to cut out the high pitched hiss of a tap, while about 3ft. were required to insulate the low pitched hum of a pump.

The results of this research are thought likely to be of value to engineers and architects who are faced with the problem of reducing the noise generated by water systems.

THE AURORA

I see an inquiry in *The Times* of January 27 as to whether anyone heard the aurora on the night of January 25. At 7.45 p.m. my wife and I were out, observing the changing lights in the sky, when we heard a peculiar noise, unlike anything we had ever heard before. It was a rushing sound, coming up the drive travelling from east to west, swirling round a copse and away in a southerly direction. There were three or four preliminary noises of three or four seconds' duration each at similar intervals of time, followed by two prolonged noises of some 10 seconds each at an interval of a few seconds.

On referring to the Encyclopaedia Britannica I see the sound is reported by some to resemble the crackling of silk, but I think a better description of the noise we heard was that given by one of the maids, who said it sounded like deer crashing through the underwood.

Enough information has now been gathered from radio workers, amateur and professional, to show that Tuesday evening's auroral-display affected wireless transmission as might have been expected from experience in high latitudes where such displays are of frequent occurrence.

Some years ago a British radio expedition, made by Professor E. V. Appleton and Mr. R. Naismith, went to Tromsö in Norway to study the effect of the northern lights on radio conditions, and there they found the remarkable result that the upper-atmospheric layers, which normally reflect wireless waves, were entirely ineffective, for medium and short wave lengths, during auroral displays. The experience of most listeners on Tuesday night was very similar, for distant reception became very difficult during the period of intense luminosity.

Auroral displays are almost always accompanied by the occurrence of magnetic storms, which are themselves caused by the circulation of intense currents in the upper atmosphere. Such currents often induce corresponding currents in the ground which may interfere with telegraphy in which a single wire and ground return is employed. It is probable that something of this kind occurred in the case of railway signalling apparatus on the L.N.E.R. Manchester-Sheffield line which gave trouble on Tuesday.

Experience in high latitudes suggests that the aurora may most probably be seen again after about 27 days (and not after a fortnight as has been widely stated). The reason for this is that after this period the sun will again be in the same position, with the same part of its disc towards us as when it caused the first disturbance.

TWO MORE SATELLITES OF JUPITER

A Harvard Announcement Card states that Professor Seares, Assistant Director of the Mount Wilson Observatory, reports the discovery of two new satellites of Jupiter (X and XI) by Professor Seth B. Nicholson on photographs taken with the 100-inch reflector at Mount Wilson. Both are about magnitude 19. Positions are given for three dates in July and August.

Jupiter's faint satellites VI and VII, of magnitudes 15 and 17½ respectively, were discovered by Perrine, of the Lick Observatory, in 1905 on photographs taken with the 3-foot Crossley reflector. Satellite VIII, magnitude 17, was found by Melotte at Greenwich in 1908 on a plate taken with the 30-inch reflector, and satellite IX, magnitude 18½, by Nicholson in 1914 on photographs taken at Lick with the Crossley reflector.

VENUS AND MARS

The atmosphere of Venus, the planet which of all most closely resembled the Earth in size, mass, and mean density, was in marked contrast with that of the Earth. Investigation had given no certain evidence of the presence of oxygen, and the most interesting fact was the abundance of carbon dioxide. Not improbably the temperature at the surface of the planet was as high as, or higher than, that of boiling water, and this high temperature, the lack of oxygen, and the abundance of carbon dioxide could be interpreted as indicating that there could not be any great amount of vegetation on Venus and suggesting that the planet was not the abode of life.

All attempts to detect oxygen in the atmosphere of Mars had been unsuccessful, but the red colour of the planet provided indirect evidence of its presence, suggesting rocks that had been completely oxydized. Nor had carbon dioxide been detected in the Martian atmosphere, but as there was some evidence of the existence of vegetation its presence might be inferred. Mars appeared to be in the state the Earth would ultimately reach when the oxygen in the atmosphere had been almost entirely exhausted by the progressive weathering and oxidation of the rocks.

IMPORTANCE OF SMOOTHING SURFACES

The report explains how drag is diminished if the airflow over the greater part of the wing can be kept free from turbulence, and how important it is to secure that the transition from smooth flow to eddying flow shall occur as far back from the leading edge as possible. It points out that turbulence may occur even in the forward portions of the wing surface as the result of roughness in that surface. It has been found that roughnesses less than one-thousandth of an inch in height are sufficient to increase measurably the drag of a high performance aeroplane.

MARS NEAREST TO THE EARTH TODAY

Travelling at a speed of four miles a second, the planet Mars will to-morrow approach to within 36,000,000 miles of the earth. This is 12,000,000 miles nearer the earth than the planet usually comes.

Astronomers throughout the world have been preparing to make use of the favourable conditions for observation in the hope that they can throw more light on the mysteries enshrouding the earth's nearest terrestrial neighbour, and South Africa is in a particularly good position to see the occurrence. Dr. Slipher, Director of the Lowell Observatory, Flagstaff, Arizona, has come to Bloemfontein, which will be his base for his observations. He will use the telescope at the Lamont-Hussey Observatory, on Naval Hill, which overlooks the town. Colour photographs of the planet will be taken with special cameras and highly sensitized photographic plates, in the hope of ascertaining whether or not vegetation exists on Mars. The astronomers here do not expect that sensational discoveries will be made.

MARS WHITE WITH SNOW

To-night Mars reaches the point at which the planet will be at its nearest to the earth. The conditions for observation last night were unfavourable, the chief difficulty experienced by Dr. Slipher, Director of the Lowell Observatory, Arizona, who has come here to make observations, being the wind. This, although scarcely perceptible to the layman, swayed the telescope, causing the image of the planet to dance. However, opportunities for observation during the next few days will still be good, because even after Mars has reached the nearest point of its approach the diminution in the size of the planet's image will be slight.

Dr. Slipher reports that there was a severe snowstorm on Mars early this week and that millions of square miles of its northern hemisphere are covered with snow. The area of the snowfields fluctuates nightly, as the temperature on Mars varies greatly from day to day.

ACCIDENT AT LICK OBSERVATORY

Through a press report we learn of an extraordinary accident to the Lick Observatory, Mount Hamilton. In a dense mist a U.S. army 'plane crashed into the Observatory building through two eighteen-inch brick walls and wrecked the store of valuable records collected over a period of thirty years. The two occupants of the machine were killed, but fortunately no one in the Observatory was injured. It is no small comfort to hear that the 36-inch refractor escaped the disaster undamaged, but the loss actually sustained is sufficient to earn the sympathy of astronomers all over the world.

STELLAR DISTANCE

One hundred years ago the first measurement of the distance of a star (61 Cygni) was successfully accomplished by the famous German astronomer Bessel. Hardly had the news been circulated among European astronomers when a second success—the measurement of the distance of Alpha Centauri, one of the brightest stars in the southern heavens — was announced (January, 1839) by Henderson, a Scotsman who had been a few years before H.M. Astronomer at the Cape of Good Hope. In the following year, the distance of Vega, one of the brightest stars in our northern sky, was measured by Struve at the Russian Observatory of Poulkowa. The distances of 61 Cygni, Alpha Centauri, and Vega, according to the best modern measures, are 64, 26, and 154 million million miles respectively, or 11, 4½, and 26 light-years. Thus, within a few months, a successful beginning had been made on three fronts in the solution of the age-long problem of the sounding of the sidereal universe.

The practical determination of the distance of an inaccessible object requires a base-line of known length and a suitable instrument for measuring the directions of the object as viewed from each end of the base-line. This is the principle inherent in ordinary terrestrial surveys, in measuring the distance of the Moon (the base-line in this instance—of a few thousands of miles—is provided by two observatories, one in the northern hemisphere, the other in the southern), and in the more difficult problem of measuring the Earth's distance from the Sun.

If the stars are at immense distances from us we require a base-line incomparably greater than that provided by any two places on the Earth's surface. As Copernicus clearly saw, the base-line that had the best claims for satisfying the exacting conditions imposed by the problem of measuring stellar distances was provided as a simple consequence of the Earth's annual motion around the Sun. At any given date the Earth is in some definite position in its orbit and six months later its position is separated from its first position, relative to the Sun, by a distance equal to the diameter of the orbit; this distance of about 186,000,000 of miles is the base-line adopted in our problem.

The practical procedure of measuring a star's distance may best be illustrated by means of a simple experiment. Hold a finger at arm's length in or near the direction of a distant object such as a church spire. With the left eye closed, the direction of the finger as seen by the right eye can be related to the direction of the spire. With the right eye closed, the new direction of the finger as seen now by the left eye can be similarly related to the direction of the spire. With suitable means we could measure the difference between the direction of the finger and of the spire in each instance, and these measures enable us to determine the difference in direction of the finger as seen by the right eye and by the left eye; the spire, it should be noted, is simply an intermediary introduced to define a constant direction. In our astronomical procedure, the star whose distance is to be measured corresponds to the finger and the intermediary (the spire) is a faint and presumably a very distant star; our astronomical base-line, the ends of which are the Earth's positions in its orbit at an interval of six months, corresponds to the base-line defined by the right and left eyes.

There are several criteria by means of which we can pick out the stars most likely to be nearest to us. One criterion, for example—although it is not always reliable—concerns the apparent brightness of the stars, for, other things being equal, the brighter a star the nearer it may be presumed to be. It was by considering all such criteria that Bessel, Henderson, and Struve selected the stars which were the first to reveal the scale of stellar distance.

To understand the merits of their achievements with the instruments of a century ago it is necessary to remember the smallness of the change in direction of the parallax star—as the near star is called—as viewed from the extremities of the base-line. If two small marks, 1in. apart, are made on the dome of St. Paul's Cathedral, and if these can be viewed from a distance of 3¼ miles, the difference in the directions of the two marks is one second of arc; in other words, the base-line of 1in. subtends an angle of one second of arc at an observer 3¼ miles away. Representing the astronomical base-line of 186,000,000 of miles by 1in., we find the distances of 61 Cygni, Alpha Centauri, and Vega correspond on this scale to about five, two, and 13 miles respectively.

It was only on the introduction of photographic methods at the beginning of the present century that astronomers were enabled to penetrate to more remote depths of space. About a dozen observatories scattered over the globe—among them the Royal Observatories at Greenwich and the Cape of Good Hope—are continuously engaged in this difficult department of observational astronomy. The photographic method, however, ceases to be effective at stellar distances beyond 1,000 light-years: this limiting distance corresponds to a distance of 500 miles from the base-line of 1in. in our illustration. The exploration of the most distant parts of the universe depends on indirect methods which are mainly based, however, on the knowledge, so laboriously acquired, of the distances of our immediate stellar neighbours.

From a newspaper viewpoint, science reporting in the Second World War was, in some ways, a re-run of what happened in the First World War. The amount of space available for reporting science decreased, both because other news received priority and because paper was rationed. The amount of reportable civilian science in progress also fell rapidly: more so if anything, than in the previous war, because scientists and engineers were recruited more efficiently to aid the war effort. In addition, secrecy concerning new military technology was better maintained in the Second World War than in the First. As a result, the end of the war saw a sudden release of information regarding the important scientific developments that had occurred during the previous five years. Many of these developments were in physics, so that the latter part of the 1940s saw a major growth in physics reporting. Historians sometimes refer to the First World War as the 'chemists' war' and the Second World War as the 'physicists' war'. In terms of newspaper coverage, this has something to be said for it (though reporting of chemistry during and after the first war never reached the level of physics reporting after the second).

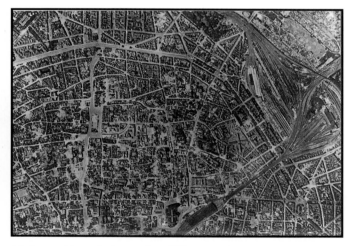

High-precision photography of German cities

The atomic bomb inescapably attracted the most attention. The discovery that heavy elements could be broken into fragments by bombardment with neutrons had been made before the war. So had the observation that such nuclear fission could be self-sustaining, given the right conditions (implying the possibility of access to large amounts of energy). The military potential of this result was recognised by a number of physicists. In the USA, Einstein was persuaded to write to President Roosevelt suggesting that effort should be put into constructing an atomic bomb. The 'Manhattan project'—the ultimate response to this letter—involved physicists from other countries besides the United States (more especially a group from the UK where parallel work had already begun). A very high level of secrecy was maintained; so the first news the public received was when atomic bombs were dropped on the Japanese cities of Hiroshima and Nagasaki.

Discussion of the bombs grew quickly. First came the question of how an atomic bomb worked, what were its effects, and who had been involved in developing it. Next came the state of German development of such a bomb. It was fear of a German bomb that had sparked Einstein's letter, but it transpired post-war that little advance had been made in Germany. Soon, the interest spread to possible applications of nuclear energy to peacetime activities. The growth of the 'Cold War' then led to discussions, often anguished, concerning the spread of atomic 'secrets'. Alongside this went reporting of the more powerful bombs being constructed and tested by those countries which already possessed, or subsequently gained, the necessary know-how. This heady mixture of science, technology and international politics meant that nuclear energy dominated the science headlines in the latter part of the 1940s.

The other major physics-based contribution to the war effort was radar. Work on radio methods for detecting aircraft had begun (especially in the UK) well before the beginning of the war. But progress had been kept secret, as were the wartime developments (e.g. of the cavity magnetron), which made radar a much more flexible and useful device. Work on radar continued after the war, but, after the first burst of publicity, received diminishing attention from the media. The important aspect from the media viewpoint was actually a side effect of wartime radar. In 1940, radar sets throughout Britain had been blacked out for a while. It was feared that the Germans had developed a new jamming mechanism, but subsequent investigation showed that the source of the radiation lay in the Sun. After the war, a number of people who had worked on radar returned to civilian life and used the skills they had acquired to develop a new subject—radio astronomy. The astronomical insights that resulted soon figured in newspaper reports, and continued to do so.

The Allied powers took the lead in developing both the atomic bomb and radar. But it was Germany who pioneered rocket propulsion. The A-4 rocket (popularly know as the V-2) was specially designed for use as a guided missile, and resulted from a German interest in rocketry which went back to the inter-war years. Bombardment of the civilian population by the V-2 overlapped in time with the more frequent firing of the V-1 (a jet-propelled missile). Consequently, public understanding confused the two. The A-4 rocket's existence was not publicised until after the end of the war, by which time both the USA and the USSR had acquired what they could of German materials and experts. In both countries, the stimulus given by this German rocket programme became the trigger for their space exploration programmes a decade later.

Not all physics reported during the 1940s related to wartime developments. The building of a bigger and better cyclotron in California, for example, represented straight-line progress from pre-war days. (Though funding for such technology became easier, since people saw such 'atom smashers' as being somehow linked to atomic bombs.) Equally, the commissioning of the largest telescope in the world at Mt Palomar represented the completion of a story that had begun well before the war. What the scale of this instrumentation underlined was the increasingly dominant role of the United States in post-war physics. Reporting of American physics advances increased markedly in newspapers everywhere. At the same time, one of the consequences of the Second World War was to create a demand for more information on science as a whole, so the amount of space devoted to science quickly expanded.

1940s CHRONOLOGY

1940
- The carbon 14 isotope is discovered

1942
- A team headed by Fermi produces the first controlled nuclear chain reaction
- Solar radio emission is detected

1943
- A team headed by Turing develops the first entirely electronic computer

1944
- The V-2 rocket is used operationally by the Germans
- Baade discovers there are two different 'populations' of stars

1945
- Second World War ends
- First two atomic bombs are dropped on Japan
- Radar signals are reflected from the Moon

1946
- The first general-purpose electronic computer (the ENIAC) is completed
- The first synchrocyclotron is built in California
- Particles—later called 'strange' particles— are detected in cosmic-ray experiments

1947
- The Lamb shift is detected in hydrogen spectra
- The pi meson is discovered
- Gabor suggests the basic idea of holography
- The first microwave oven goes on sale

1948
- Feynman and others develop quantum electrodynamics
- The shell theory of the nucleus is proposed
- The first transistor is developed
- The 200-inch telescope on Mt Palomar comes into operation
- The Big-Bang theory of the origin of the Universe is proposed
- The atomic clock is introduced

1949
- Shannon publishes his work on information theory
- The polarisation of starlight by interstellar dust is observed

1940s

1940

SCIENCE AND WAR

In an embattled world few men might appear more enviable than the pure mathematicians,—those whose whole contemplation is fixed upon number as it is in itself and not as it is made manifest in the objects of sense, to whom 1,000 is the cubed radix of the scale of notation, not the weight in kilograms of a ton bomb, and 1 the mystic unity whence all multiplicity is generated, rather than the one individual with his mission to enslave mankind. Yet world war, it seems, has power to make even these intellectual ascetics question their vocation. Dr. G. H. Hardy, from whom so many generations of undergraduates at both universities have had their first introduction to the esoteric world of mathematics, has been moved to consider the justification of his life's work, and to publish the result in a little book called "A Mathematician's Apology." He concludes that what he calls the "real" mathematics—which is what the Greeks called arithmetic, the abstract study of number, as opposed to logistic, the apparatus of calculation required by tradesmen for their traffic—has its worth of the same high order as music or any other of the arts; and that, although its ultimate value is only to be apprehended by the few who have been trained to perceive its beauties, yet it may be sufficiently defended to the multitude as "a harmless and innocent occupation."

With his own conscience thus clear, Dr. Hardy nevertheless cannot ignore the indictment brought against the "trivial" mathematics, which by throwing out branches such as aerodynamics and ballistics has acquired so grim a significance in the daily life—and death—of unmathematical men. Is the scientist, whom our grandfathers so unquestioningly hailed as the leader of humanity into the promised land, really marching us blindfold into the ambush of the powers of evil? The evidence for the prosecution needs no emphasis: it is scored into the earth's surface from Warsaw to Coventry and from Barcelona to Canton. For the defence Dr. Hardy calls Dr. J. B. S. Haldane to testify "that bombs "are probably more merciful than "bayonets; that lachrymatory gas and "mustard gas are perhaps the most "humane weapons yet devised by military "science; and that the orthodox view "rests solely on loose-thinking senti-"mentalism." Without committing himself to this view he suggests that it might be reinforced with the argument "that the "equalization of risks which science was "expected to bring would be in the long "run salutary; that a civilian's life is not "worth more than a soldier's, nor a "woman's than a man's; that anything is "better than the concentration of savagery "on one particular class; and that, in "short, the sooner war comes 'all out' "the better."

AERIAL PHOTOGRAPHS BY NIGHT

The United States Army Research Staff, in experiments over Rochester and New York on Tuesday night with an explosive light-bomb of 1,000,000,000 candle-power, illuminated an area of five square miles so brightly that it was possible to take extraordinarily detailed pictures of the whole terrain. The camera used was equipped with a photo-electric cell which opened the shutter at the moment of greatest illumination.

RADIUM STORED IN A CAVE

The transference of radium to a cave in the Derbyshire hills was among one of the war measures taken by the Christie Hospital and Holt Radium Institute, whose annual meeting took place here to-day.

The annual report stated that the board was informed that anything approaching a direct hit where radium was present would make the immediate neighbourhood unliveable, so the radium store was lowered into a shaft dug 50ft. deep in the hospital grounds. Later, it became necessary to provide special protection in a non-vulnerable area for the reception and treatment of patients by radium. A member of the board loaned his country house for the duration of the war, and patients from a wide area were now able to have radium treatment under conditions of safety.

To extend still further the radium treatment, the radium in solution was transferred to a deep subterranean cave in the Derbyshire hills where a laboratory was fitted up and equipped.

HYDRODYNAMICS ON A TRAIN

IF an excuse be needed for recording the experience described below in these pages, it may be pointed out that hydrodynamical principles are being applied to many modern astrophysical problems, so that some day we may find an analogous phenomenon in the external universe.

A passing shower had left some raindrops on the window of a train travelling southwards. A passenger looking out of the window on the western side noticed a peculiar eddy-motion of dust-specks in the drops; perhaps because it was war-time there was more grime than usual on the window-pane, and each drop had a fair ration of dust-specks suspended in it. *All the specks were rotating anticlockwise.* Other windows were examined on the same side of the train; whenever motion could be observed (which was not always), the rotation was in the same sense. On examining the eastern windows, the only rotations that could be observed were *clockwise.* Judging by the smoke of the train the west side was more sheltered from wind than the east. It seems probable that the direction of rotation is determined by the asymmetrical shape of the drops produced by gravity, and that the undercutting action of the wind tends to initiate motion in the bulge of each drop. A further phenomenon was noticed when the train began moving after a stop. Although the motion of the train was quite smooth the rotation in the drops began in pronounced jerks as though checked by viscosity.

GIANT CYCLOTRON AND TELESCOPE

The current Rockefeller Foundation Review draws attention to the work proceeding in the United States on two of the mightiest instruments the world has seen for the peaceful exploration of the universe. One is the 200in. telescope nearing completion on Mount Palomar, in California; the other is the giant cyclotron under construction at Berkeley, California.

The new telescope will explore the outer reaches of the universe, the realm of the infinite; the new cyclotron will probe the inner reaches of the universe, the realm of the infinitesimal. The construction of the telescope was made possible by an appropriation in 1928 of $6,000,000 by the International Education Board, established by Mr. John D. Rockefeller, jun. Last year the Foundation appropriated $1,150,000 to the University of California for the construction and housing of the cyclotron.

The new giant cyclotron will contain over 4,000 tons of steel and copper in its magnet alone, and will produce a beam whose voltage will range from 100,000,000 to perhaps 300,000,000. The beam of the largest cyclotron now operating penetrates, in air, about 5ft; the beam of this giant instrument will penetrate 140ft. The most powerful cyclotrons now in existence produce particles whose speeds, when fired at atoms, enable them to knock off only the external and more loosely bound features of the atoms under attack. It is at this point that the new giant cyclotron, now under construction, is of critical importance; for it is designed to produce projectiles so powerful that they can penetrate and explore the nucleus itself.

WAR RESEARCH IN CANADA

Research Enterprises, Limited, a Government corporation, had been set up by the Canadian Ministry of Supply primarily to produce for the armed forces and for industry inventions and apparatus developed in the council's laboratories. Already optical instruments, including gun-sights and range-finders, radio gear and similar articles, were in production in a large new factory, and very shortly the company would be turning out its own supplies of optical glass in quantity. Thus a small pre-war nucleus had been developed into a key industry of essential importance for our war effort.

In the field of radiology special attention had been paid in the council's laboratories for many years to the examination of castings, particularly those in light alloys required to carry stress in aircraft construction. The knowledge of the highly developed technique of making sound castings was proving of great value to the Canadian aircraft industry. X-ray photographs could be taken by the penetration of several inches of steel. For greater thicknesses a plentiful supply of radium, tested by the council, was available.

MICROLETTERS

The first batch of photo-letters under the newly instituted airgraph service from the Middle East arrived in London yesterday morning. The mail contained 50,000 microfilm letters which had been flown from the Middle East. They weighed only 13lb., whereas 50,000 ordinary letters would have weighed nearly three-quarters of a ton.

As soon as they were received the films were sent to the Kodak works, where the staff began to enlarge the letters from the size of a man's finger-nail to 4in. by 5in. As fast as this was done the enlargements were rushed back to the post office to be placed in special envelopes, sorted, and sent to all parts of the country for delivery.

The airgraph service was begun last month, and the first batch of letters has taken about a fortnight to arrive. It is hoped that this time will be shortened as the scheme develops. At present airgraph letters can only be sent from the Middle East to this country, but it is hoped to establish a similar service in the outward direction.

THE SUN'S DISTANCE FROM THE EARTH

The Astronomer Royal, Dr. H. Spencer Jones, has recently completed the most accurate determination yet made of the sun's distance from the earth. The distance proves to be 93,005,000 miles, with an uncertainty either way of only 9,000 miles. The uncertainty in the new computation of the sun's parallax—which determines its distance—corresponds to the apparent breadth of a human hair at 10 miles, or of a halfpenny at 3,250 miles.

The facts from which the new figure is derived were secured in 1931 by cooperation between observatories in England, Belgium, Germany, Italy, Spain, Czechoslovakia, Russia, India, China, Japan, Algeria, South Africa, the United States, Argentina, and Australia. The method was to photograph Eros at a time when that small asteroid (which is only 15 miles in diameter) had come within 16,200,000 miles of the earth.

By careful measurements on each photographic plate the exact position of Eros among the stars was determined. As that position differed slightly according to the situation of the observer on the earth's surface, comparison of photographs taken at different stations enabled the relative displacement of Eros to be determined and the distance of the sun to be inferred.

STAR-GAZING

But probably never before this present hour has astrology been so pertinaciously broadcast (the B.B.C. has long since dropped it, and the word has its pre-wireless sense here) and brought to bear upon a future of such immeasurable import to the whole of the human race. That, no doubt, was what moved MR. KEELING to ask the MINISTER of INFORMATION whether he would stop astrological predictions about the war in order to counteract the risk that addicts of astrology would relax their efforts.

"Addicts" is a good word; it carries the requisite taint of surrender to a drug or some other means of escape from the hard truths of fact. The irresponsibility induced by this kind of self-protection is not to be lightly dismissed as of no public concern. The stars say that England cannot be invaded; the stars say that England can and will be invaded—you buy your newspaper and you take your choice. The authorities are, no doubt, satisfied that there is no possible danger from unlucky accident. Suppose, for instance, that one of the prophets should read in his stars that a certain act of war would take place on a date which had been actually, and very secretly, chosen for it, that would be a very unhappy coincidence. We may take it for granted that the Government has glanced at that possibility and has not thought it worth notice.

WATSON-WATT, THE BRITISH RADAR PIONEER, SPEAKING AT A WAR-TIME MEETING

A NEW SOLAR PERIODICITY

Lord Horder was a lively and amusing chairman at the Poetry Society's meeting yesterday afternoon in Portman-square. He presided at a lecture by Dr. Douglas Webster, the Harley-street specialist, on periodic inspiration in poetry and music.

Dr. Webster believes that the stimulus on life comes from solar activity. This revolves in a period of 33 weeks, and its rhythm can be traced in diseases—such as tuberculosis—as well as in artistic creation.

A LAKELAND PHENONMENON

A meteorological phenomenon shortly before sunset on August 20 fascinated thousands of Lakelanders. In an otherwise clear sky a spiral of cloud like a gigantic waterspout was seen to rise in the south. Within 10 minutes it had passed overhead and extended to the hilltops in the north, forming a complete arc from horizon to horizon. The setting sun tinged it with red, and as the sun passed from sight the narrow belt of cloud vanished in a few moments.

SCIENCE AFTER THE WAR

Sir Lawrence Bragg, speaking on the exposition of science, said scientists were often accused of being too technical in their presentation of their subject. The main trouble was that science had not been regarded as an essential part of a balanced education, and the great public schools were the main sinners in this respect. Looking through a list of the leaders of science he found that Harrow, Clifton, Malvern, and Westminster contributed one each; all the rest came from secondary schools. However, a knowledge of science was rapidly becoming more widespread. It was much easier to give a popular scientific lecture to young people than to their parents; they had a background and both understood the points and appreciated the interest.

Professor J. A. Lauwerys said that in education the museum had been neglected. Our great national museums were second to none in the world, and 200 others were good, but the remaining 600 were junk shops. He suggested that as part of the post-war rebuilding there should be put up in each city a civic centre in which one of the most important buildings would be the museum of science. There one would find a microcosm of the city, showing its relationship to its surroundings and to the State; illustrating, too, the relationship between industry and society.

Mr. H. G. Wells, winding up the discussion, said he had a profound conviction that the newspaper was dead. It was perfectly well known that no one got news that was worth reading from the newspapers. When one wanted to know the time one dialled TIM. Why not, when things got into order again, dial NEWS and listen to a summary of what had happened in the last two or three hours? It would be a more reasonable way of getting news than making desperate attempts by buying three or four newspapers to find out what was being concealed from us. He believed that we were entering a period of pamphlets and that they would do what newspapers had utterly failed to do, and provide material for discussion.

US ENDS BAN ON WEATHER REPORTS

Washington, Oct. 11.—The United States censorship office announces that restrictions on the publication and broadcasting of official weather reports are being removed at midnight. The director of the office, Mr. Byron Price, said that this action was warranted by improved defence and other conditions.—

ANGLO-AMERICAN COOPERATION

In explaining the objects of his four months' visit to India in response to an invitation from the Government to the Royal Society, Professor A. V. Hill described the close cooperation that has existed, ever since the outbreak of the war, between British and American scientists.

Early in 1940 he went to Washington, and found that American scientists were quite clear in their minds about the war and wished to help in every way possible. In Professor Hill's words, the cat of radio-location peeped from his bag and from their bag, but it could not be let out until President Roosevelt had agreed to a complete pooling of knowledge. By November of that year, 13 months before Pearl Harbour, they were sharing information and American scientists were in the war. There was a British scientific office in Washington and Americans had a similar branch at their Embassy in London.

1944

IMPROVED OSCILLATIONS

To tell the time, astronomers instead of counting the swings of a pendulum now count the vibrations of a quartz crystal. . . . . The pendulum vibrates once a second, the quartz crystal 100,000 times a second. That is the measure of improvement between the old and the new clock.

D-DAY WEATHER

A cartoonist went so far as to draw the Clerk of the Weather taking off the mask of elderly, Puritanical benevolence with which the imagination for some reason endows him and disclosing beneath the hideous features of the fifth columnist. And yet all the time, had we but known it, the clouds we so much dreaded were big with opportunities Supreme Headquarters were quick to grasp. The weather, indeed, while we were able to probe its secrets, bluffed the enemy completely for, in the words of the report:

The German commanders were advised by their meteorological service that there could be no invasion in the period including June 6 because of continuous stormy weather. That is why D Day forces landing during a brief break in the windiest month in Normandy for at least 20 years found so many German troops without officers, and why other enemy coastal units were having exercises at the time of the landings.

A LONG VIEW OF V-2

Recently an occasion of extraordinary clarity in the much abused London atmosphere permitted the view of a V 2 rocket being fired from its distant base on the far side of the North Sea. Before dawn on that morning I chanced to glance eastward from a small London window and was astonished to see, through a gap between the surrounding buildings, the bright trace of a rocket against the still dusky sky. I had previously seen Press photographs of such trails taken from a point nearer their source, and I readily recognized what I was seeing. A visual range of 200 miles or more may seem surprising, but although London was hidden from the direct rays of the sun, this would not apply to a rocket ascending from a point much nearer the sunrise.

A LIFE SAVING PLUG

The United States raids deep into the heart of Germany had been made possible by detachable auxiliary petrol tanks made from waste paper in Britain. These tanks enabled Thunderbolt and Mustang fighters to fly the distance to give protection to the Fortresses. Tanks were made of old newspapers and other waste paper pressed and treated under a formula first developed by British technicians.

" Operational needs of our air force for these detachable tanks increased so fast early last spring," the report says, " that for a few weeks tanks that were delivered from factories one day would be in use over Berlin or some other German city the next day. Since then there have been plenty of them. Already deliveries are approaching the 300,000 mark.

" These paper tanks have been made in over a score of small plants scattered throughout Britain, plants that used to turn out such products as ice-cream containers. To find enough labour to meet our needs the British went out and recruited women, men in their fifties and sixties, and boys in their early teens to get production up to the required levels. By their devoted efforts this pickup labour force, working principally with old newspapers and glue, has contributed much to bringing destruction to the heart of German war industry, to shortening the war, and to saving the lives of American bomber crews."

HOW THE BOMB
WAS MADE

From the cruiser Augusta, in mid-Atlantic, President Truman has announced that the United States Air Forces have used, for the first time, against a Japanese target an atomic bomb which has an explosive power equal to 20,000 tons of T.N.T.

The statement, which was issued through the White House, said that on Sunday an American aircraft dropped one of the new bombs on Hiroshima, an important Japanese army base, west of Kobe. The bomb had more power than 20,000 tons of T.N.T. and more than 2,000 times the blast power of the British " grand slam " (22,000lb.) bomb, which until now was the largest bomb ever used in the history of warfare.

The new bomb, said Mr. Truman, meant a new and revolutionary increase in destruction to supplement the growing power of the United States against Japan. " It is an atomic bomb. It is the harnessing of the basic power of the universe." He added that the new bomb was now in production and that even more powerful forms were being developed.

RESEARCH "POOL"

The use of the new bomb meant victory in a feverish race with German scientists to find some way to harness and release atomic energy. In 1942 it was known that the Germans were working hard to find a way to use such energy to make engines of war with which to enslave the world. The battle of the laboratories went on and had now been won, as had other battles.

Before Pearl Harbour the United States and Britain had pooled their scientific knowledge that could be useful in war, and many priceless aids to victory had come from that arrangement. With American and British scientists working together, the allies had entered a race of discovery against the Germans. There were now two great plants and many lesser works devoted to the production of atomic power.

" We spent $2,000,000,000 on the greatest scientific gamble in history—and won." As a result of these discoveries the United States was now prepared to obliterate more rapidly and completely every productive enterprise the Japanese had above the ground.

" We shall destroy their docks, their factories, and their communications. Let there be no mistake: we shall completely destroy Japan's power to make war. It was to spare the Japanese people from utter destruction that the ultimatum of July 25 was issued from Potsdam. Their leaders promptly rejected that ultimatum. If they do not now accept our terms they may expect a rain of ruin from the air the like of which has never been seen on this earth. Behind this air attack will follow by sea and land forces in such numbers and power as they have not yet seen, but with fighting skill of which they are already aware."

Mr. Truman went on to say he would recommend that Congress should consider the establishment of an appropriate commission to control the production and use of atomic power so that it could become a powerful influence for the maintenance of peace. Atomic energy might be harnessed to supplement the power that came from coal, oil, and waterfalls, though at present it could not be produced on a basis to compete with them commercially.

The centres of production were at Oakridge (Tennessee), Richland (Washington), and Santa Fe (New Mexico). The number of people employed at one time on these projects was 125,000, and more than 65,000 were still engaged in operating plants. Only a few of these people had known what they were producing, and, although they were making materials for the greatest destructive force ever known, they had been in no more danger than attended many other occupations.

A message from the Augusta late this afternoon states that Mr. Truman personally made the announcement about the bomb to officers of the Augusta in the wardroom. He said " the experiment has been an overwhelming success."

IMPROVED VERSION SOON

The Secretary of War, Mr. Stimson, said later that an improved bomb would be forthcoming soon. It would increase by " several-fold " the present effectiveness of the new weapon. Uranium was the essential ore in the production of the new bombs, and steps had been taken, and would continue to be taken, to ensure adequate supplies of this mineral.

Mr. Stimson was convinced that Japan would be unable to use such a bomb during this war, and it was abundantly clear, he said, that possession of this weapon by the United States, even in its present form, should prove a tremendous aid in shortening the war. He added that no praise was too great for the unstinted efforts, courage, and devotion to national interests of the scientists whose contributions were behind the concrete achievements the bomb represented.

Mr. Stimson said substantial patent control had been accomplished in the United States, the United Kingdom, and Canada to make certain that the weapon would not fall into the hands of the enemy. In each country the scientific and industrial personnel engaged on the work were required to assign their entire rights to any inventions in this field to their respective Governments.

Mr. Stimson also stated that scientists had already discovered means of releasing atomic energy other than explosively, that is in regulated amounts, but the necessities of war had so far precluded a full exploration of the peace-time application of these new discoveries.

A TRIAL BOMB

The War Department, describing the first test firing of the atomic bomb, said that the weapon immediately vaporized a steel tower from which it was suspended and sent a massive cloud billowing into the stratosphere with tremendous power. At the appointed time there was a blinding flash which lit up the whole area brighter than the brightest daylight, with a mountain range three miles away from the observation point standing out in bold relief. Then came a tremendous sustained roar and a heavy pressure wave, which knocked down two men standing outside the control tower 10,000 yards from the scene of the explosion.

GERMAN DEVELOPMENTS

Many of the German secrets, at the time the Combined Intelligence Objectives Sub-Committee, as the body carrying out the investigations was called, discovered them, were considered of great potential value in the war against Japan, and many of them were being adopted for use in the Pacific when the Japanese surrendered.

Some of the secrets which may now be disclosed show that not only had the Germans made significant progress in the development of an atomic bomb and on the production of "heavy water" but they had contemplated a piloted missile with a possible range of 3,000 miles. The designer thought it could be adapted to carrying passengers across the Atlantic in 17 minutes.

VIEW OF HIROSHIMA AFTER THE ATOMIC BOMB EXPLOSION

FOG DISPERSAL

The first operational use of Fido took place on November 19, 1943, when four Halifaxes landed successfully after an operational flight to the Ruhr, though the surrounding visibility was only 100 yards. Ten minutes after Fido had been lit visibility on the runway increased to the equivalent of two to four miles.

Since that day more than 2,500 allied aircraft have been safely landed—many of them in conditions of dense fog—with their crews of over 10,000 airmen. Later 15 operational airfields in this country (14 on the east coast and one in Cornwall) and one on the Continent were equipped.

It is of interest that in the early days, when the experts were pessimistic about the outcome of their researches, Fido was said to stand for " Fog investigation dispersal operations," but when success was achieved the R.A.F. altered it to " Fog, intensive dispersal of."

Fido provides the necessary heat on airfields by a continuous line of burners installed parallel to and some distance from each side of the main runway. A standard fog dispersal installation consists of three main portions—burner lines, pumping and distribution, and storage.

SCIENTISTS AID
IN THE WAR

The plans for the famous attack on the Ruhr Dam were first worked out, he said, with models at the Road Research Laboratory. At the Building Research Station the study made of the effect of enemy incendiary bombs on our buildings was adapted to the study of the effect of allied bombs on enemy methods of construction, models of which were put up at the station.

The Fuel Research Station successfully investigated methods of treating fuel for flame-throwers, and also produced a simple and effective means of preventing the emission of smoke by ships at sea. The help of all departments of the Physical Laboratory had been sought on many problems. Models of assault craft of all descriptions were tested in the tanks of the Ship Division, including parts of the Mulberry harbours.

The Cotton Research Association had produced a waterproof cotton cloth from which garments were being made for airmen immersed in sea water. The fibres were so twisted as to swell on coming in contact with moisture and to block up the interstices of the cloth without other treatment. A hose-pipe had been made from these cloths which would stand up to Home Office tests. The Forest Products Research Association had cooperated in the study of plywood and adhesives for wooden aircraft.

RESEARCH ON
THE ATOM BOMB

At the beginning of 1940 Professor Peierls, of Birmingham University, Dr. O. Frisch, and Professor Sir James Chadwick, both of Liverpool University, independently called attention to the possibility that a fast neutron fission chain reaction could be obtained if pure, or nearly pure, uranium 235, were available.

A committee of scientists, with Professor Sir George Thomson as chairman, was set up in April, 1940, originally under the Air Ministry, and later under the Ministry of Aircraft Production to examine the whole problem.

By the early summer of 1941 the committee decided that the feasibility of a military weapon based on atomic energy was definitely established, and that this weapon had unprecedented powers of destruction, that a method of producing the amounts of material required was in view, and that a fair estimate of the industrial effort needed to accomplish the project could be given.

Work on finding conditions under which a mixture of uranium and some suitable " slowing-down " medium might give a neutron chain reaction in which the release of energy was obtained in a controlled way was carried out at Cambridge by Drs. Halban and Kowarski, two French physicists who had been sent by Professor Joliot to this country at the time of the fall of France in June, 1940. They brought with them the 165 litres of heavy-water—practically the whole world stock of this material—which the French Government had bought from the Norsk Hydro Company just before the invasion of Norway. Facilities were provided at the Cavendish Laboratory, Cambridge, and by December, 1940, they produced strong evidence that, in a system composed of uranium oxide (as actually used) or uranium metal, with heavy-water as the slowing-down medium, a divergent slow neutron fission chain reaction would be realized if the system were of sufficient size.

ATOMIC ENERGY
FOR CIVILIAN USE

In a discussion here yesterday by nine scientific and industrial experts, all of whom played important parts in the development of the atomic bomb, it was agreed that in anywhere from three to 25 years the development of atomic energy for peace-time uses in competition with coal at $15 a ton would be commercially practicable.

1946

LONG TERM PLANNING

German scientists had their eyes on plans 50 to 100 years ahead, said Lt.-Col. John A. Keck, chief of the Technical Intelligence Branch of the United States Army, today. He was talking to reporters. . . . .

They also had the idea of using sun rays to scorch nations and cities out of existence. The power of the sun was to be focussed on the target with huge reflectors, two miles square, which were to be placed on "space platforms" 5000 miles above the earth, where gravity would be neutralised.

His statement was greeted with incredulity, but he declared that United States officials were taking the matter seriously.

RADAR AND THE WEATHER

Soon after the installation of the first centimetric radar sets certain peculiarities in their performance were noticed. The principal advantage of radar is that it enables objects to be detected in thick weather or at night, and, of course, it also gives an accurate measurement of range under all conditions. The normal range of such equipment is approximately as far as the optical horizon. It was therefore a matter of great interest and importance when reports began to come in from various parts of the coast that surface warning sets had been plotting shipping out to distances far beyond the horizon, or that echoes had been received from land at many times the normal range.

It was obviously desirable to determine the underlying cause of this phenomenon. The operators had already begun to notice that these extraordinary variations in the performance of their sets appeared to be related in some way to the prevailing weather conditions. "Anomalous propagation," as it came to be called, was associated with fine weather. When the weather was rough, rainy, or cloudy performance was nearly always orthodox, but on fine, calm, warm days ranges usually began to go up to above normal, particularly in the afternoon and early evening.

Later, when radar equipment had been installed in other parts of the world, it was found that the degree of intensity of "abnormal propagation" varied greatly in different regions. The normal rules were found to be violated on a stupendous scale on the west coast of India during the hot season. A radar set situated near Bombay which during the monsoon season is only able to plot ships out to the expected range of 20 miles, commonly during the hot season plotted them out to 200 miles, and has on occasion plotted them out to 500 miles. Land echoes during the hot season have been identified as coming from ranges up to 1,500 miles.

although radar does not affect the weather, it has the exceedingly important property of being able to give warning of the approach of bad weather. This is because centimetric radar sets are able to detect echoes from concentrations of water drops in the atmosphere either in the form of moderate to heavy rain, or even in the interior of a thunder cloud. On the screen of the Plan Position Indicator (P.P.I.) which gives the radar operator a kind of bird's-eye view of his surroundings up to a radius of about 50 to 60 miles, belts of rain, isolated rain storms, and storm clouds from which rain is liable to fall, show up as bright patches which are easily recognized by experienced operators. The motion of such rain areas can be plotted and accurate predictions of the time of onset of rain at a given place may be made. The radar screen picture also enables an estimate to be made of whether a given storm is likely to develop into something serious or die away.

THE ATOMIC BOMB TESTS

The object of the first Bikini test is to ascertain the effect of an atomic bomb exploding in mid-air over warships. In a second test, planned to take place in August, a bomb will be exploded just below the surface. Next year a third test is to take place in deep water, which means away from Bikini, with a bomb exploded some thousands of feet below the surface.

The physical tests consist of measurements of the various pressure, shock and blast effects, and of the strain produced in various structural features of the ships. A simple device for measuring one kind of blast effect is an empty five-gallon petrol tin: the crushing, measured by the change in the amount of water the tin will hold, indicates the blast pressure.

Some of the instruments are simple, too, such as the so-called crusher gauge, in which a piston, pushed in by the blast, squeezes a ball of soft copper, but others are elaborate devices which utilize the instantaneous electrical effects produced when certain crystals are squeezed—the so-called piezoelectric effect.

What will be the scientific value of this massive test, in the way of new information, on fundamental points, either of atomic theory or of the behaviour of shock waves? So far as can be seen, very little. Apparently the test has been organized without the brilliant expert opinion that lay behind the measurements made in New Mexico, and no points of major importance are likely to be settled. No doubt interesting details about the propagation of the blast wave in air, the length of duration of the radioactive contamination in different circumstances, and other similar points will be obtained, but the Bikini tests are extremely unlikely to furnish a signpost at any scientific crossroads.

ATOMIC ENERGY AND DISEASE

The first peace-time products of the United States Government's atomic energy plants have been given to research institutions for use in the study of cancer, diabetes, and other diseases. These products are minute units of radio-active carbon called Carbon 14, which were produced in the same chain-reacting uranium ovens which produced the atomic bomb.

It was stated in June that some hundreds of varieties of radio-active materials, to be produced at the plant at Oakridge, Tennessee, and other centres, will be made available soon to accredited hospitals, universities, and industrial research laboratories. To-day's action was the first in that programme.

CONTROLLING THE ATOM

Unless sanity prevails, the atomic bomb carries with it the doom of cities and citizens as we know them, if not of all mankind. Many far-sighted men of science foresaw the implications that are clear to all to-day. Dr. Vannevar Bush, chairman of the National Defence Research Council of the United States, was not speaking for himself alone when he expressed, a few years ago, the hope that science would never succeed in tapping the energy of the atom, with the words: "It will be a hell of a thing for civilization." Ironically enough, Dr. Bush became, by force of circumstances, a leader in the project that led to the appalling success which now poses the gravest problem of the day—that is, how can we make sure that no atomic war is ever waged? Coupled with this problem is the necessity of permitting the vigorous pursuit of those aspects of atomic research and technology which bid fair to be useful to society.

Before the question of control can be profitably raised something must be said of the scientific aspect of the release of atomic energy, or, as it is better called, nuclear energy, since even the energy of ordinary combustion involves a re-arrangement of the outer parts of the atomic structure, and so may be called atomic, while the new energy is derived from a rearrangement and redistribution of the component parts of the atomic nucleus.

What are the social, as distinct from warlike, uses of nuclear energy? The development of power installations—for instance, the generation of electricity through steam produced by the heat of nuclear reactions—is an obvious objective. It does not follow that the power, even after considerable further research and experience, will be particularly cheap, for the materials are never likely to be easy to prepare, and, in any case, the cost of fuel is only a comparatively small part of the cost of generating electricity. The easy transport of atomic "fuel" will, however, offer great advantages in certain cases - one pound of uranium is capable of producing the same heat as 10 tons of coal.

COMETARY CONFUSION AND CONFUSED DISTANCES

A new comet, which should easily be visible to the naked eye at about 10 p.m. to-day, has been discovered in the constellation of Saturn, Aachen astronomical observatory announced yesterday, Hamburg Radio reported.

. . . As meteors shed by the Giacobini-Zinner comet burned across the skies . . . thousands of persons clambered on to rooftops or stood in the streets to watch the celestial pyrotechnics.

Dr. J. A. Pearce, Director of the Dominion Astrophysical Observatory, Victoria, described the show as "the greatest fireworks display of the century".

The comet itself passed about 24 million miles away, but the visible meteorites were within 135,000 to 500,000 miles, Dr. Pearce said.

TRITIUM PRODUCTION

It was announced yesterday to the American Chemical Society, at its spring meeting at Atlantic City, that laboratories operated for the Atomic Energy Commission at Oak Ridge, Tennessee, were now producing tritium, a super-heavy water containing three times the mass of hydrogen that ordinary water does.

REPORTED THEFT OF ATOMIC SECRETS

A staff correspondent of the *New York Sun* in Washington reported to that newspaper to-day that several files of highly secret data on the atomic bomb had been stolen " by unknown agents working from within the atomic energy plant at Oakridge, Tennessee." He said his " alarmed informants " had told him that the full story, whenever it was disclosed, would rival that of the atomic bomb spies in Canada.

MOUNT PALOMAR TELESCOPE

Polishing of the reflector for the Mount Palomar telescope—begun in 1936, discontinued in 1942, and resumed in 1945—is now virtually completed; the surface is within one-millionth of an inch of perfection. However, it will be some months yet before the reflector is sent to the observatory in preparation for installation in the telescope. The California Institute of Technology expects this to be about September 1.

The 10-ton reflector, or mirror, will rest in a cradle at the bottom of the telescopic tube, 55ft. from the photographic plate at its upper end. That plate and the astronomer using the telescope will be enclosed in a cartridge-shaped house which will project slightly beyond the top of the tube.

The telescope, with the reflector and all its other parts, will weigh 1,000,000lb., but will be so delicately balanced and so nearly free of friction that a pressure of only 2lb. will be enough to move it.

The 200in. instrument will be able to photograph light sources 1,000,000,000 light-years distant, as against the 500,000,000 light-years within the range of the 100in. Mount Wilson telescope, and will have four times the latter's light-gathering capacity. Among other things, it is possible that it will yield evidence to prove or disprove that the universe is expanding, and there are some astronomers who believe that it will reveal the nature of the so-called canals on Mars.

THREE-DIMENSIONAL PHOTOGRAPHS

The photographs, which are shown by Deep Pictures, Limited, are primarily of still-life objects and portraits, though a few early experimental photographs of moving objects are included.

SOVIET OFFER TO ATOM EXPERTS

Dr. Werner Heisenberg, German winner of the Nobel prize in physics for 1932, told a correspondent of the Associated Press at Göttingen to-day that Russia has a standing offer of 6,000 roubles a month to any German atomic expert who will engage in atomic research for the Soviet Government. He said that he was promised, in addition, to his pay, 50lb. of fresh meat a month, 3,500 calories of food a day for each of his six young children, and a comfortable house.

Although he declined the offer, three other German scientists, he said, accepted it. They were Professor Gustav Hertz, who built the German cyclotron for smashing the atom, and is an authority on the separation of the explosive isotope U-235 from uranium; Dr. Robert Doepel, a Leipzig physicist; and Dr. Ludwig Bevilogua, who was Heisenberg's assistant during the war. They went to Russia soon after the end of the war, he said. Bevilogua for a while worked in the Crimea and then was suddenly moved. Heisenberg said that he was not certain where any of the three were now.

He told the correspondent that the production of atom bombs " is no longer a problem of science in any country, but a problem of engineering."

NUCLEAR REACTORS

The natural uranium nuclear reactor could be used in its present form as a power generator, and it was probable that experimental plants of this type would be built during the next five years to obtain this country's first operating experience of nuclear power generators. If nature were exceptionally kind a reactor containing 100 tons of uranium would generate 100,000kw. of thermal energy for 20 years without fuel replacement. This would yield about 20,000kw. of electrical energy.

It was expected, however, that the formation of fission products would lead to poisoning of the chain reaction, due to their additional absorption of the neutrons produced. Difficulties might be experienced due to deterioration of materials under the intense neutron bombardment of the pile. Professor Cockcroft pointed out that upon these, and perhaps other unknown factors concerning the repurification of the elements, would depend the ultimate cost of nuclear power.

One other major uncertainty stood in the way. The safe disposal of the very large scale production of radio-active materials in the generation of nuclear power was a major problem, and one which would require a cautious approach. While nuclear power presented a rather dazzling prospect and was worth a determined attack its future was by no means clear.

1948

ATOMIC ENERGY FOR SHIP PROPULSION

The United States Atomic Energy Commission, in the hope of developing an atomic engine for ship propulsion, has given a contract to the Westinghouse Electric Corporation to construct an experimental pile. Requirements for the pile have been set by the Navy Department's bureau of ships. The work will be carried on in cooperation with the Argonne National Laboratory, which is the centre of the commission's pile development programme.

NEW SUBSTITUTE FOR RADIUM

Mr. D. E. Lilienthal, chairman of the Atomic Energy Commission, announced to-day the discovery of a new, comparatively cheap and plentiful radioactive substitute for radium in the treatment of cancer. He said the commission was undertaking to put the production of radioactive cobalt under way now. It would be safer to handle but had a limited life compared with radium.

ACTINIUM ISOLATED

The isolation from radium of actinium—a radioactive element long known to exist but occurring in such small quantities that it has never before been seen—was announced to the American Chemical Society at Chicago yesterday by Dr. Hagemann, of the Argonne National Laboratories.

ATOMIC ENERGY RESEARCH

Mr. David Lilienthal, chairman of the United States Atomic Energy Commission, speaking here on Saturday, said that the United States, the United Kingdom, and Canada were expanding their technical cooperation in atomic energy research and development. They were working more closely together than hitherto to prevent duplication of effort and, as during the war, were keeping under review the problem of raw materials for the production of atomic energy.

Mr. Lilienthal referred to his disappointment that the Soviet Union had blocked an effective plan for international control which a majority of the members of the United Nations commission had evolved. In the circumstances the United States was pressing on with its own programme of research and development, which was one of the greatest magnitude and intensity. This undertaking was yielding, even at this early date, knowledge of basic value to human life, knowledge strikingly new and far-reaching in its consequences.

NEW URANIUM ISOTOPE

The new fissionable element is uranium of the atomic weight of 233. Some milligrams of it were created by subjecting the non-fissionable element thorium to an intense bombardment with neutrons.

Like its relative Uranium 235, U233 can be used for nuclear explosions in bombs and for the release of nuclear energy for power. At present only the natural substance U235 and the man-made element plutonium, created by the transmutation of Uranium 238, can be utilized for the release of nuclear energy by a self-sustaining chain reaction. Thorium, an element three times more abundant in the earth's crust than uranium, will thus theoretically triple the availability of nuclear fuels or explosives.

ATTEMPT TO SELL HEAVY WATER

Sentences of six months' imprisonment for possessing and attempting to sell "heavy water"—used for making atomic bombs—were to-day imposed by a Military Government court on two Germans, Count Angus von Douglas and Stephen Schmidt, and on a Yugoslav, Alexander Babisch-Armstrong.

BLUE SUNS

After the eruption of Krakatoa in 1883 not only was the moon seen green and blue but the sun also, and even Venus. Coloured suns were mostly, though not entirely, confined to low latitudes, but blue and green moons were more widely observed. The colours were due to the dust and vapour thrown into the high atmosphere by the explosion, dust which produced the extraordinary sunsets that some of us can remember.

INDUCING RAIN

Having read of attempts to bring down rain by various scientific methods, I was interested to observe in Italy last year what seemed to be a very elementary means of inducing rain. While travelling in the train over the Ruscada Pass from Domodossola to Locarno I saw in one of the fields by the railway line a peasant touching off, by means of a lighted taper at the end of a long pole, some six antiquated mortars which were set up in line, and directed at the heavy rain-clouds hanging over the pass. I was unable to ascertain the nature of the projectiles, if any; it was only too apparent that the charge must have been considerable, for the noise as each mortar was fired was deafening. Very little time had elapsed before it began to rain quite heavily.

ARTIFICIAL MESONS

The University of California announced yesterday that artificial production of the meson, a cosmic ray constituent, had been achieved in the Berkeley radiation laboratory there. Credit for this belongs largely to Dr. Cesare Mansueto Giulio Lattes, a Brazilian physicist, aged 23, who first found the track of a meson on a photographic plate, and to Dr. Eugene Gardner, aged 35, who had been studying atomic disintegrations for more than a year by bombarding photographic emulsions in the big cyclotron.

SETBACK ON MT PALOMAR

Trials of the 200in. mirror—the largest in the world—of the astronomical telescope, or camera, on Mount Palomar in California have revealed that its outer rim is 20 millionths of an inch too high. Because the correction of this fault will take time observations with the instrument will probably not be possible until next autumn.

RADIOACTIVE MINERALS IN CANADA

Samples of uranium ore discovered in Hastings County, Ontario, have been brought here for analysis by a mining engineer, whose company has already staked 32 claims covering a 2,000-acre area in the region where the ore was found.

The discovery adds another area to the growing number of uranium sources in the country which can be exploited only under the supervision of the Federal Government. Exploitation of radioactive minerals is proceeding in Saskatchewan, Ontario, the North-West Territories, and British Columbia

1948

US ARMY'S NEW WIND TUNNEL

The United States Army is using a wind tunnel which can produce speeds 10 times that of sound. It is being used to help to develop types of rockets which could cross the Atlantic from New York to London in less than 30 minutes.

The Defence Department, announcing this to-night, said that the tunnel was at the California Institute of Technology at Pasadena. It was the first time, the announcement added, that American scientists had produced speeds of more than 10 times that of sound. The highest previous known speed of air flow in supersonic wind tunnels was about seven times the speed of sound, which is 760 miles an hour at sea level.

PHOTOGRAPHS TAKEN WITH NEW TELESCOPE

Astronomers on Mount Palomar in California, after several weeks' study of photographs taken there on February 1 with the world's most powerful telescope-camera, which has a 200-inch reflector, have announced that for the first time there have been brought into sight objects 1,000m. light years away. This is twice as far as the longest penetration into space ever made by the 100-inch telescope of Mount Wilson Observatory, but even greater results are thought to be possible.

The photographs, made by Dr. Edwin Hubble, showed objects appearing only as large as pin-points, but the astronomers said they were pictures of nebulae, each one a galaxy of stars like the Milky Way. The photographs were made as a preliminary to further polishing of the reflector, which is soon to be dismounted for that purpose.

EINSTEIN'S NEW UNIFIED THEORY

A spokesman of the university said that Professor Einstein's new theory set forth in a series of equations the laws governing the two fundamental forces of the universe—gravitation and electro-magnetism. Physicists were not yet willing to try to talk publicly about it. The theory had yet to be tested to establish whether it covered all the known laws or observations about the way that matter, energy, and gravitation behaved.

Mr. William Laurence said in his *New York Times* article to-day that the new theory "attempts to inter-relate all known physical phenomena into one all-embracing intellectual concept, thus providing one major master key to all the multiple phenomena and forces in which the material universe manifests itself to man. . . . Einstein's latest work promises to bridge the last gap that now separates the infinite universe of the stars and galaxies and the equally infinite universe of the atom. These at present are widely separated, one being explained by relativity, while our knowledge of the other rests on the quantum theory, of which Einstein was also one of the major architects." It was known that Professor Einstein intended to crown his life's work as a "universe-builder" by bringing these two theories into one all-embracing, comprehensive system. They were the two major pillars on which man's basic understanding of the universe rested, Mr. Laurence added. Professor Einstein's present work was regarded as a major step towards his objective.—*Reuter.*

PALOMAR SCHMIDT TELESCOPE

Astronomers at the Palomar observatory in California will begin on Tuesday night a unique undertaking which is expected to extend over four years and to open the way to great new discoveries in the skies. It is the photographing of three-quarters of the heavens with a 48in. Schmidt telescope—a comparatively new instrument which has a corrective lens bending light rays in such a way as to produce clearer photographs of stars than any other telescope.

Although the instrument was invented 20 years ago, by the German optician Bernard Schmidt, its virtues were apparently not appreciated in Europe, but when some of Schmidt's photographs were brought to this country they roused great enthusiasm among astronomers and at once they set to work to build one of his instruments here.

ATOMIC ENERGY EXCHANGES

The talks between Britain, the United States, and Canada on atomic energy were resumed to-day with Mr. James Webb, Under-Secretary of State, and Mr. David Lilienthal, the retiring chairman of the Atomic Energy Commission, representing the United States and the respective ambassadors representing the other two countries. The present agreement on the exchange of information expires on December 31, after which time there will be no method under American law for even the limited exchange permitted at present.

At a Press conference this morning Mr. Lilienthal said that he personally was in favour of continuing the war-time partnership with Canada and Britain, and in answer to questions said that both countries had scientists and information which would be useful to the development of the American atomic energy programme.

This was not the only point on which he disagreed with Senator Johnson, who yesterday accused him of being in a " nefarious plot " to disclose secret processes to England. He also disagreed about the secrecy of the information which the Senator broadcast over a television programme earlier this month. At that time, the Senator said, among other things, that the United States already had a bomb six times as powerful as the one dropped on Nagasaki, and was working on one a thousand times as powerful.

A HISTORIC THIMBLE

A small silver thimble which has been given to the Science Museum, South Kensington, is one which has a curious place in scientific history. When in 1866 two cables were at last successfully laid between Ireland and Newfoundland, the battery used for the first experimental transmission of current across the Atlantic consisted of this thimble, a few drops of acid, and a zinc wire. The thimble had been borrowed for the purpose from the late Miss Emily FitzGerald, daughter of the Knight of Kerry, upon whose land stood the building in which the cables ended. It has now been given to the Science Museum by her nephew, Mr. R. B. FitzGerald.

PROPULSION OF SHIPS

The Argonne laboratory is also working on a land-based prototype of a reactor suitable for ship propulsion. On this, research and development work is well advanced, and detailed engineering designing will begin in about a year. It is hoped that construction will be under way by 1952. The ship-propulsion reactor, according to the commission, will be a single-purpose machine designed specifically for the purpose of producing large amounts of heat under conditions that will permit conversion into power for the propulsion of warships.

A 'BLACK' MOON

The total eclipse of the moon showed watchers here last night a rare phenomenon—a moon at moments completely " black." This, astronomers said, was caused by an unusual cloud blanket around part of the earth which kept the refracted rays from the sun from reaching the moon.

Probably for the first time the eclipse was televised, and thousands of persons chose to see it by this means. Theirs was an unimaginative choice, for they missed far more than they gained.

TESTS AT BIKINI

Scientists who for three summers have been studying water, sea life, birds, plants, and animals at the Bikini atoll, in the Pacific, where the atom bomb tests were made in 1946, reported yesterday at Seattle that radioactivity was still present there in measurable amounts, but that it was too early yet to determine whether or not people could subsist without harm on food produced in the lagoon. They said that the answer to that question might not be known " in our generation."

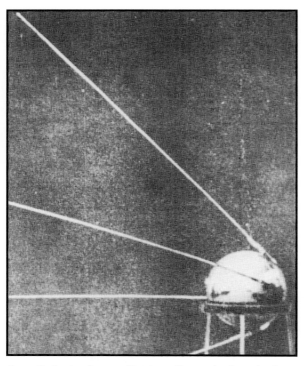
The
1950s

Physics reporting, having mushroomed immediately post-war, continued to expand in the 1950s. It is reasonable, from this period on, to talk of real 'front page physics', because major items of physics news now typically appeared on the front page. In the first half of the decade, nuclear physics—usually, if confusingly, referred to as 'atomic' physics—still dominated the reporting of science. Most media attention was devoted to nuclear weapons—more especially, the transition from fission-based bombs to the more powerful H-bomb, based on fusion. Tests of atomic bombs continued throughout the period, the best remembered probably being the test at Bikini atoll since it gave its name to a swim-suit. As the tests continued, however, public concern about radioactive fallout, and its possible effect on human beings, began to grow. Yet the main debate in the press concerned not this, but the speed shown by the USSR in catching up with the US development of atomic weapons.

Sputnik 1—the first artificial satellite to be launched

By the beginning of the decade, nuclear energy was being publicised as a potentially cheap and clean way of producing power. A number of countries became interested in the possibility of producing commercial amounts of electricity via nuclear reactors. In 1952, there was the first public announcement of an accident at a reactor (at Chalk River in Canada), but this had little effect on the general acceptance of nuclear power. Indeed, its apparent economic viability was enhanced in the same year when the first breeder reactor came into operation in the USA. (Such reactors provide not only power, but also a replenishment of the fissile material.) Some public concern was expressed about the disposal of radioactive waste from reactors. This latter problem was one reason for the considerable interest in thermonuclear fusion, when news of research on this was released in the latter part of the decade. At the time, it seemed that such fusion might be achieved quite quickly, and it offered access to nuclear power without the production of dangerous side-products.

It sometimes seemed in the 1950s as though any activity with the word 'nuclear' in it was considered worthy of financial support. But already the large particle accelerators were beyond the means of any but the wealthiest nations. This was illustrated in 1954 when CERN (Centre Européen de Recherche Nucléaire) was established, on the basis of joint funding from several West European nations. In the popular press, the violence of nuclear reactions using these 'atom smashers' was often emphasised, perhaps in an attempt to tie the image of high-energy nuclear physics to that of nuclear weapons. Discoveries of new particles became increasingly common. For example, the existence of both anti-protons and anti-neutrons was confirmed experimentally. Theory developed along with experiment, but was typically less reported. One exception was the theoretical prediction (and subsequent experimental confirmation) of the non-conservation of parity. The idea that God was 'a weak left-hander' (as Pauli put it) intrigued readers.

The latter part of the 1950s was dominated not by nuclear physics, but by the onset to the space age. The IGY (International Geophysical Year) started in 1957 (and actually ran for more than a year). It represented the largest co-operative effort ever mounted to study the physical characteristics of the Earth and of its environs. The IGY attracted a good deal of press coverage in its own right, but the aspect that caught the public interest was the launching of the first artificial Earth satellites.

At the end of the Second World War, both the USA and the USSR had imported German scientists to help expand their already existing interests in rocket development. By the early 1950s, publicity was already being given to the first ultraviolet observations of the Sun made with US rockets. During the 1950s, the Russians actually released a good deal of information about their intentions for launching satellites. But the greater publicity surrounding US efforts and, perhaps, suspicion generated by other Russian claims to historic 'firsts' in science, led to some discounting of the Soviet plans. This may explain the shock, and the consequent publicity, that greeted the launch of the first Russian Sputnik in October 1957. To experts, however, the main shock was the mass of the satellite, which was much greater than anything planned by the Americans. This aroused immediate fears regarding US defence policy, since the Russian rockets were clearly capable of launching inter-continental ballistic missiles. Such American questioning was exacerbated by the failure of the first attempts to launch a US satellite. (It was the team organised by Wernher von Braun, the major German rocket expert in the Second World War, which finally succeeded.) The shock administered by Sputnik had a significant influence in the USA both on the funding of research, and on science education.

The Russians remained ahead in terms of rocketry into the following decade, which their Lunik series of space probes at the end of the 1950s either crash-landing on the Moon, or surveying its hitherto unseen rear face. This Russian advantage was somewhat offset by the more sophisticated US instrumentation, which led, for example, to the discovery of the Earth's radiation belts.

In astronomy, the 1950s were marked by the growing importance of radio observations. Early in the decade, the discovery of the 21 cm hydrogen line started off a programme to map the structure of our Galaxy (impossible at light wavelengths). At the same time detection of new radio sources went on apace. Large, specially designed radio telescopes were now coming into use, as, for example, at Jodrell Bank in England. As though to demonstrate that large optical telescopes were still essential, observations in California showed that the accepted distance scale for the Universe was out by a factor of two. This cleared up an embarrassing problem. Estimates of the age of the Earth had previously suggested that our planet was older than the Universe. The new adjustment to the expansion rate of the Universe removed this discrepancy.

Particle accelerators, spacecraft and large telescopes all required large amounts of funding for their creation and use. The 1950s thus saw the origins of 'big science' (which might equally well have been labelled 'big physics'). Underlying all these advances was a need to handle the large amount of scientific data generated by such facilities. Early in the decade, commercially produced transistors became available. Their importance for the manufacture of computers was obvious. As computer power grew, so the first signs of 'user-friendliness' appeared—notably in the implementation of programming languages, such as FORTRAN and COBOL. However, at this stage, computers were mentioned in the press mainly for their games-playing ability.

1950s CHRONOLOGY

1951
- 21 cm radiation is detected from interstellar hydrogen
- Ultraviolet observations of the Sun are made from rockets
- Mueller develops the field ion microscope

1952
- Baade revises the distance scale for the Universe
- The breeder reactor is built
- The bubble chamber is developed
- The first fusion bomb is exploded by the USA

1953
- The concept of 'strangeness' is introduced into particle physics
- The first maser is developed
- The double-helix structure of DNA is determined

1954
- CERN (Centre Européen de Recherche Nucléaire) is set up

1955
- Neutrinos are detected experimentally
- Anti-protons are produced

1956
- The programming language FORTRAN is introduced

1957
- The International Geophysical Year (IGY) starts
- The USSR launches Sputnik, the first artificial Earth satellite
- Non-conservation of parity is observed
- Bardeen, Cooper and Schrieffer formulate the BCS theory of superconductivity

1959
- The Russian Lunik observes the rear face of the Moon
- The Xerox photocopier is introduced

1950s

ACCELERATION OF PARTICLES

The interest of such fast particles to the nuclear physicist, he said, resulted first of all from the fact that while a good deal was known about the pieces of which atomic nuclei were built, little was known of the nature of the forces holding them together. More could be learned about the nature of forces by studying collisions of such particles at high speeds. It was already known that new particles called mesons were produced, just as at lower energies light quanta were produced in the collision of electrically charged particles.

A close knowledge of the emission of light was, in fact, essential to a clear understanding of the electric forces that governed the motion of electrons in atoms, and similarly it was expected that more information about the properties of mesons would help in the understanding of the phenomena that took place when fragments of nuclei collided and lead ultimately to a better understanding of the structure of the atomic nucleus.

Faster particles would also help to solve the problem of whether the proton—that is, the hydrogen nucleus which was one of the important nuclear constituents—possessed, like the electron, a counterpart of opposite charge. The least energy required to produce " negative " protons which in Nature always held a positive charge was about 1,000m. electron volts, but in practice rather more would be needed.

NEUTRAL MESONS

" Strong evidence " of the existence of neutral mesons was reported to-day by the University of Rochester (New York State) to have been found as the result of a study of photographic plates carried 19 miles into the air by balloons released near Harrisburg, Pennsylvania.

The short-lived particles called mesons are sometimes described as the " glue " of the atomic nucleus. So far scientists have identified two kinds of charged mesons but have only suspected the existence of neutral mesons. Charged mesons are said to be responsible for attraction between dissimilar nuclear elements, and neutral mesons, it is assumed, are a similar force between identical elements.

THE HYDROGEN BOMB

the problem which has come to the fore in recent discussions of the hydrogen bomb is how to achieve the conditions which will allow hydrogen, which, unlike the materials at present used in atomic bombs, is cheap and plentiful, to combine to form helium. The task is simplified by using heavy hydrogen, the atoms of which have each a mass of two units (with, of course, the small percentage excess which has been already mentioned as the operative condition) and a positive charge of one unit, so that two heavy hydrogen nuclei contain the component parts of one helium nucleus. It is easier to incite two particles to combine than the four which we have if we start with ordinary hydrogen. It remains to ensure that the concentration of heavy hydrogen is sufficient, that the necessary temperature is attained, and that the transmutation—or combination—is effected before the hydrogen can be dispersed by the explosion necessary to attain the required temperature.

Heavy hydrogen, like ordinary hydrogen, is normally a gas, which, even when compressed as much as would be feasible, is still not sufficiently dense, quite apart from the complications caused by the massive container that would be necessary. In the liquid form the concentration of atoms is, of course, very much increased, but hydrogen will remain liquid only at extremely low temperatures, which again would introduce many complications. The obvious thing to do would seem to be to use the heavy hydrogen in the form of a solid or liquid compound with some other element. The particular compound used, if this is the method adopted, is clearly one of the secrets of the hydrogen bomb.

MILES OUT

Yesterday the National Physical Laboratory announced that the speed of light was wrong and that Dr. Essen was right. Dr. Essen has made Professor Einstein alter his calculations by 11 miles.

FIRST USE OF ATOMIC BOMB

Twelve of this country's leading physicists, all of whom took important parts in the development of the atomic bomb, joined yesterday in urging the United States to make a solemn declaration that it would never use the hydrogen bomb in warfare unless an enemy used it first against this nation or against its allies.

In a statement issued at the annual meeting here of the American Physical Society they confirmed, for the first time here, that there could be developed a bomb with a thousand times the destructive power of the present type of atomic bomb and that a single one of these bombs could destroy New York or any other of the great cities of the world. In fact, they indicated that a bomb could be made of a destructive power even greater than this. " The thermo-nuclear reaction on which the H-bomb is based," they said, " is limited in its power only by the amount of hydrogen which can be carried in the bomb."

ELECTRONIC COMPUTERS AND DEFENCE

The United States Navy exhibited to-day at Princeton, New Jersey, a " project typhoon," said to be the largest and most accurate electronic computer ever built, which was designed to help develop guided missiles such as rocket-borne atomic bombs. It cost $1,400,000, but is expected by its builders, the Radio Corporation of America Laboratories, to save many millions of dollars in the designing of guided missiles, and also to solve many of the problems encountered in air defence. It was said that eventually it might also cut the time for the design and construction of an aircraft from five years to six months.

In a demonstration of its capacities the computer solved a simulated air defence problem in which a high-speed bomber was successfully attacked by a radar-controlled supersonic rocket-propelled guided missile. The solution involved 20 additions, 67 multiplications, 30 integrations, and 20 aero-dynamic functions, all carried out simultaneously with continuously variable factors. The answer appeared in less than 60 seconds.

BOHR'S LETTER TO THE UN

While the possibility still existed that negotiations within the United Nations might result in arrangements for the use of atomic energy guaranteeing common security, I was reluctant to take part in the public debate on this question. In the present critical situation, however, I have felt that an account of my views and experiences might perhaps contribute to renewed discussion about these matters, which so deeply influence international relationships. I am acting entirely on my own responsibility and without consultation with the Government of any country. International scientific cooperation continued as a decisive factor in the development which, shortly before the outbreak of the war, raised the prospect of releasing atomic energy on a vast scale. The fear of being left behind was a strong incentive in various countries to explore in secrecy the possibilities of using such sources of energy for military purposes. The joint American-British project remained unknown to me until, after my escape from occupied Denmark in the autumn of 1943, I came to England at the invitation of the British Government. At that time I was taken into confidence about the great enterprise, which had already reached an advanced stage.

Professor Bohr quoted from a memorandum which he submitted to President Roosevelt as a basis for a long conversation which he had with him in August, 1944, and then continued: When the war ended . . . the creation of new barriers restricting the free flow of information between countries further increased distrust and anxiety. In the field of science, especially in the domain of atomic physics, the continued secrecy and restrictions deemed necessary for security reasons hampered international cooperation to an extent which split the world community of scientists into separate camps.

METEORITE CRATER FOUND IN CANADA

The *Globe and Mail* of Toronto reports to-day that what is thought to be the world's biggest meteoritic crater has been discovered in the barren lands of Ungava, northern Quebec, by a scientific expedition from Toronto.

The crater is situated about 60 miles from Hudson Strait, has a rim of about 7¼ miles in circumference and contains a small lake which was covered with ice 3ft. thick when it was viewed on July 26.

FAMILY FAVOURITE

"I really have nothing to say", she told the enquirer from the broadcasting corporation. "I have hardly any time to listen to the wireless".

"But have you any criticisms to make of the items you do listen to", he asked.

"No."

"You do not feel they could be improved in any way"?

"No."

"Could you tell me what you do listen to"?

"Chiefly the time signals".

POWER FROM ATOMIC PLANTS

For the first time useful electric power has been produced from atomic energy. A successful experiment, made on the night of December 21 at the national reactor testing station near Arco in south-eastern Idaho, was reported yesterday at Idaho Falls by a spokesman from the Atomic Energy Commission.

By means of heat withdrawn from a breeder reactor by a liquid metal enough steam pressure was produced to drive a turbine which generated more than 100 kilowatts of electric power. During the hour in which the experiment lasted the reactor and equipment in the building housing it, including lights, pumps, and other devices, were operated from this source.

ATOMIC AID IN CANCER TREATMENT

The research establishment at Chalk River, 120 miles west of Ottawa, has been developed since the war by the National Research Council to promote the peaceful application of atomic energy and has produced isotopes of great importance to metallurgy, agriculture, and biological research. Cobalt 60 is regarded as probably the most important isotope produced there and the most powerful radioactive source ever applied in peace-time, more powerful than the conventional radium bombs now in use.

Cobalt is not naturally radioactive, occurring in nature as an element with an atomic weight of 59. After exposure to bombardment by neutrons in the reactor its atomic weight is increased to 60. Cobalt 60 gives off energetic gamma radiation of about 1,200,000 volts.

The cobalt 60 deep therapy unit costs about $50,000 to produce. Radium producing an equivalent radiation would cost about $50m., if such an amount of refined radium existed.

EXPLODING UNIVERSE

The theory that the universe is exploding is reported from Los Angeles to have found strong support in observations made with the 200-in. Hale telescope on Mount Palomar of clusters of galaxies as distant from the earth as 360 million light-years. This is said to be farther by 50 per cent. than man has ever seen before.

Dr. Humason found, as other astronomers had concluded, that stars were moving outward in space and that their runaway speed was increasing in proportion to the distance of the stars involved. Those 250 million light-years away had a runaway speed of 25,000 miles a second, those at 300 million light-years away a speed of 31,000 miles a second, and so on—an increase of 100 miles a second for every million light-years.

NEW PLANETOLOGY

He said that he had calculated the existence of four new planets after long study. His observatory does not have powerful equipment, but if he had he could photograph the planets which he described as 'tremendous'. He claimed that he had learned of the planets' existence through physical laws and by applying 'intuitive radar'.

HIGH VACUUM

By a high vacuum we meant that so little air, or other gas, was left in a vessel that the pressure was about a millionth of atmospheric or less. This meant that if one imagined the ordinary barometric mercury column magnified until it appeared as high as Mount Everest, the pressures dealt with in high vacuum work would be represented by a mercury column varying from half an inch down to a small fraction of the thickness of a cigarette paper.

The modern pump for producing a rough vacuum was no longer a cylinder pump but a pump in which, by rotation of a specially devised shaft, a space is alternatively opened and closed, and with this special oils have to be used. But the speedy and easy production of very high vacua depended upon pumps based on a totally new principle, actively developed after the first world war. In this a rushing stream of mercury vapour or of oil vapour entangled the atoms of gas and carried them away.

DIAMONDS AS DETECTORS OF RADIOACTIVITY

Certain diamonds, he said, were found to be highly sensitive to gamma rays and could be used as detectors.

When a gamma ray was absorbed by a diamond atom a photo-electron was emitted and the freed electron was accelerated through the inter-atomic space toward the positive electrode of the current supply in contact with the diamond. Within a very short distance the freed electron acquired such a high velocity that other atoms in its path were ionized by collision with additional electrons released, which in turn were accelerated in the same direction.

The multiplication of charges was repeated in rapid succession, producing an avalanche of electrons equivalent to a small pulse of current. The larger the diamond the more electrons involved, but in practice small diamonds capable of amplifying the original signal a million times had been sensitive enough to operate the associated electronic equipments satisfactorily.

TOWN SHAKEN BY ATOMIC EXPLOSION

An atomic explosion in Nevada this morning —by general agreement of distant observers the most violent of the four set off this week— shook Las Vegas, 40 to 70 miles away, like an earthquake. Buildings were rocked and the plate glass window of a store was shattered.

The flash, at first white, then orange, and finally yellow, was seen farther away than any of its predecessors—according to one report as far as Caldwell, in Utah, 530 miles from Las Vegas. In Los Angeles several hundred persons claimed to have seen it.

The Atomic Energy Commission announced to-day that it was investigating reports, that snow which fell in Rochester, New York, this week was " measurably radio-active."

NEW JODRELL BANK TELESCOPE

At Jodrell Bank a fixed radio telescope— a paraboloid aerial, 220ft. in diameter—has been used to pick up radio waves reaching the earth from sources as far distant as the great spiral nebula of Andromeda, 750,000 light-years away; but, though the largest in the world, it is inadequate to carry this research to all regions of the sky.

The new, completely steerable telescope, with a paraboloid aerial 250ft. in diameter will be free to scan the whole of the sky, and to transmit signals or to receive them from any part of it. The diameter of the platform on which the radio telescope will rotate will be 310ft.; the height to the top of the horizontal axis will be 185ft., and when the beam from the aerial is horizontal the total height will be 300ft. The total weight of the telescope will be 1,270 tons.

The instrument, which will have an aperture a little larger than that of the fixed one at Jodrell Bank, will be used for all aspects of radio astronomy; priority being given to the study of the galactic and extra-galactic radio emissions, with particular reference to the number and nature of the dark "radio stars." It will also be used to plot the intensity of radiation, particularly from those important regions of the Milky Way systems which are obscured from normal vision by the great dust clouds in interstellar space. It is confidently expected that the great instrument will do for radio astronomy what the large visual telescopes in America have done for classical astronomy.

THE JODRELL BANK RADIO TELESCOPE, ENGLAND

DR PICCARD'S PLANS FOR MARS

Dr. Jean Piccard, the first stratosphere explorer, who is holidaying near Duncan, Vancouver Island, states that he is planning to explore Mars in 1954 in a free balloon. His plans, he said, are already advanced and clusters of balloons have been tested and a gondola model built. The planet in 1954 is in the most favourable position for astronomic study.

SPACE TRAVEL IN OUR TIME

The lecturer recounted the difficulties to be overcome in space travel. The rocket motor, he said, was capable of functioning in the almost complete vacuum between the spheres and of overcoming the forces of gravitation. In plotting the probable courses of interplanetary flight, he said that the rocket-driven craft would travel vertically through the earth's atmosphere from a point near the equator to take best advantage of the globe's speed. They would then incline in an orbit to gain the speed of 25,000 miles an hour needed to escape from the world's gravitational pull.

When the escape had been accomplished, craft would cruise until they came within the influence of some other body, and would then check speed with parachutes, by use of wings for gliding, or by braking with motors to make a gentle descent.

"Australians will soon be able to arrive in New Zealand by air before they leave Australia according to Air Vice Marshal Sir Keith Park."

ALL COMES TO HIM WHO WAITS

"Early arrivals at Dundee Mills Observatory had a long wait before Jupiter, showpiece of the evening, was high enough in the heavens for the telescope to be trained on it. Their patience was well rewarded by an excellent view of Jupiter and his five suns.

ARTIFICIAL 'COSMIC RAYS'

" A stream of protons, cores of hydrogen atoms, already accelerated in a preliminary operation to energies of 3,600,000 volts, was introduced into the cosmotron's giant circular track. Round and round the hydrogen nuclei travelled through the machine's circular magnet, being accelerated by giant electric pulses from a generator which delivers a maximum of 40 million watts a second.

" In seven-tenths of a second, the awestruck scientists observed, the atomic bullets had made two million trips around the 204ft. circular track of the magnet, attaining speeds of 175,000 miles a second, or about 95 per cent. of the velocity of light, highest possible velocity in the universe.

" In that incredibly short time the protons had been accelerated to energies of 1,360 million volts, an energy three times greater than that produced by the most powerful atom machine in existence. It was the first time that atomic bullets had reached, and even exceeded, the 1,000 million-volt range. In fact, it was the first time on record that even the 500 million-volt range had been achieved by a man-made device.

GUIDES FOR ATOMIC PARTICLES

A discovery at the Brookhaven laboratory of the Atomic Energy Commission on Long Island, New York, of an entirely new principle for focusing strong magnetic fields which are used as guides for atomic particles when they are being accelerated within a giant ring-shaped magnet, has made possible the construction of an atomic machine capable of firing atomic " bullets " with energies far in excess of any that can be developed by existing machines—energies as high as 100 milliard electron volts.

USE OF ISOTOPES IN AGRICULTURE

When a phosphate fertilizer containing the radioisotope is added to the soil analysis of the crop discloses how much of the phosphorus in plants has been derived from the fertilizer and how much from the soil.

Much of the pioneering work in that and similar problems was undertaken in the United States, where isotopes were available earlier in the quantities needed for field-scale experiments. It was quickly realized that the results recorded would be misleading or erroneous if plants used were affected by radioactivity during growth.

There is proof that plants are damaged by large amounts of radioactive material, but the quantities used in pot and field-scale experiments do not appear to affect plant growth, and controls are being used extensively to confirm the conclusion. There is no evidence that isotopes stimulate plant growth or increase yields.

US HYDROGEN BOMB

Press reports that the first American-made hydrogen bomb had been exploded earlier this month in the South Pacific in the course of the Eniwetok tests. Eyewitness accounts of a reported hydrogen bomb explosion have been published during the last few days in several American newspapers, most of them being in the form of letters home from members of the naval task force concerned with the South Pacific test operation

RUSSIAN RESEARCH

Mr. Malenkov's statement that the hydrogen bomb was no American monopoly was preceded by other claims to Russian achievements in science, engineering, and technology, many of them of an unusual nature.

Observations conducted in the Kazakh mountains by Professor Fesenkov were said to have confirmed the hypothesis of Soviet astronomers that stars were still being formed. Professor Fesenkov's observations were said to have been made at Alma Ata with a big telescope designed by the Russian scientist Maksutov. Professor Fesenkov had found that many nebulae had a fibrous structure and a vortex motion. Photographs of the fibrous formations had shown them to be chains of condensation areas. Further photographs taken through various filters had confirmed that the condensation areas were very young stars. These findings, according to Russian scientists, had "completely refuted the senseless suppositions about the existence of some special moment at which the universe began and at which all stars were allegedly created."

Professor Tikhov, of Alma Ata, the founder of Russian astrobotany, claims to have discovered that in the cold climatic conditions prevalent on Mars vegetation had a much greater need of solar heat than vegetation on earth. It therefore almost completely absorbed the rays having the highest calorific value— i.e., red, yellow, and green. The blue and violet rays were not absorbed by Martian vegetation, which therefore had a blue or violet appearance. He claims as supporting evidence of this hypothesis the fact that in the Pamirs or on the tundras of the far north, where climatic conditions are similar to those said to prevail on Mars, the vegetation frequently has a blue or violet appearance.

THE BREEDER REACTOR

The achievement of "breeding" in a nuclear reactor is indeed a landmark in the development of atomic energy for peaceful use. What has not been stated, and may not yet be known, is the size of the credit margin.

The attainment of breeding, as such, means that for every atom of fissile material destroyed in the breeding reactor, one atom, or a little more, of new fissile material has been formed to replace it. The crux of the matter is the size of "the little more," for on this depends the rate at which fissile material can be accumulated.

CRYSTAL DETECTORS

The crystal detector, with its cat's-whisker, was among the earliest devices used to identify a radio signal; and now, after being in eclipse for many years, the crystal is again to the fore. Fresh interest in it as a detector device was aroused when it was found to be a convenient and indeed more efficient means than the thermionic valve for identifying very short radio waves, such as are used in radar systems. Still more recently it has shown great promise in other directions, but always within the field of electronics.

Much still has to be done before these new valves can be produced cheaply and in large numbers with uniform characteristics—an essential need if the crystal is ever to supersede the thermionic valve in its multiplicity of uses. Fundamental problems remain. The crystal material, silicon—or more likely germanium—has yet to be more fully understood.

Germanium crystals are now being grown in quantity out of flue dust; into the refined metal certain " impurities " have to be put, perhaps amounting to only one part in a hundred million, and great skill is needed if reproduceable results are to be obtained. The mounting of the crystal and the contact element is at the moment " a watchmaker's task " and means would have to be found to do it mechanically before large-scale factory production could be organized.

NEVADA'S BIGGEST BOMB

The most powerful atomic bomb ever exploded in the United States was dropped over the Nevada desert at dawn to-day, and its flash was observed, it is stated, more than 1,000 miles away in Canada and Mexico. Hundreds of residents in Los Angeles, 250 miles away, reported that they felt the shock about 20 minutes after the detonation; windows rattled, cats howled, and dogs barked.

The spring series of nuclear tests was to have ended last week with the firing of the first atomic gun, and to-day's explosion, estimated to be equal in power to 40,000 tons of T.N.T., was apparently staged at the request of the scientists to test a device not previously used. One of them called it a refinement embodying a "discovery" made in one of the recent tests.

THE COBALT BOMB

The idea behind the "cobalt bomb" is that cobalt, when exposed to irradiation by neutrons, is converted into a radio-active form (cobalt 60), which is in use already as a substitute for radium in medical treatment. Cobalt 60 is normally prepared by the exposure of cobalt in a nuclear reactor. However, if the structural parts of an atomic bomb were made of cobalt they would be exposed during the brief instant of explosion to an extremely high intensity of neutrons and would become intensely radio-active. At the same time the cobalt would be vaporized and dispersed, along with the mushroom shaped cloud of gases from the explosion.

Cobalt 60 has a relatively long "half-life" —5.3 years. Large quantities of radio-active material are produced in any case during the explosion of an atomic bomb. In a "cobalt bomb" not only would more radio-active material be produced, the proportion of long lived radio-activity would be greater.

ATOMIC WEATHER

BRIGADIER MEDLICOTT (Norfolk, Central, Nat. L. & C.) asked the Under-Secretary for Air what examination and research was taking place into the possible connexion between atomic explosions and the abnormal weather conditions which have prevailed throughout the world since the first atom bomb was exploded.

MR. WARD, in a written reply, states.— No research on this question has been done by the Meteorological Office. The available information, however, does not suggest that the weather has been more abnormal since the first atom bomb exploded.

DR EINSTEIN'S 74th BIRTHDAY

The occasion also led Dr. Einstein to converse indulgently for a few moments about implications for medical science of splitting the atom, and about current researches into his "unified field theory," by which he is seeking to establish a natural law capable of explaining all the physical phenomena of the universe. Discussing these latter studies, he said: "I have finished work on the structure of the equations, but I have not been able to find out if there is any physical truth in it. There is a mathematical way of showing which laws are mathematically correct. These equations define the general (unified) field theory. The mathematical difficulty is to find out if the whole theory has to do with nature—if it is true in the ordinary sense of the word."

JEFFREYS ON DATA HANDLING

Geophysics originally arose from the need to know the size of the earth in order to coordinate survey instruments with differences of latitude and longitude ; from the use of the mariner's compass, and from the destructive effects of earthquakes. They would still like to know what the earth was made of, but direct chemical analysis was possible only for surface rocks. The elastic properties of actual rock specimens in the laboratory suggested that far the most likely material for most of the shell below a depth of 30 km. (about 18.6 miles) or so (less under the oceans) was olivine, possibly with a crystallographic change about 300 km. (about 186 miles) down.

He thought the subject worth the attention of workers on crystal structure, it should be possible to work out the behaviour of a silicate at great pressures and find out at what pressure olivine acquired the properties of a metal.

STRANGE OBJECT IN SKY

An Air Ministry spokesman said yesterday that there was insufficient evidence upon which to base an opinion as to the existence or non-existence of so-called "flying saucers." For about four years, whenever reports were received of the sighting of unusual objects, they were investigated. In 95 per cent. of the cases a simple explanation had been found. This might well apply to the other 5 per cent., but proof could not be obtained. Some reports undoubtedly referred to natural phenomena. "Sightings" had proved to have been meteorological balloons, light reflected from the wing of an aircraft as it banked, or unusual cloud or other weather.

OPENHEIMER CASE

General Eisenhower disclosed a little later that last year he had received a report on Dr. Oppenheimer which he had found very disturbing, and that he had forwarded it to the Atomic Energy Commission for investigation.

Dr. Oppenheimer's lawyers issued a statement in New York to-day saying that the dissenting opinion of one member of the Atomic Energy Commission on his case had said all that needed to be said. Dr. Oppenheimer himself issued a statement this evening saying that Mr. Smyth's statement was fair and considered, and had been made with full knowledge of the facts. He added: " Our country is fortunate in its scientists and in their high skill and devotion. I know they will work faithfully to preserve and strengthen this country. I hope that the fruit of their work will be used with humanity, with wisdom, and with courage. I know that their counsel when sought will be given honestly and freely. I hope that it will be heard."

RUSSIAN ROCKETS

He said that a Russian, Professor Konstantin Tsiolkovsky, was the first to place the problem of interplanetary flight on a scientific basis. " He has designed a rocket flying apparatus for interplanetary travel, and has worked out the principles of its flight," Professor Stanyukovich said.

" Soviet scientists are working not only on the construction of rocket ships but are also working out a number of other problems linked with interplanetary travel," he said. Mr. Sternfeld, an engineer, had defined the " most convenient " trajectories along which cosmic ships would be able to fly to the moon, Mars, and Venus.

ANTI-PROTON?

The particle, it was said, was caught in a pack of photographic films carried in a high-altitude balloon over Texas last winter through which " it tore like a bullet through a deck of cards." In so doing, the dispatch reported, " it struck squarely an ordinary proton in the aluminium covering the pack and produced a scientifically thrilling picture.

" Moving at incredible speed and energy of 10 million billion volts," the account continued, " the particle converted matter into energy and the energy turned back into another form of matter." The account went on to say that although final proof is lacking, the supposition is that the particle is an anti-proton.

APPLICATIONS OF TRANSISTORS

The most immediate and significant applications for crystal valves are likely to be in the telecommunications field. For domestic radio and television receivers the earliest use is likely to be in personal or small portable radio sets, which will give performances equal to present valve receivers but will require only one-tenth of the power to work them. A complete change of design in television receivers to crystal valve operation appears to be unlikely for some time.

The main uses of the present semi-conductor devices are in electronic equipments where small size, light weight, long life, strength, reliability, and low power consumption are important considerations.

To demonstrate some of these qualities a miniature radio transmitter, using transistors in place of conventional valves, was dropped from a height of about 200ft. Signals were received on an equally small transitor receiver as the transmitter fell through the air, and the set continued to function just as well after hitting the ground with considerable force—an impressive endurance test.

WEATHER NOT 'ATOMIC'

Dr. Sutton, who was giving a broadcast talk, said that there was only one way he could think of in which atomic explosions could affect the climate of the world as a whole. That was by throwing up so much dust that the heat balance of the atmosphere was upset.

" Many meteorologists think this happens sometimes with volcanic eruptions but, although I do not know how much débris was thrown into the atmosphere by the recent hydrogen bomb explosions, I find it difficult to believe it was anything comparable with, say, the Krakatoa eruption of 1883, and it would have to be of that sort of magnitude to produce a noticeable effect.

" In 1903, the worst year in the Kew record, there were no atom bombs, and the Wright brothers had only just built the first aeroplane. . . . In the twenties broadcasting was blamed for both rain and drought—and long before that railways were accused of making rain by blowing steam into the air.

" We can point to no one feature as the cause of persistent wet weather, and it is probable that many influences are at work."

RAINBOWS

The learned tell us that it is only when the sun is high in the heavens that we can see rainbows below us in this manner and that the great arcs that span the dark clouds and are the most impressive form of the bow become visible only when the sun is rather low: the greater the arc the lower the sun, in fact.

The glory, even of solar rainbows, is sometimes eclipsed by the unusual beauty of lunar rainbows. On the night of September 23, 1918, when I was returning to camp along the road from Matringhem to Senlis, in the Pas-de-Calais—the moon, three days past the full, just rising on my left hand, after heavy rain—a wonderful bow appeared on my right, brilliant but of palest opalescent colours against a wall of black sky that looked solid—a breathtaking and unforgettable sight in its immensity. And since the air was at that time full of hopeful rumours—hopeful for the allies at least and almost too good to be realized—I accepted this as a portent of dawning peace.

ROCKETS TO PHOTOGRAPH THE MOON

A rocket, in order to go up to the moon and round the other side, take photographs, and return, would have to get up the necessary speed. Once its motor had attained that speed it would go on its course again without any propulsion. It might take three minutes to achieve the right speed. It would take an elliptic course, and from a speed of about 25,000 m.p.h. would slow down gradually to about 400 to 500 m.p.h. For an unmanned rocket to make such a trip and return, with photographs, would take about 119 hours, or five days. If it was desired to land on the moon the journey could be done more quickly.

It would not be possible to send one rocket direct to the moon. The general idea was to have three rockets—two contained in the largest, the smallest inner rocket to be the one with the instruments. The whole thing would be about 200ft. long. At stages two of the rockets would leave.

ODD OCCURRENCES ON JUPITER

She went along to the London Refraction Hospital. She looked at Jupiter and said: "I can see two of its moons."

"But Jupiter has nine moons," said the experts, beginning to doubt her.

Telescopes and delicate instruments were brought—and it was then found that at this time of the year there *are* only two moons visible on Jupiter

MARS STUDIED FROM BALLOON

M. Charles Dollfuss, director of the Musée de l'Air at Paris, and his son, M. Audouin Dollfuss, who is an astronomer at the Meudon Observatory, spent more than half an hour early yesterday morning studying the planet Mars through a specially built 11in. telescope from a balloon 22,500ft. up.

The balloon, which was launched at Meudon at 1.50 a.m., reached its maximum height soon after 4 a.m., where the astronomical observations were carried out in 47deg. of frost. It afterwards came down at 6.35 a.m. in a woodland clearing near Jaucourt, 120 miles south-east of Paris.

ANOTHER NEW ELEMENT

The new substance is radio-active and short-lived, and in a few minutes changes into element 97. It is not likely to be useful for atom bombs or power, the commission said: it is not fissile. As with all the new elements heavier than uranium, which is number 92 and the heaviest of the conventional elements, number 99 was made by adding nuclear particles to uranium; it has an atomic weight of 247. It was first made in secrecy in the University of California cyclotron at Berkeley as well as in other laboratories of the Atomic Energy Commission.

The new element has no name yet, but is known tentatively as ekaholmium, because it is related to the traditional element number 67, called holmium.

A NEW ATOMIC PARTICLE

The Atomic Energy Commission and the University of California have jointly announced the discovery of a new atomic particle, the anti-proton, which they say may inaugurate a new era of nuclear research.

Welcoming the news, Dr. Libby, acting chairman of the A.E.C., said that although the discovery had no immediate practical application it opened the way towards fuller understanding of the basic nuclear processes; and the director of the University of California radiation laboratory, Dr. E. O. Lawrence, wondered whether it would be comparable to the discovery of the positive electron, a quarter of a century ago, as "a milestone on the road to a whole new realm of discoveries in high energy physics that are coming in the days or years ahead."

The anti-proton, or negative proton, is described by the A.E.C. as "a nuclear ghost which has haunted the world's physicists for a generation." It is not a part of the atomic nucleus, which consists of only protons and neutrons, but is created after some event such as a high energy collision of nuclear particles. Its destination concerns a part of atomic theory that had been generally accepted by theoretical physicists but had sometimes been questioned by experimental workers and the elimination of any uncertainty about the anti-proton's existence is said to be one of the most valuable features of its discovery.

The team of scientists responsible said that the main reason why the anti-proton had not been discovered earlier was that it occurs only at high energy and until the construction of the great "bevatron" on the University of California's campus at Berkeley it was impossible to create nuclear bombardments of sufficient energy.

The "bevatron," which was built and operated by the A.E.C. at a cost of $9,500,000, enabled protons to be accelerated to 6,200 million electron volts, at which point they were directed at a target of copper inside the "bevatron" chamber. The collision of one of the protons with a neutron of copper produced not only the original proton and neutron but also a new set of heavy particles—another proton and an anti-proton. "In the collision," said the A.E.C. announcement, "a part of the bombarding proton energy is converted into mass according to Einstein's theory."

In a vacuum the anti-proton is stable and will not disappear spontaneously, the announcement continued, but in contact with a proton both particles immediately decay into mesons and disappear; this has led some reports to describe the anti-proton as the "annihilator of matter." To identify the anti-proton the scientists devised a "maze" through which only anti-protons could pass. The first observations were made on September 21, and the experiment was confirmed on October 17.

THERMONUCLEAR ENERGY

The Atomic Energy Commission announced to-day that its research into the possibility of producing a controlled thermonuclear reaction had now reached the status of a "major project" and more money would be spent on it this year than in the previous two years combined.

At present, it was said, research is concentrated on the use of deuterium—a heavy form of hydrogen—which is present in water in the ratio of one to every 6,400 nuclei of ordinary hydrogen. There was enough deuterium in the seas, said Mr. Strauss, to supply 1,000 times the present power needs of the world for the next one million years.

Mr. Strauss described himself as "a great optimist" about the project, and said he would not be surprised to see a "breakthrough" achieved within 20 years—an estimate that recalls that of Dr. Bhabha at Geneva recently. On the other hand, he added, he would not be astonished if it took much longer than that, for the problems involved were "gigantic," including that of how to control temperatures of several hundred million degrees. He did not expect the development of controlled fusion energy to have any effect on investment in ordinary atomic power installations, which would, he believed, pay for themselves long before hydrogen power was proved feasible or not, because many years of intensive effort might be required to produce the first operating thermonuclear machine

HEAVIEST ATOMIC ELEMENT

The new element 101, named mendelevium after a nineteenth century Russian chemist, is a synthetic matter heavier than plutonium or natural uranium. Dr. Glenn Seaborg, a Nobel prize winner, and other university scientists created it by bombarding nuclei in their cyclotron. Intensely radioactive, the element has no practical place in atomic energy, for bombs or power, but will help broaden man's understanding of atomic matter, the university announcement said.

Identification was completed two months ago during work being conducted under the United States Atomic Energy Commission contract.

SPACEFLIGHT

Professor Krafft-Ehriche, a German-born American scientist, in a lecture on satellites, said that he was working on the construction of an artificial satellite which should be able to move around the world for six days and nights on a minimum of fuel. It would fly at an altitude of 100 miles and would, at this altitude, use about 33lb. of fuel—a mixture of oxygen and petrol—to circle the earth once. Altitude would determine not only the shape of the satellite, but the period it was desired to keep it flying. To keep a ball-shaped satellite in space for one year it would have to start 200 miles up, but at a low altitude, for instance two miles, the satellite would be able to fly for only a little more than an hour.

RADIOACTIVE ATMOSPHERE

There is no evidence that there has been, during recent years, any significant increase in the radioactivity of the atmosphere, which remains mainly due to natural causes. Nor has there so far been any evidence that radioactivity is a contributory factor in the causation of bronchial cancer but investigations are continuing into this possibility.

VEGETATION ON MARS

An announcement said the discovery of the area, 200,000 square miles in extent, had produced the greatest change in Martian geography since the planet was first mapped 125 years ago. The discovery is credited to Dr. E. C. Slipher, leader of an expedition which photographed Mars from Africa in 1954. Dr. Slipher alone made more than 20,000 photographs.

The announcement said the discovery was "totally unexpected." It added: "Never before has such a new dark splotch appeared, except as an increase of an existing dark area. The remarkable transformation indicates that the division between Martian (orange-red) desert and dark areas is not necessarily fixed or permanent; one may change to the other at any time. It helps support the conclusion that Mars is not a dead world, that the darkening is due to the growth of plant life. Biologists suggest that this life may be akin to lichens that grow on the earth's barren rocks and mountain-tops."

SOVIET FORECAST OF PLANETARY TRAVEL

A Russian professor said:—

"Modern technology is building rockets capable of rising 400km. [approximately 250 miles] or more into space. The further perfection of these rockets will lead to the conclusion of the first stage of the development of astronautics—the creation of an automatic artificial satellite of the earth—a rocket which will rotate round the earth at a distance of several hundred kilometres.

"The rocket will be equipped with instruments for transmitting to the earth the results of scientific observations, and scientists believe that the creation of such an artificial satellite can be expected within a very few years. The next stage will be the construction of a further artificial satellite on which people will stay. Step by step stations will be created which will rotate round the earth. On these it will be possible to assemble small rockets. These remote-controlled rockets will be able to reach the moon, travel round it, and return.

"Finally, the flight of a manned rocket to the moon will be carried out. The crew will be able to land on the moon, undertake scientific research there, and then return to the satellite stations or to the earth."

KEEPING FOOD FRESH INDEFINITELY

Dr. Seligman said that it looked as if radiations from the many fission products could be turned into something very useful. Food could be treated by passing it on a moving belt, for instance, and exposing it to radiation in such a way that things like meat could be made to last indefinitely.

"It is possible that widespread use of radiation sterilization may bring about the reduction, or perhaps even the elimination, of the necessity for refrigeration to keep food fresh," he said.

Radiation from fission products could also be used to sterilize materials and there was an obvious future for its use in the sterilization of pharmaceuticals, including antibiotics such as penicillin and streptomycin, where heating was not permissible. A tenth of the people who had stitches inserted into wounds might be spared the agony of having stitches broken because of weakened fibres.

The most successful fission product was Caesium 137, and radiations from this in compact sources were equivalent in strength to thousands of grams of radium. Radiations from large sources of caesium were envisaged for use in place of radium or large scale X-ray machines in the treatment of disease. Bones could be sterilized so that a bone "bank" could be formed; similarly, sections of artery required for grafting could be preserved.

1956

ANOTHER NEW COSMOLOGY

Dr. Simpson of Chicago University recently made a number of observations which show that the cosmic ray equator, which may be expected to coincide with the geomagnetic equator, is displaced at some points as much as 40° . . . If these results substantiate the previous findings, a point of fundamental importance to cosmologists and philosophers, as well as to physicists, is suggested: that is, that the Earth, instead of being relatively close-lying to the Sun and in a vacuum—as was once thought—is in fact at the edge of the solar system and in no vacuum. Much else follows from this once it has become established

LAUNCHING SITE FOR SATELLITE

The Air Force and the Navy issued a joint announcement to-day saying that the American earth satellite for the International Geophysical Year would be launched from Patrick Air Force base on the Banana River in Florida.

This site was chosen, the announcement states, " on the basis of operational requirements for large rocket launchings." It will be recalled that Patrick Air Force base is the northern terminus of the Bahamas long range missile proving ground

ATOMIC CLOCK

The possibility of an atomic standard depends in principle on the fact that the electrons surrounding an atom exist in a variety of configurations or states, each with a definite energy. A change from a higher level of energy to a lower one, or the other way about, is balanced by the sending out or absorption of radiation; and the frequency of this radiation is determined by the difference in energy to be balanced. What is new in the situation is that the frequencies of microwave radio can now be measured with very great accuracy. Therefore, if the properties of the atoms are taken to be constant, the radiations which they emit or absorb can be used as a standard of time.

The atoms which have been chosen are those of caesium. There were several reasons for the choice; one of them was merely that they offered a convenient frequency to measure (one with an equivalent wavelength in free space in the neighbourhood of 3.25 centimetres). Such measurements can be made quickly, and, through an atomic clock, as now demonstrated, can serve as a permanent source of time, independent of the earth. It seems certain at the least that they will be used as the future working standard of time. Whether the final arbiter should, in the long run, be earth or atom is a question for later decision. And it will be of interest to know from observation whether astronomers' time and atomic time keep in step over long periods, or diverge perceptibly, one from the other, as some theorists have suggested.

THE ANTI-NEUTRON

What is particularly striking about the anti-neutron is what happens when it comes close to an ordinary neutron in the heart of an atom—the two particles annihilate each other. With their destruction there is released several hundred times more energy than is produced in the reaction of a hydrogen bomb.

Commenting on the discovery Professor Ernest O. Lawrence, director of the university's radiation laboratory, said that it had no practical application yet. " The value of this work lies in the expanding of an understanding of the nature of matter," he said. " This is part of a programme in basic nuclear research, and past experience has shown that even fundamental discoveries of this kind bring material progress in ways that are unpredictable." Other scientists noted, in this connexion, that when uranium atoms were first split by ordinary neutrons many years ago no practical application of that discovery was immediately foreseen, though eventually it led to the creation of the atomic bomb and to the harnessing of atomic energy for power.

SEEING INTO SPACE

A photoelectric recording surface is 100 times as sensitive as the photographic plate. It emits electrons when light falls on it, and the emission of the electrons leaves a pattern of electric charge on the surface. Two quite distinct methods of use are thus possible; either the electrons emitted, or the charge left behind them, can be used to give the record of the picture. Methods which use the electrons are analogous to the wartime "image converter." Their result is the formation of one image from another—in this case a brighter image from one that is less bright. They are thus rather an extra stage between telescope and photographic plate than a wholly independent method of recording.

Nor is there an accumulation or build-up of the picture with time in the photoelectric part of the equipment; accumulation is wholly photographic as before. The second main method—that using the charge left on the surface—is like having a television camera adapted for long-exposure "stills." Accumulation of the picture with time is now on the light-sensitive surface—or after projection on to another surface. The following stage is electronic, as in a television camera, and tricks can be played with the "picture." For example, the brightness of the background sky could be subtracted, so that longer exposure times would be possible. Finally, the "pictures" could be recorded electronically, if so wanted, and only reproduced visually when required. But more probably they would be photographed at once, and the plates used for storage.

More specific forecasts would be rash. But it can be said now that methods analogous to the "image converter" could, in principle, enable a 20-in. telescope to take pictures which can now be taken only by the 200-in.; but that there are difficulties to be overcome which may make the actual gain less. Also, the limitation of sky background would remain. For this reason, the performance of the 200-in. telescope would not be increased in the same proportion by this method. The second method—that of the television camera—offers the possibility, in principle, of a gain of 16-fold in sensitivity. This would apply even to the largest telescopes, so that astronomers should be able to see four times as far into space as they now can.

RUSSIAN PLANS FOR AN EARTH SATALLITE

Professor G. Pokrosky, a member of the Soviet Commission for Interplanetary Communication, writing in *Trud*, the trade union newspaper, said that Russia could launch an artificial earth satellite this year. "The problem of interplanetary communication has passed from the project stage to that of practical implementation," he said.

A three-stage rocket could be manufactured which could circle round the earth in 103 minutes. Soviet scientists, however, had evolved another method. "A rocket, containing an explosive charge of a special design, rises to the required height, and then, when the charge explodes, a small mass of metal is expelled from the rocket and becomes an artificial celestial body. The manufacture of a satellite by means of a rocket with a special explosive charge is simple and cheaper. We have all possibilities to launch such a satellite in 1956."

THE CERN ACCELERATOR

What is to be seen on the main site is rather more than a half-circle of a concrete tunnel roughly big enough to carry a double-track railway. Round the middle of this there will pass eventually the race-track proper, an oval-shaped vacuum tube, some 3in. deep by 6in. wide. The protons accelerated have to be kept within this small cross-section while they travel a total of 180,000 miles, making in round figures 500,000 circuits of the race-track. As with other types of cyclic accelerator, the particles accelerated are held to the required course by large magnets while they are given repeated small accelerations by electrical means.

The new element in this type of machine is the more precise limitation of their course. This means that the magnets, as well as the vacuum chamber, can be of smaller cross-section, though each of the 100 magnet units will contain 32 tons of steel. They will be lowered into position from a small high-level rail track on one side of the tunnel. Part of this already exists.

The precise control of the course of the accelerated particles has been obtained by making use of a new principle which was first suggested in the United States. To secure the promised precision in practice, the steel of the magnet must be uniform in its magnetic properties to a degree which has raised new problems in production, and the placing of the magnets must be accurate to within about one-75th part of an inch vertically, and only slightly less radially. To ensure stability, it has been found necessary to carry the weight of the magnets on concrete pillars which are being taken down to bedrock.

1956

THE FIRST ARTIFICIAL EARTH SATELLITE

The Russian satellite soaring over the United States seven times a day has made an enormous impression on American minds, and that impression is by no means reassuring.

The first surge of awe and admiration at man's first successful venture into outer space has been swiftly coloured in political and military quarters, by a distinct feeling of uneasiness, if not alarm, at some of the underlying implications of such an achievement by a potential enemy; and, almost inevitably, calls have gone out for an urgent review of American defences, which in various ways are being subjected to large-scale retrenchment.

The United States has a special concern, having been the only other country to undertake a satellite programme as part of the International Geophysical Year. The first launchings had not been expected until the spring, and naturally no one finds pleasure in having lost the race, for all the generous tributes with which the Russian success is greeted.

What strikes a note of fear is the American scientists' estimate that it must have taken something like an inter-continental ballistic rocket to launch a satellite of nearly double the expected weight—and hitherto the official tendency has been to belittle Russian advances towards the "ultimate weapon." This tangible evidence that the United States is not ahead in the race to conquer space may well be salutary, though President Eisenhower, taking the announcement calmly during a week-end's golf at Gettysburg, let it be known that it would not entail any acceleration of plans to launch an American satellite.

But all such considerations, and the recognition of a great "propaganda victory," pale in the contemplation of what has been achieved. The first statement from Moscow that the satellite was encircling the earth made an impact such as few events have done since the end of the war, and every move was closely followed, while an array of tracking devices was rushed into service to pick up the satellite's signals and calculate its course.

The rotation of the earth causes its north-south orbit to move progressively westwards across the United States, which will have seven "visitations" a day. In addition to the official listening posts, thousands of short-wave amateurs all over the country have picked up the satellite's signals.

Some disappointment is expressed that the secrecy of the firing precluded the mobilization of "Operation moon-watch" established at Cambridge, Massachusetts, by the Smithsonian Institution, but scores of observers were rapidly mustered in the expectation that the satellite would be visible through binoculars at dusk and dawn. So far, however, it has not been in the American sky during these twilight hours, and some doubt is expressed whether it will be visible until its orbit is nearer the earth.

This is estimated by American observers to be only slightly elliptical, and calculations so far invariably place the satellite at a height of more than 500 miles. According to some experts, it could remain in flight for as long as 30 years before it began to encounter "drag"—an unknown factor—from the earth's atmosphere.

A description was recently given in *The Times* of the extraordinary computing equipment installed in Washington by International Business Machines as part of the American programme and, if the claims are justified, this should be able to translate the tracking data into predictions of the satellite's course over the United States.

According to these calculations, it was launched in a north-easterly direction from a point north of the Caspian Sea at 9 p.m. (G.M.T.) on Friday; and, according to Russian statements here, the satellite contains no instruments other than wireless transmitters. Such refinements, it is stated, are being reserved for later launchings.

It happens—perhaps it was no coincidence—that a group of Russian scientists have been in Washington to attend a I.G.Y. conference on rockets and missiles. At the time of the Moscow announcement they and their colleagues from other countries were attending a cocktail party at the Russian Embassy, which was suddenly transformed into one of the most hospitable places in the capital, with exchanges of toasts and congratulations.

CONSERVATION OF PARITY

The point at issue is about the nature of the restrictions which are considered to be imposed on the behaviour of nuclear particles—and hence on the makers of theories designed to explain that behaviour. There are several such restrictions; they apply to quantities which, in a closed system, remained unchanged in total amount and are said therefore to be conserved. Examples relevant to the properties of elementary particles are energy—which in this context means the mass of the particle when at rest—and angular momentum, which can be defined loosely as quantity of spin.

The characteristic now in question distinguishes, analytically, between ability, or not, to undergo reflection in a mirror without being changed in the sense that a mirror image differs from the original. Pictorially it can be thought of as corresponding to the presence or absence of " screwiness "—the association of a direction of rotation (that of tightening the screw) with a direction of movement (that in which the screw advances).

It is not quite correct to say that the conservation of parity—and hence of " screwiness-or-not "—has been taken for granted. Some theoretical physicists have been ready to question it, in spite of the apparent success with which the rule has been applied. But there has been no substitute for it; experimental physicists have taken it for granted; and until a paper by Lee and Yang, published last October, there had been no suggestion from the theoretical side of experiments which might be done to test it.

The consequences of these experiments are likely to be far-reaching. The removal of a false starting-point for theory must open the way for new ones, which will need time to consider, but must in the long run be more hopeful. Secondly, the circumstances in which " parity conservation " has been found to be broken suggest to phycicists a new distinction between nuclear reactions which proceed on a fast time scale and more slowly—and this, too, may be important.

US ROCKET EXPLODES

A terse signal, " Exploded—no casualties," said everything at Cape Canaveral to-day of the failure of the first spectacular effort to launch an American test satellite. A naval spokesman said it would take weeks before another launching could be attempted.

After days of delay for the elimination of minor defects, the Vanguard rocket carrying the 6in. sphere lifted off the launching platform but fell after two seconds and burst. It was officially explained both at the Florida launching site and in Washington that the failure was due to loss of chamber pressure, or thrust, in the first stage of the 72ft. projectile, which blew up in a billowing cloud of flame and smoke. The ensuing fire was quickly extinguished and no one at the Patrick Air Force base was hurt

Such are the bald facts about the failure of the accelerated efforts to " catch up with the Russians," and its impact will be the greater for the immense publicity to the week's activities at Cape Canaveral. The news, described by the Secretary of Defence, Mr. McElroy, as disappointing but " not too surprising," was telephoned to President Eisenhower at Gettysburg, and strong doubt must now be cast on his prediction that American test satellites would be shot into orbit this month.

THE INTERNATIONAL GEOPHYSICAL YEAR

The I.G.Y. organization is an association of international scientific bodies and of national scientific academies. It is non-political. Scientists come together at its meetings, in harmonious cooperation, from countries between which, unfortunately, there are deep political divisions. Yet the scientific academies of these countries have all succeeded in getting their Governments to provide the money for their respective parts in the I.G.Y. programme. Many of them have also been enabled to contribute to the expenses of the central organization. At first these were supplied by I.C.S.U., with special grants from the United Nations Educational, Scientific and Cultural Organization (U.N.E.S.C.O.). But in 1956 the expenses grew beyond the scope of these grants. The central expenses are a very small " overhead " item in the whole I.G.Y. budget. It has been estimated that the whole sum spent and to be spent on the I.G.Y. will exceed £100m. The total central expenditure is estimated at £115,000—less than a tenth of 1 per cent. of the whole cost.

HEISENBERG AND UNIFIED FIELD THEORY

Professor Heisenberg, the leading German physicist and Nobel Prize winner, said in a lecture at the university here that he and his associates had developed a mathematical equation by which the whole structure of the universe might be explained.

His research is aimed at achieving a uniform field theory, a problem on which Einstein worked for 30 years. Professor Heisenberg said that at the present stage he could not finally prove that his formula was right but when proof was attained Einstein's search for a unified field theory would be realized.

US TO TRY FOR THE MOON

The United States to-day announced plans to send space vehicles to the moon, and perhaps even land on it. They will carry a "primitive television" device to scan the moon, including the side which cannot be seen from the earth.

The American plans, which have been approved by President Eisenhower, call for at least five attempts to send unmanned space vehicles to the lunar area. Two of these attempts will be made by the Army, three by the Air Force. A Defence Department spokesman said it was not the present intention to hit the moon, but it was possible that one of the five vehicles could do so.

GRAVITY AND THE NEUTRINO

The question "Will physics ever again be easy?" must have been asked many times. It is prompted this week by the impossibility of explaining in a few sentences what anyone should want with a theory of gravitation based on a particle, the neutrino, which has neither mass nor electric charge. Such a theory has just been advanced by Dr. L. I. Schiff, of Stanford University, California.

INTENSE COSMIC RAY ACTIVITY

American satellites in orbit around the earth experienced cosmic ray activity hundreds of times more intense than had been expected, according to scientists on the American committee of the International Geophysical Year.

At the apogee of their orbit, more than 1,000 miles from the earth, the radiation met by the Army's Explorer satellite was so intense as to "overwhelm the cosmic ray counters" contained in the cylindrical satellite; the count of particle pulses reached 35,000 a second.

LAUNCH OF EXPLORER

Relief and joy—eloquently reflected in President Eisenhower's "well done!"—has surged through the United States this weekend over the launching from Florida of the American satellite Explorer which, having soared into a predictable orbit, is circling the earth 12 times a day in company with the Russian sputnik. It is, of course, this latter contemplation that lends such strong political emphasis to a great scientific achievement; and its psychological impact on American military planning and diplomacy will clearly be no less profound than that made by Russia's initial advances in outer space.

The Army, for instance, has become the toast of the hour for having brought its modified Jupiter rocket to the rescue of the Navy's ill-starred Vanguard, which has twice failed to take off from the launching site at Cape Canaveral.

The Army's success, achieved with all the aplomb of a polished golfer driving straight and long from the first tee, will certainly give weight to its claim that it could have put a satellite into orbit a year or two ago; and not the least of these ironies—the master mind behind the Jupiter was Dr. Wernher von Braun, one of the foremost designers during the war of the German V-weapons.

MAN-MADE AURORA

An abnormal display of aurora, seen at Apia, Samoa, well inside the tropics, has been connected by Dr. A. L. Cullington, of the Geophysics Division of the New Zealand Department of Scientific and Industrial Research, with a high-altitude nuclear explosion some 2,000 miles to the north. He describes it as a man-made aurora—the first on record.

The date was August 1, when a nuclear warhead was said to have been exploded above Johnston Island, in the central Pacific, as part of the American test programme. From the effects now reported, it appears that the explosion was not only of great magnitude but that it must have taken place at a very great altitude.

As well as the auroral display there was a magnetic storm, which appeared to be localized in the central Pacific. The storm began at about the same time as the nuclear explosion: the auroral display was first seen about a minute later.

TOWARDS POWER FROM HYDROGEN

Mr. Kurchatov said that several groups of scientists and engineers were designing hydrogen power reactors by means of which the fuel problem would be solved practically for ever.

Rarefied deuterium (hydrogen with an atomic weight of two instead of one) had been heated to a million degrees Centigrade by a two-million amperes current and the emission of neutrons had been registered in the process.

Kurchatov favoured the idea of using a magnetic field for isolating the "plasma" of ionized gas, to prevent the heat from affecting the chamber. He added that in future reactors it should be possible for the magnetic field of the electric current passing through the gas to provide this heat insulation in a thick-walled chamber for a period of about one second. "The means to produce this are technically feasible," he said.

RUSSIA LAUNCHES GIANT SPUTNIK

Russia to-day launched a third sputnik, a giant of one ton 679 lb., three times the size of its predecessor. Moscow radio gave the news of the launching to the Russian people and also broadcast signals emitted by the sputnik—a series of dots and dashes.

The sputnik is in orbit at an angle of 65 degrees to the Equator, the same orbit as was followed by the two earlier sputniks. An announcement by the Tass news agency said it completes a revolution of the earth in 106 minutes and is about 1,175 miles above the earth at the highest point of its orbit.

Tass said it carries apparatus for measuring the pressure and composition of the atmosphere, the concentration of positive ions, the size of electrical charges, the intensity of gravity, and the intensity of different kinds of radiation from the sun, and thermometers to measure interior and exterior temperatures.

Powered by solar energy batteries, its radio transmitter sends out uninterrupted signals at 20.005 megacycles with a duration of 150-300 milli-seconds. The total weight of instruments and radio equipment is about 2,133 lb.

VOLCANIC ERUPTION ON MOON

A volcanic eruption on the moon was observed by a Soviet scientist on November 3, the Soviet news agency Tass reported to-day.

Mr. Nikolai Kozyrev, observing the moon with a 50in. mirror telescope at the Crimea observatory, noted a reddish outline around the Alfons volcano.

Professor Alexander Mikhailov, chairman of the Astronomical Council of the Soviet Academy of Sciences, said there was no doubt that Mr. Kozyrev had really seen a volcanic eruption on the moon. Professor Mikhailov said: "We can now regard as completely unfounded the existing view on the origins of characteristic features of the lunar landscape, which ascribes them to the fall of meteorites. The volcanic eruption shows that the moon shares similar processes to the earth, which are responsible for the formation of mountain contours."

THE MOON AND KHRUSHCHEV

The Russian probe to the moon shares to-day's headlines, as was clearly intended in Moscow, with the imminent arrival of Mr. Khrushchev in a jet aircraft—and for many he might almost be a man from the moon as the United States gets ready for the first visit ever made by a Russian ruler. Experts point out that conditions for sending a satellite into orbit round the moon—deemed to be even more difficult at this time than making a direct hit—were at their best five days ago, and they conclude that Russian scientists have been required to bow to political expediency.

This further manifestation of Russian prowess in outer space is obviously calculated to reinforce the views and proposals which Mr. Khrushchev will be expounding to President Eisenhower and others in the next fortnight. Indeed, it may be doubted whether Mr. Khrushchev, accorded full honours as a head of State, would now be leaving for Washington had not the first sputnik two years ago shocked the United States into a dire, if grudging, admission that Russia in some respects held a commanding lead in the conquest of space.

THE FIRST RUSSIAN LUNIK

NEW ATOM SMASHER

The President told his audience that plans for the construction of a huge new " electron linear accelerator " (or atom smasher) had been recommended to him by his science advisory committee. The project had been proposed by Stanford University, California, and it was planned to construct the accelerator there. The university already has in operation the largest machine of its kind, a 220ft. tube. The new machine will be roughly 50 times the length of the existing one, and its initial operation is expected to produce between 10,000 million and 15,000 million electron volts.

EARTH'S RADIATION ZONES

Professor James Van Allen, a leading physicist engaged in the Vanguard satellite programme, said here that America's Pioneer IV satellite, now orbiting the sun, had revealed that radiation round the earth extended much farther than was previously thought.

Professor Van Allen told space scientists here last night that, of the two known radiation bands, the second has been found to extend from 8,000 to 55,000 miles into space. The inner band has been calculated as extending from 1,500 to 3,000 miles around the earth.

Pioneer IV showed, he said, that the inner band was by far the more deadly. Men who tried to pass through it by rocket would have to be heavily shielded, or leave the earth over the poles, where radiation thinned out.

RADAR MESSAGE TO VENUS AND BACK

Scientists at the Lincoln laboratory project at Westford, Massachusetts, announced to-day that they had received an echo of a radar message sent to the planet Venus and so had established a new record—52,000,000 miles—for a message into space and its return. This was by far the longest man-made radio transmission ever achieved.

They said they had struck Venus with 10 cents' worth of electricity in February of last year, but only after a year of effort and the expenditure of thousands of dollars had they been able to confirm the radar echo received from Venus at that time. It had taken 2½ minutes for the signal to reach the planet and the same time for it to return.

THE MOONS OF MARS

Dr. Shklovsky said that Phobos and Demios differed from the satellites of other planets by their insignificant size and their extreme closeness to their planet. Phobos, moreover, showed another striking dissimilarity from all other satellites in the solar system in that it had deviated in the past few decades from its calculated orbit by two and a half degrees and its movement had accelerated. This meant it had gone closer to the surface of Mars.

"The same thing happens to satellites launched from the earth; they are slowed down by the resistance of the earth's atmosphere and consequently come down and in doing so accelerate," Dr. Shklovsky said. He had concluded that Phobos was hollow, and that "as no natural body can be hollow, it must be an artificial satellite of Mars." He suggested that Demios might have had a similar origin.

CLAIMING THE MOON

The committee was in some geometrical difficulty about what would happen if national boundaries were extended into space. Senator Keating (Republican, New York) argued that, because of the curved face of the earth, extended boundaries, in the form of an "inverted cone," would overlap, with the result that more than one nation would be occupying the same air space.

Another member felt, on the contrary, that, if boundaries were projected upwards like columns, gaps would be left between them. But, asked Senator Keating, would not the rotation of the earth cause the columns to get mixed up?

A Republican colleague gave him "the full weight of my scientific ignorance," and the committee was about to vote on "gaps or overlaps" when someone thought of approaching the National Geographic Society for a ruling.

WEATHER SATELLITE

The United States to-day fired into orbit another satellite, specially instrumented to gather meteorological data and relay it to earth in an experiment that holds much promise for long-range weather forecasting. The "weather satellite" was carried in a Vanguard

This rocket was of the type that has had many failures in the past, but this time it seemed to function smoothly from the moment of its ignition on the firing pad at Cape Canaveral, Florida. A little more than two hours later the National Aeronautics and Space Administration reported that the 21lb. sphere contained in the last of the rocket's three stages had gone into orbit.

The science headlines for the decade told predominantly of the growth, now rapid and continuous, of spaceflight. Whereas the International Geophysical Year had acted as an important focus for the launching of the first satellites, the corresponding International Year of the Quiet Sun in the mid-1960s had little influence on space developments. The logic of spaceflight had become political: based on President Kennedy's vision of landing a man on the Moon by the end of the decade, and on the accompanying US–Soviet competition for leadership in space research.

During the 1960s, the Americans moved ahead of the Russians in most areas of both manned and unmanned space exploration. In 1961, Gagarin became the first person to orbit the Earth, whilst, in the same year, the Americans only made their first sub-orbital hop. Yet, in the following year, John Glenn also circled the Earth. By the mid-decade, both Russians and Americans were investigating what tasks human astronauts could carry out in space—space walks, space rendezvous and so on. In 1967, there was a hiccup: three US astronauts died in a fire during a ground test of the Apollo spacecraft designed to go to the Moon. Three months later, a Soviet astronaut died during the final descent of his space capsule to Earth. This increased the glare of publicity as, in the following year, the first US manned flight orbited the Moon. Then, finally (and within the timescale set by President Kennedy), Neil Armstrong became the first person to walk on the surface of the Moon. A few months later, the Americans ended the decade triumphantly with a second

Man on the Moon (Apollo lander)

lunar landing.

In terms of unmanned spaceflight, much effort naturally went into examination of the lunar surface. Here, too, the Russians were in the lead initially, but the more sensitive American instrumentation proved to be a major advantage. By 1966, when both Russians and Americans sent their first orbiting and soft-landing space craft to the Moon, the original Russian advantage had disappeared. The same was true of planetary probes. The Russians took the lead in exploring Venus, but the Americans proved more successful in exploring Mars. The results from both planets cast doubt on hopes that either might harbour life. Venus proved to be too hot, whilst the Martian atmosphere was found to be much thinner than expected, and the first pictures of the planet's surface indicated a highly hostile environment.

If hopes of life on other planets now diminished, other types of investigation brought forward well-publicised new ideas of life elsewhere. One excitement concerned a certain type of meteorite which was found to have traces of chemicals normally associated with life processes, and which was even claimed to contain 'fossil-like' structures. Another controversial study was Project Ozma, which entailed looking for radio signals from intelligent life elsewhere in our Galaxy. The telescopic discovery of a plant orbiting a nearby star (an interpretation that was later disputed) increased interest in the possibility of life outside the solar system.

In terms of value, the most important launches continued to be those of unmanned Earth satellites. Increasingly, satellites for weather detection and for communications were put into orbit (some, for the first time, into geosynchronous orbit), along with a much larger, but much less discussed, number of spy satellites. The chance of seeing international television programmes in real time naturally made a major impact on the rapidly growing TV audience. But the interesting research development lay elsewhere—in the use of rockets, balloons and satellites for examining the Universe in new regions of the spectrum. During the 1960s, cosmic x-ray sources were discovered in this way. At the same time, ground-based astronomy continued to surge ahead. Probably the most important discovery of the decade was the detection of a universal background radiation in the microwave region (corresponding to that from a low-temperature black body). It was realised that such radiation could be interpreted as convincing support for a 'Big-Bang' origin to the Universe, and this theory now came to dominate popular accounts of cosmology. The 1960s also saw the discovery of pulsars. Their short-duration radio pulses were attributed to a variety of sources initially (including little green men) until they were finally tied down to a new type of star—the neutron star.

Quasars (quasi-stellar objects) were initially detected via their radio emission. However, the real excitement came when optical measurements showed that these objects were very bright distant galaxies. Radio waves, in the form of radar were also used actively within the solar system to examine nearby planets: the rotation rates of Venus and Mercury were first determined in this way. One non-radio form of observation that caught public attention was the detection of neutrinos. This required a very large detector to be buried deep in a mine. By the latter part of the decade, monitoring of neutrinos emitted by the Sun had become routine.

The value of creating major projects in order to attract funding, along the lines of nuclear and space research, was now widely recognised by researchers. Thus geophysics in the 1960s was beginning to develop a more comprehensive theoretical base—the idea of seafloor spreading became widely accepted during the decade—and a number of leading American geophysicists grouped together to propose a technologically difficult new experiment in Earth exploration. This was the Mohole project—the idea of drilling through the Earth's crust in order to examine the nature of the top of the mantle. Though the project produced some interesting results, it eventually ran out of money; but it helped direct more media attention to studies of the Earth.

Within the accepted 'big science' field of high-energy physics, the decade saw significant advances in providing a unifying theoretical base for the continuing experimental advances. The problem was the rapidly increasing number of 'fundamental' particles that were being detected. Early in the decade a method of classifying heavy sub-atomic particles—the 'eightfold way'—was introduced. This led during the next few years to development of quark theory. It was envisaged that a small number of these supposed quarks, brought together in various combinations, could explain the variety of 'elementary' particles that had been found. Indeed, the theory predicted the existence of other possible particles that might be found with more powerful accelerators. The latter part of the 1960s saw the first step towards a unified theory of the four fundamental forces known to physicists—in this case, of the weak nuclear force and the electromagnetic force. Again, this theory predicted the existence of particles which could not be detected with existing instrumentation. The debate engendered by these theories kept nuclear physics before public attention despite the complexity of the issues involved.

The idea of a laser (an acronym for 'light amplification by stimulated emission of radiation') had emerged in the 1950s, but the first useful versions appeared in the 1960s. Initially, practical uses for them seemed to be limited (apart from their potential in holography), but gradually applications appeared. One of the first was in eye surgery. As the range of applications increased, so did popular references to lasers in the press.

1960s CHRONOLOGY

1960
- The Mossbauer effect is discovered
- The first laser is developed
- The idea of seafloor spreading is proposed

1961
- Gagarin becomes the first human to orbit the Earth
- The 'eightfold way' is introduced as a way of classifying heavy sub-atomic particles

1962
- US space-probe Mariner II reaches Venus
- The existence of more than one type of neutrino is established
- The first active communications satellite, Telstar, is launched

1963
- The first survey of celestial x-ray sources
- The existence of quasars is established
- The concept of seafloor spreading receives support from magnetic measurements
- Launch of the first satellite into geosynchronous orbit

1964
- The idea of quarks (and of 'charm') is introduced
- The Mohole project gets under way

1965
- Penzias and Wilson discover radio radiation apparently remaining from the 'Big Bang'
- Cosmic masers are discovered
- Mariner IV reaches Mars
- Neutrino astronomy gets under way

1966
- A Soviet space-probe makes the first soft landing on the Moon
- The Soviet Venera III space-probe makes the first landing on another planet, Venus

1967
- The first pulsar is detected
- Electroweak theory is developed

1968
- US astronauts make the first manned landing on the Moon in Apollo 11

THE ORIGIN OF THE SUN

Professor Agrest said in February he believed Sodom and Gomorrah were destroyed by an atomic explosion. He believed a gigantic spaceship came to earth at a speed close to that of light, and went into orbit while its occupants studied the earth. They possibly landed in the region of the Baalbek terrace, a platform of huge stone slabs in the Anti-Lebanon mountains which has defied explanation.

His theory was that the space travellers could have exploded their surplus nuclear fuel before leaving earth, after warning the local population of the danger.

UNUSUAL SPACE THEORY

But in the last few years physicists have been driven to take seriously the idea that a small proportion of cosmic ray particles might have such high energies that, even if originating in the Galaxy, they could not be retained in it. The mechanism by which (if formed in the Galaxy) they are supposed to be held in it depends, again, on the existence of interstellar magnetic fields. There is reason, from optical astronomy, to think that such magnetic fields exist.

At their Albuquerque observing station in New Mexico Professor Rossi and his colleagues have been working with a collecting array which extends over some three square kilometres. The reason for this large area is that cosmic ray particles of the highest energies can be detected—and their energies estimated—from the number of secondary particles, up to a million or even many millions, that they give rise to in the atmosphere.

Within a few months of beginning work with his new array, he detected a shower of secondary particles from a primary cosmic ray particle of higher energy (about $2 \times 10^{19} \text{eV}$) than has before been observed.

the direction of arrival of the particle was determined as well as its energy; and that account was taken of this in arriving at the conclusion that the particle was of extragalactic origin.

US LAUNCHES VENUS SATELLITE

Officials of the National Aeronautics and Space Administration said the satellite, known as Pioneer V, was going successfully into orbit. By noon it was 42,460 miles from the earth.

The orbit will take 311 days to complete, and the greatest distance from the earth will be about 186 million miles. The satellite will approach within 74,700,000 miles of the sun, following a path inside that of the earth and outside that of Venus, but touching the orbit of each once in every trip. It is expected to last 100,000 years before being burned up in the earth's atmosphere.

This is the first known attempt to establish communication at such ranges. The radio transmitter is powered by solar cells in four " paddlewheel " vanes, and will transmit its information only when within a range of 50 million miles, and in response to a signal. The measuring equipment includes a high-energy radiation counter and an ionization chamber and Geiger-Müller tubes to measure medium-energy radiation, a micro-meteorite counter to measure dust particles and a magnetometer, to determine the strength and direction of magnetic fields in space, and an aspect indicator which triggers an electrical impulse when it faces the sun.

Measurements of magnetic fields in almost any part of the solar system are of interest to astronomers, and the same point applies to observations of particle streams encountered. In both cases the value of information will be much increased if the ambitious arrangements for the storage of information and its radio transmission back to earth are successful, so that changing conditions are observed.

UNCERTAINTY

From one point of view Heisenberg's "uncertainty principle" amounts to the fact that it is impossible to make an accurate survey of a particle, or set of particles, of atomic or sub-atomic sizes. The more accurately the position of a particle is known, the less is known about its velocity and vice versa; although, for a particle of given mass, for example an electron, the product of the two uncertainties can be specified precisely.

But this is not the only—or the most informative—way of putting it. It can also be said that one is not entitled to look on an electron as a billiard ball; and that, if one insists on so doing, one will get into this kind of difficulty. In other words, the world is not such that questions about the positions and velocities of electrons or other atomic particles can properly be asked.

A further point is that the uncertainty principle follows directly from a theory—that of wave mechanics—which enables more precise predictions to be made than before about the wavelengths of radiation that different kinds of atom can emit or absorb. In return for not asking meaningless questions, physicists have gained an ability to answer other questions, often of practical interest—for example, about one class of materials used in transistors, and also those that provide the viewing screens of television sets

ORBITER PICTURE OF THE LUNAR CRATER COPERNICUS

PROJECT OZMA

The first systematic attempt to search for signals from intelligent beings on other planets was begun last night at the National Radio Astronomy Observatory, Green Bank, West Virginia. For the next month the observatory's 85 radio telescopes will be focused alternately on two stars that might have planets able to sustain life. They are Tau Ceti and Epsilon Erindi, each about 66 million million miles from the sun. It would take radio signals 11 years to bridge the distance.

The project is known as Ozma, after the queen of the mythical land of Oz, "a place very far away, difficult to reach and populated by strange and exotic beings," as a spokesman explained.

MAN-MADE NUCLEAR MOLECULE

the nuclear molecule was created in an "atom smasher", known as the tandem accelerator. The molecule that was created consisted of two nuclei of carbon 12 instead of atoms and it existed for less than one billionth of a second

This, if substantiated, will be an entirely new kind of effect. Since any two nuclei are positively charged it follows that the carbon nuclei have approached close enough together—in what can only be described as a temporarily illicit union—for the short-range forces which hold together the individual particles which make up a nucleus to balance temporarily the force of electrical repulsion driving them apart. The usual results of such a collision are either the formation of a compound nucleus, which is unstable and shortlived, but while it lasts is a single nucleus or that the bombarding nucleus is deflected so that the two nuclei go their own ways.

1961

NEW MESON DISCOVERED

The particle, called the omega meson, was discovered by scientists in the University of California's Lawrence Radiation Laboratory at Berkeley.

The particle, whose existence was predicted several years ago on a theoretical basis, has a lifetime of a hundred billionth of a second. It plays an important part in the structure of protons and neutrons, the basic "building blocks" of atomic nuclei. Like the neutron, it has no electrical charge.

SPACEMAN SAFE AFTER 15 MIN FLIGHT

Exactly according to plan, Commander Alan Shepard, of the United States Navy, was launched 115 miles into space from Cape Canaveral this morning and was successfully recovered from the Atlantic 15 minutes later, 302 miles down the missile range. He was in constant communication with the ground during his flight, and was in good physical condition, having already climbed out of the space capsule when he was picked up by helicopter. His first comment was: "Boy, what a ride".

MAN ON THE MOON BY 1967

The tentative timetable now being prepared for the landing on the moon begins this year with two unmanned rockets carrying instruments which will survive the impact on the moon's surface and be able to transmit information to help to decide on a suitable area for the manned space craft to land.

The plan to make a "soft landing" on the moon will be advanced so that during 1963 at the latest there should be mechanical instruments on the moon able to prepare samples of the moon's surface and return them to earth.

In the meantime other satellites and instrument-carrying rockets will continue to record information about radiation, meteorite activity, and other hazards that will be found during the journey from earth to the moon. Work on the three-men Apollo space craft will also be continuing, and perhaps in 1964 this capsule will be tested by being sent into orbit round the earth, with one or more men on board, for more than a day.

After the Apollo has successfully completed this test it will be sent with its full crew of three men for a journey round the moon without landing. It may orbit the moon a number of times to let the astronauts study its surface and check the selected landing areas.

Finally, during 1967, if all goes well, but by 1969 at the latest, the landing of men on the moon will take place. The journey is expected to take about five days. The men will stay on the moon for a few days and then make the return trip, which will also take about five days.

THE SOLAR WIND

The Massachusetts Institute of Technology said here today that a space probe wh ch was sent up last week from Cape Canaveral, Florida, disclosed the existence of a "solar wind" sweeping into space from the sun at a speed of millions of miles an hour.

Dr. Bruno Rossi, of the institute, who worked with the National Aeronautics and Space Administration on the probe, said that the "solar wind", believed to be the result of evaporation of burning gases from the sun, could provide the power for future space vehicles.

NEW SATELLITE LAUNCHES

The United States successfully put the Tiros weather-detection satellite into orbit this morning and followed it a few hours later with an experimental Midas satellite whose infra-red sensors are supposed to pick up the heat of missile exhaust flames. The two wide-angle television cameras in the Tiros were reported to be sending back excellent pictures of cloud formations in the lower atmosphere, which will be of considerable help to meteorologists, particularly in the early detection of hurricanes.

FOSSIL REMAINS FOUND IN METEORITE

The meteorites from which samples were examined were the Orgueil meteorite, which fell in southern France in 1864, and the Ivuna meteorite, which fell in an arid region of central Africa in 1938. Both remained out of doors for only a few hours after they had fallen.

Interest in carbonaceous meteorites as a possible source of evidence of the existence of life other than on the earth has increased greatly in recent years.

Of five types of "organized structure" which they describe—and in three cases illustrate—four are said to be similar to known terrestial species, but not identical. The species they resemble are small single-celled animals—Dinoflagellates or Chrysomonads, which live only in sea or lake water. The fifth is unlike any known terrestrial organism.

TEKTITES

The origin of the small glassy objects known as tektites that are found strewn on various parts of the earth's surface presents a problem in detection, with enough clues to provide scope for argument and enough uncertainty to sustain it. What can be said definitely about tektites is that they have passed at least once through the atmosphere and been long enough outside to have been solid when they began their descent. They are therefore in some sense space-travellers. Some 30 theories of their origin have been advanced. But in spite of some new arguments in favour of a terrestrial origin that were put forward at a recent meeting of the American Geophysical Union in Washington, the balance of evidence seems to favour the moon.

REACTOR EXPLOSION IN IDAHO

Radioactive forms of gold and copper found at the scene of the atomic reactor explosion at the Idaho atomic research centre on January 3 have established that the blast was caused by a "runaway" nuclear chain-reaction, the Atomic Energy Commission reports.

COLLIDING PARTICLE BEAMS

The Brookhaven National Laboratory in the United States has already a machine which is marginally bigger than the Cern one (at Meyrin near Geneva)—though, when it began operation early last year, this was the biggest in the world. The Russians are building a 50,000 MeV machine, and still bigger ones are projected: 100,000 or 300,000 MeV in the United States, and 70,000 MeV and possibly more in the Soviet Union. Moreover, Cern is proposing to set up a joint study group with the United States and the Soviet Union to look into the possibilities of machines that will accelerate particles to energies of half a million or even a million MeV. Apart from cost, there seems to be no insuperable difficulty in going to at least 10 times the present size.

Mere scaling up, though already certain to be carried some way, is open to another objection besides cost. Progressively less benefit is obtained, because of an effect due to relativity. Even at lower energies than those discussed, the particles accelerated are already moving at close to the speed of light and most of the extra energy imparted to them in the later stages of "acceleration" is spent in increasing the mass of the particles. Their use in experiment is, however, in collisions with other nuclear particles, in fixed targets, which the accelerated particles are caused to hit. From the point of view of the particles it is irrelevant which of them is moving and which fixed. What matters is the speed of each of them relative to their common centre of gravity.

For this reason there has been increasing interest for a number of years in the possibility of bringing about collisions between opposing beams of particles. If particles from, for example, the present Cern accelerator could be stored in a ring system of comparable size, and caused to make head-on collisions with newly accelerated particles, the effective energy of collision between particles in the two beams would be greater than could be got with a fixed target from the biggest machine yet dreamed of.

1962

NEW RADIOACTIVE DATING METHOD

The new method of dating iron meteorites was outlined at the meeting by a German and a Swiss scientist. By measuring rhenium-187, an isotope which loses one-half of its radioactivity in 48,000 million years, they had confirmed with a great degree of reliability the assumption that stone and iron meteorites, as well as the crust of the earth, had been formed "approximately at the same time, some 4,000 million years ago."

GLOBAL WEATHER WATCH

In a report issued today the World Meteorological Organization recommends to the United Nations the establishment of an international weather watch, based on meteorological satellites and a system of world and regional centres, to develop existing capabilities of weather forecasting. The report has been made in response to a resolution on the peaceful uses of outer space, unanimously adopted by the United Nations General Assembly last December.

Satellites now make it possible, the report says, to observe clouds and other atmospheric features over the whole world, and the observations, combined with those obtained on land, on board ships and aircraft, and with rocket and balloon observations, should enable the detection and tracking of every major storm.

Satellites, the report points out, provide an observational platform outside the atmosphere from which the following data are obtainable: photographs of the earth's surface and of cloud cover, solar and terrestrial radiation measurements, and, in the future, radar and other observations.

The proposals provide for cooperative efforts by member countries directed towards the rational use of artificial earth satellites. Coordination of types, scope of observations, and orbits is recommended.

COL. GLENN IN SPACE

Colonel Glenn, whose food on the flight consisted of two toothpaste-type tubes filled with apple sauce and beef stew, said today: "I believe that any type of food can be eaten, as long as it does not come apart easily or make crumbs." This was one of the lighter moments in a day devoted largely to a mass of scientific data about the capsule, the pilot's training, reactions, and physical condition, and the roles played by others.

The theme of the day was that Colonel Glenn had proved the value of having a man in space, and it became clear that had the spacecraft been entirely automatic he might never have returned safely to earth. The fact that he was able to take over control, turning what might have been serious failures into minor nuisances, established beyond question that man's role is important in space, and augured well for more ambitious two-man or three-man flights in the future.

US SPACECRAFT HITS MOON

A few hours earlier it had been announced that the United States had hit the moon for the first time with its Ranger IV spacecraft, which was launched last Monday. In addition two satellites were launched from Point Arguello, California, by the United States Air Force, making it one of the busiest weeks in the field of space research.

The Ranger IV hit the far side of the moon, and, although its instruments were not working properly, it was described as a tremendous guidance success. " It is the first step towards man eventually walking on the moon ", said Mr. Clifford Cummings, of the joint plotting laboratories at Goldstone, California.

A NEW FUNDAMENTAL PARTICLE

Brookhaven National Laboratory at Upton has discovered one of the last remaining undiscovered particles in nuclear physics, the anti-Xi⁻. The particle is one of the 30-odd known fundamental particles.

Its life is only one 10,000-millionth of a second and the scientists had to examine 34,000 photographs before they were sure they had identified it. The announcement of the discovery, which physicists consider important because it gives further proof of the basic theory that every particle has its anti-particle, was made simultaneously in the United States, Switzerland, and France. Teams working at Cern, the nuclear research centre in Geneva, at the French centre at Saclay, and at the Ecole Polytechnique in Paris all cooperated in the research.

COMMUNICATIONS SATELLITE

Telstar, the communications satellite by means of which it is hoped to transmit the first live transatlantic television pictures, went into orbit successfully this morning from Cape Canaveral, Florida.

The satellite, which weighs 170lb., was placed in orbit by a three-stage Delta rocket and was reported a few hours later to be travelling once round the earth every 157.8 minutes at heights of between 593 and 3,502 miles. It is the first commercially owned satellite ever to have been put into orbit

It is expected to be rather more than a week before the first formal transatlantic programme is conducted. The American television networks are planning a 12-minute transmission of news events from various parts of the country to Europe and a similar programme will be sent in the reverse direction. The United States Information Agency has made arrangements for telephone conversations between people in 20 American and 20 European cities.

THE OPTICAL MASER

Physicists in two British laboratories have talked over a short distance by means of a beam of infra red radiation—using it in the same kind of way as radio waves as a means of communication—so repeating a demonstration given rather more than a year earlier at the Bell Telephone Laboratories in the United States.

This was reported at a discussion meeting held by the Royal Society in London yesterday by Dr. P. I. R. King, of the Services Electronics Research Laboratory at Baldock, as a late addition to a contribution which he made jointly with Dr. J. H. Sanders, of the Clarendon Laboratory, Oxford.

The device used is known as an "optical maser". American scientists have also operated a device giving visible light waves instead of infra red waves in a form that could be used in the same way.

Practical applications—apart from possible space communications over great distances—appear a little remote at the present stage, although ability to produce light waves which are all in step with each other as are waves from a radio transmitter has many possible applications in research, some of which are already being studied.

DOLPHINS TO AID SPACE RESEARCH

The United States Space Agency today awarded a contract for investigation of the dolphin as part of a study in communication between humans and "other species" which may live on distant planets.

1963

Objectives outside the Solar System were quite beyond present possibility, since the relative speed of sound waves, our only form of communication at present, gave information of conditions just one hundred thousand years out of date.

US DISCOVERY OF NEW PLANET

The existence of a planet revolving round a star is determined by recording the path of the star. If its track is uniform the star does not have any satellite. If the star wobbles from its expected course this indicates the presence of some unseen mass accompanying it.

The Sproul observatory has been photographing Barnard's star every year since 1938, and also from 1916 to 1919, and the series of more than 2,000 plates thus accumulated has enabled the observatory to make a precise determination of the motion and distance of the star. By measuring the variation of the star from its predicted path the observatory has calculated that its accompanying planet is half as large again as Jupiter, which is the largest planet in the solar system and has a diameter of 88,000 miles, 11 times as large as the earth.

LASER OBSERVATIONS OF THE MOON

Russian scientists have bounced a beam of concentrated laser (Light Amplification by Stimulated Emission of Radiation) light off the moon from a Crimean observatory, Tass claimed today. The report said the laser generator, which emits powerful, narrow pulses of light, was fitted at the focal point of a 2.6 metre (about 102in.) telescope. After several attempts, a special receiver on the telescope picked up the light pulses reflected back " greatly weakened " off the lunar surface.

ANTI-MATTER

The discovery of a missing link in matter—named Anti-xi-Zero—has given scientists another clue to the make-up of galaxies in space. The discovery of a long-missing member of the family of anti-particles—the direct opposite of ordinary matter known on earth—was announced by the Atomic Energy Commission here.

Anti-particles are created by bombarding particles of matter with other kinds of particles, duplicating the work of cosmic rays in the atmosphere. Virtually all anti-particles have a life measured in a fraction of a second. Scientists believe the discovery lends new weight to the theory that elsewhere in the universe whole galaxies may be made up of anti-matter, the A.E.C. said.

One scientist outside the A.E.C. said that powerful radiation from distant galaxies that had long puzzled astro-physicists could conceivably be explained on the basis of clouds of matter colliding with clouds of anti-matter.

FIRST WOMAN IN SPACE

Miss Tereshkova, a junior lieutenant, was probably launched from Baikonour, in Soviet Central Asia, like the other cosmonauts. Her constant training with the cosmonaut squad since last year pulled her through the important minutes of entering into the state of weightlessness and going into orbit.

Mr. Alexei Lenotyev, a Soviet psychologist, quoted by Tass, said the orbiting was not only remarkable because a woman had been put into space. She was younger than the other cosmonauts, had less experience and lacked the training and the familiarity with danger of the pilots who became cosmonauts.

Moscow television, showing pictures of her before she began her space training, said she had always been " full of the joy of living ". In most of the pictures—in the back row of the class at school, with her brother and married sister, or among parachute friends—she was smiling

FIRING A CANNON ON THE MOON

The cannon, which will be landed on the moon from a spacecraft when the United States space programme has mastered the art of getting one there, has already been tested in the Mohave desert. It is designed to fire a projectile that would travel for about a mile, carrying behind it a series of small explosive charges with time fuses. Miniature seismometers would then record details of surface and underground tremors and transmit them to earth.

THE ROLE OF THE BREEDER REACTOR

While the advent of the commercial fast breeder cannot be expected before the middle seventies both countries are pursuing nuclear energy competitiveness with conventional power, against the odds of high building costs which still add up to about two-thirds of the cost of the unit sent out, while the same item represents roughly one-third of the cost of conventional power. The latter is further favoured by the collapse of conventional fuel prices in world markets—not entirely unconnected with the long term prospects of atomic energy.

Demonstrations of relative competitiveness in Britain and the United States are bound to have a substantial impact on the use of nuclear energy for civilian purposes throughout the world. The next step forward is expected to come from the operation of such advanced " converter " reactors as the British Advanced Gas Reactor, the Canadian heavy moderate reactor, or any of the many projects now in the experimental stage in the United States. They are all meant to bring about further reductions in electricity costs and set the scene for the full commercial development of the fast breeder power plants. In the presence of such ultimate targets, arguing about one-half of a thousandth of a dollar or about one-tenth of a penny in the present cost of nuclear power may seem a purely academic exercise.

UNSTABLE HYPERNUCLEI

In nuclei which show this kind of instability, the particle whose place is taken is a neutron. The particle that takes it is known as a lambda particle, first discovered in cosmic radiation by Dr. G. D. Rochester and Dr. C. C. Butler at Manchester University in 1947. More specifically, it can be described as a " lambda nought hyperon "; the nought denotes that, like the neutron, it has no electric charge, while " hyperon " is a general name for the heavier unstable particles, those with masses greater than those of the proton and neutron. Whereas the mass of the neutron is 1,878 times that of an electron, that of the lambda is some 2,180 electron-masses.

Nuclei containing a lambda particle are known for obvious reasons, as hypernuclei. They are usually produced by the breaking up or fragmentation of a larger nucleus after interaction with a suitable high-energy particle—and so are referred to also as " hyperfragments ". They are unstable in a way which differs radically from the more familiar kind of instability shown by the nuclei of the radioactive elements, for the latter is determined by the composition of the nucleus as a whole.

Much of the interest of hypernuclei derives from the fact that they contain one, distinguishable, particle which differs from any other in the nucleus. That the " lambda nought " does, in fact, differ in important respects from the neutron (i.e. not only in mass) is most clearly shown by the existence of hypernuclei analogous to a non-existent isotope of hydrogen— hydrogen-4. If hydrogen-4 existed, it would consist of one proton and three neutrons. But there is a theoretical rule which makes this an impossible combination of particles, except in so unstable a form as to be undetectable. If, instead of three neutrons, there are two neutrons and some further particle that is different, this theoretical ban does not apply.

SPACE NEEDLES RELEASED

Radar contact confirmed today the successful release in space of about 400 million tiny copper needles, thinner than a human hair, in a United States Air Force experiment to study a new method for radio communication.

The needles are expected to spread along a circular orbital path, about 40,000 miles in circumference, to form a thin, narrow ring or belt round the earth. Fifty pounds of wire were used.

PECULIAR GALAXIES

Roughly 99 per cent of the galaxies in the universe fit neatly into one of the two main groups of elliptical and spiral galaxies that form the basis of Hubble's classification system. Those in the remaining 1 per cent were at first an embarrassment. They are of much less symmetrical shape, or show freakishly distorted spiral arms; others are of more normal appearance but show abnormal concentrations of interstellar dust, or of hot interstellar gas in the very centre of the galaxy.

Recently these " freak " galaxies have attracted much interest. If, during the many millions of years in which a galaxy slowly changes from one form to another, one evolutionary stage is passed through rather rapidly, the number of galaxies we see in that rapidly changing state at the present instant in time will be very small. Rare types of galaxies may thus be the rapidly changing ones, though even these changes are too slow to be seen during a human lifetime. Peculiar galaxies are, therefore, of interest because they may tell us how galaxies evolve.

It was at first accepted that these objects must be true stars in the Galaxy, though there were serious difficulties in this explanation: in particular the spectra were unlike spectra of any known type of star, and defied satisfactory interpretation. It was later found that the features in the spectrum of some of the objects could be explained if they were supposed to be known emission lines displaced towards the red end of the spectrum by an amount proportional to the wavelength. There are various possible causes of such a displacement; the most plausible at present seems to be that it is the same phenomenon that causes the redshift in the spectra of the galaxies. If this is correct, then the redshift gives the distance of the objects.

Their inferred intrinsic brightnesses and sizes seem fantastic even to an astronomer accustomed to extreme sizes and energies: the objects are optically brighter than the brightest known galaxies by factors of 10 or even 100 times, yet they are no bigger than a few thousand light years across—only a tenth the size of a normal galaxy. This size is an upper limit; there is evidence that some of the objects vary in brightness slightly in the course of a year or so, and this would require a significant fraction of the whole light to come from volumes of matter that are not more than a light year in diameter.

Even a name for these surprising objects has not yet been agreed upon: quasi-stellar radio sources (or QSRSs) seems clumsy, and the derived word *quasars* unattractive.

VENUS 'SPINNING BACKWARDS'

Radar observations from two observatories using three different techniques have all shown that Venus is slowly spinning backward, it was reported here today. Consequently, after several years of debate, astronomers are beginning to accept the proposition that Venus is really spinning against the traffic pattern of the entire solar system. Why it should do so is a mystery. It has not been possible to observe its spin visually, as its surface is obscured by clouds.

'SIGNALS FROM A PLANET'

Two Soviet writers claim that light signals from highly developed beings on another planet had reached earth at least three times, the Tass news agency reported today. According to the writers, Generikh Altov and Valentina Zhuravleva, the signals came in 1882, 1894 and 1908.

The last was in answer to the eruption of the volcano Krakatao, on an island in the Sunda strait between Sumatra and Java, on August 27, 1883, which the beings mistook for a signal from earth, they said in an article in the Leningrad magazine *Zvezda* (Star).

Altov and Zhuravleva think the signals came from one of the planets of the sixty-first star of the constellation of Cygnus (The Swan). They believe that plasma erupting from Krakatao sent into space a powerful radio, and perhaps light, signal which was received on the star 11 to 12 years later.

CLASSIFYING FUNDAMENTAL PARTICLES

An intensive search for the Omega-minus particle started several months ago at a number of high energy physics laboratories. At Brookhaven, where the particle was first demonstrated, the 33,000 million electron volt alternating gradient synchrotron was used by a team of 33 physicists, including three from Syracuse University and one from the University of Rochester.

The search for a theoretical way in which these particles could be linked together in atoms has occupied mathematicians and physicists around the world in recent years. Physicists know the physical characteristics of a ' zoo " of elementary particles, but they cannot classify the particles and their relations to each other as animals have been classified.

Three years ago, Dr. Gell-Mann and another scientist, working independently, proposed to classify some of the well-known particles in groups of eight. The attempt was to find an orderly arrangement of particles just as the elements are classified into an atomic table. With the atomic table at hand, chemists can know in advance some of the ways in which chemical elements are going to react with each other and know some of their properties. In a similar way, physicists could use a classification of the elementary particles.

INTERNATIONAL YEAR OF THE QUIET SUN

Yesterday saw the start of the International Years of the Quiet Sun, an international cooperative enterprise in which at least 64 countries are taking part. The aim of the scientific programme of the years is to extend knowledge of solar-terrestrial relations and the way in which radiation from the sun interacts with the earth's atmosphere and its magnetic field.

This interaction leads to the formation of the ionosphere, the electrically charged layer which makes long-distance short-wave radio communication possible; it also causes auroral displays in high latitudes and disturbances in the earth's field frequently referred to as magnetic storms.

As the name of the enterprise implies, the 11-year sunspot cycle is reaching the point of minimum activity; during the time before, during and after the actual minimum, studies will be made of the properties and behaviour of the neutral atmosphere and of the charged ionosphere in relation to particular events in the solar chromosphere and on the sun's surface.

ROCKET SEARCH FOR EXPLODED STARS

Dr. Herbert Friedman, an astronomer from the Naval Research Laboratory, said today that the experiment was being conducted to try to find out whether mysterious celestial X-rays discovered in space last year during another rocket flight, were generated by neutron stars or the ultimate remnants of super-novae.

Dr. Friedman said that the X-rays were discovered emanating near the Constellation Scorpio and the Crab nebula in widely separated areas of the Milky Way. He said it was believed that they might be the invisible fingerprints of small but extremely hot stars made up entirely of neutrons—stars that are the final remnants of huge stars that exploded more than 1,000 years ago. It is believed that such neutron stars might have a diameter of no more than 10 miles, but have more than one thousand million tons of matter a cubic inch of volume, and burn with temperatures exceeding one thousand million degrees.

RADIOACTIVE FUEL LOST IN SPACE

High up in space off the West African coast the United States has lost nearly $1m. worth of plutonium 238, one of the most toxic commodities known to man, but the Atomic Energy Commission said this weekend that it posed no health hazard to the population of the world. The plutonium was being used for a S.N.A.P-9A generator, designed to provide power for a navigational satellite, but an operator failed to flick the right switch during launching, and the satellite did not go into orbit.

1965

ORIGIN OF THE MOHOLE PROJECT

With some obvious exceptions—among them Newton, Einstein and Charles Darwin—it is largely chance that determines whether the name of a former scientist shall become widely known at some later date—and, if so, whether as a person, a name, or part of a name. One who seems well on the way to the last is Dr. Andrija Mohorovičić, a former director of the Meteorological Observatory at Zagreb in Croatia. The first part of his name is assured of wide circulation by the Mohole project—the scheme for drilling a hole through the crust of the earth to the mantle beneath it. Even should the first attempt fail the experiment will be done sooner or later, and the name of Mohorovičić will become familiar in whole or part to many more millions of people than have heard of the man or his work.

An opposite example is provided by the late Sir Edward Appleton. He will remain known to scientists for all time as the discoverer, first, of the Kennelly-Heaviside region of the ionosphere—whose existence Kennelly and Heaviside had predicted—and later of a region above it, which was known for a time (but never officially) as the Appleton layer. In the event, the first became known for scientific purposes as the E region and the second was found to be divided into two regions—F_1 and F_2—while yet another region—the D region—could be distinguished at times at a lower level. The result, predictably, is that, whereas the term, Heaviside layer, is still sometimes used the term Appleton layer has dropped out.

The case of Mohorovičić is to some extent parallel. From an intensive study of records of an earthquake which occurred on October 8, 1909, some 25 miles to the south of Zagreb, he discovered a discontinuity in the properties of the rocks of the earth which is now recognized as a major boundary, that between crust and mantle. Earthquakes must have seemed a natural subject for interest: Zagreb had suffered severely on two, then recent, occasions—in 1880 and again in 1901.

Usable records of the 1909 earthquake were obtained at observing stations throughout Europe, and Mohorovičić undertook their analysis. This led him to the conclusion that waves of alternate compression and expansion in the direction of propagation had travelled by more than one route from earthquake to observatories. At short distances the first wave to arrive travels at a speed of around 3½ miles a

second. At a distance of about 105 miles, this wave is overtaken by another which is of the same type but has travelled by a longer route at higher speed—a little over five miles a second. A critical point in this discovery was the fact that over a considerable range of distances, both waves could be distinguished in records obtained at individual observatories.

The depth of this discontinuity—the boundary between the crust and the mantle—varies between about 20 and 25 miles beneath the continents, somewhat less than Mohorovičić thought, and it is less again beneath the oceans, where the Mohole is to be drilled. Before the Mohole project, the boundary was known to geologists and geophysicists as the Mohorovičić discontinuity. Since the Mohole idea was first mooted in 1957, it has tended to become the Moho.

OBSERVING NEUTRINOS

Scientists attached to the Case Institute of Technology in Ohio have confirmed predictions that the earth is being bombarded by a mysterious particle, known as a neutrino, that weighs nothing, travels at the speed of light, and could pass through 100 million miles of lead.

The existence of the particle was surmised by the late Enrico Fermi, and some evidence was discerned earlier in the tracks of sub-atomic particles emerging from an atomic reactor. Yesterday Dr. Frederick Reines announced that over a nine-month period seven natural neutrinos were detected by a device at the bottom of a gold mine in South Africa.

BLUE GALAXIES

Observations in California have identified what seemed to be paradoxically young stars in the halo of the Milky Way as enormous objects of unmatched brilliance which occur all over the universe. Mount Wilson and Mount Palomar observatories announced the discovery of this "major new constituent" yesterday. The objects resemble the recently discovered "quasars", except that they do not emit strong radio waves. They are much more numerous, outnumbering the quasars by about 500 to one; there may be as many as 100,000 of them within the range of the largest telescopes.

EXPERIMENTS RULE OUT NEW FORCE

The law to which it now seems that exceptions must be admitted asserts, in effect that if all the velocities in a system were reversed in an instant the previous history of the system would be retroversed, as in a film that is run backwards. The Brookhaven experiment left the position (slightly simplified) that either this law did not hold in the particular interactions of particles that the experiment was designed to study or, if the law did hold, there must be some new kind of interaction—that is, a new kind of force. The Cern experiment confirms the main result of the one done at Brookhaven, but is said to exclude the second alternative.

ANTI-PROTONS

In an all-night session at the west German Electron Synchrotron Research Institute in Hamburg last night, the 11-member team for the first time produced "anti-protons" by bombarding protons with light, confirming a 20-year-old theory, they claimed.

A proton is a particle of the nucleus, or central core, of an atom. An anti-proton is a particle with an opposite electrical charge to the proton. Until now, anti-protons have been created only by bombarding protons with protons—that is, bombarding matter with matter.

INFRARED ASTRONOMY

A new kind of telescope whose mirror "jiggles" 20 times a second has disclosed that the sky, when viewed in the deep infra-red, dazzles with a multitude of strange bodies. Some, at these invisible wavelengths, are brighter than anything in the sky except the sun, the moon, and perhaps a few planets. Yet with the world's largest telescope, the 200in. reflector on Mount Palomar in California, most of them are invisible.

The new objects have both excited and perplexed the astronomers. Some of them seem to be as cool, on their surfaces, as a planet like Venus. The new-type telescope on top of Mount Wilson, overlooking Pasadena, California, has already surveyed more than half the sky visible from that site, and from 400 to 1,000 objects cooler than 1,500° Fahrenheit have been found.

MARINER SPACECRAFT ON MARS

What one of Mariner's experiments indicated was that the surface pressure on Mars is something between 10 and 20 millibars, or equivalent to pressures at between 102,000 and 93,000 feet above the earth. It also reported that the ionosphere of Mars was less dense than the earth's.

Among the conclusions which may be drawn are that the lower the atmospheric pressure, the more difficult it is to imagine life, such as we recognize on earth, existing on Mars. It is also possible that the high winds which have been assumed to blow over Mars because of dust storms observed from earth do indeed blow very strongly.

By mid afternoon here, scientists at Pasadena had received the first three pictures of the Martian surface, the first one of which, published last night, showed a broad, generally featureless desert with a few low hills round the edge.

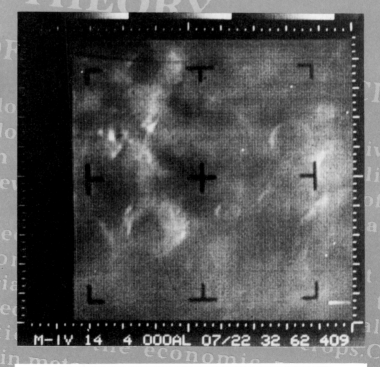

M-IV 14 4 000AL 07/22 32 62 409

MARTIAN SURFACE FROM MARS FLY-BY

MATTER AND ANTI-MATTER

The recent discovery at Columbia of the antideuteron, Dr. Paolo Franzini said at a press conference, confirmed that entire systems of planets and stars composed of antimatter might exist. If antimatter had been the mirror image of matter, there would have been no way of identifying such heavenly bodies short of annihilatory contact.

But now, if antimatter does exist, it would be possible to ascertain whether a cosmic body is a world or antiworld by comparing its eta meson with any of its substances and communicating the results to earth. If the energetic particles of its eta mesons have the same sign as the heavy particles of its substances, it is matter. If not, it is antimatter.

PLANETARY INTERIORS

It is with the object of drilling deep enough in an ocean area to sample the mantle that the American Mohole project—now in its final stage near Hawaii—has been planned. Geochemists will argue at the conference that particularly dense silicates, brought up through cracks in the crust, are just such a sample, in the form of olivine nodules, and one much more cheaply available.

both compression and extension occur and this has led to the hypothesis that the mantle is not as solid as seismologists had thought, but that great convection currents move within it—transporting heat from the deep interior, compressing the crust where they descend, and causing tension where they rise. These currents, moving a few centimetres a year, can explain continental drift. Mars and the moon give no indications of such vigorous internal stirring, but their interiors may not be so dead as used to be thought. Careful study of telescopic and Ranger photographs shows that small volcanic cones and faults exist, and opinion is hardening that the maria are lava which could have issued from the interior.

SOFT LANDING ON THE MOON

Moscow television tonight displayed the circular panoramic picture of the moon's surface, showing what it would look like to gaze round from Luna 9 space ship. The picture was apparently made up of several photographs taken by the television camera of Luna 9 on Saturday evening.

A commentator said that the camera was inclined at an angle of 30° so that in one direction only sky was visible and in another only the pock-marked lunar surface close to the ship.

As the camera revolved, parts of the moon craft appeared in the picture, including aerials and two small objects which showed white against the black lunar sky and appeared to be suspended from a higher part of the ship. The picture also showed two dark objects, described as parts of the ship jettisoned as it settled gently on the moon's surface.

The new picture showed that Luna 9 had settled in a relatively flat area near the lunar equator, in the east of the waterless Ocean of Storms.

SAFEGUARDS FOR NUCLEAR REACTORS

Mr. Foster said the United States was offering the International Atomic Energy Agency access to a commercial nuclear fuel processing plant at West Valley, New York.

This would allow scientists to test various safeguard techniques so that they would be ready for the day when supervision was applied universally.

It was highly pertinent to the objective of seeking to ensure that what was peaceful today remained peaceful tomorrow.

He explained that such plants reprocess the by-products of a nuclear power reactor, to recover uranium and plutonium. The need for international safeguards on all peaceful nuclear installations was becoming more and more evident, if the parallel development of nuclear weapons was to be checked.

'COHERENT' RADIATION

The use of all-in-step—or, as it is usually described, " coherent "—radiation to carry information can be effected in principle at any wavelength. Its use to provide a new means of taking pictures is restricted to wavelengths that are short compared with detail in the object to be depicted. This illustrates the general point that ability to produce coherent light waves with lasers opens the way to applications that could not be envisaged when the only coherent radiation available was, for practical purposes, that from radio transmitters.

The advent of the laser changed the whole outlook. The pioneer experiments have been done in the United States, the home of the laser. The methods used to snatch an image, as it were, from the empty air can be described adequately only in terms of optical theory. It is enough here merely to assert again that they depend on timing —in the sense used in this article—which is why an abundant source of coherent light waves was necessary for progress. The new age in image construction by " holography ", as it is called, began with a report published in 1963 by E. N. Leith and J. Upatnieks of the University of Michigan. A year later they achieved a further and striking improvement. Now more than 100 laboratories in the United States are working on the method.

One advantage is that, in principle, more detail can be recorded on a given area of fine-grain photographic plate, than is possible with the best lenses. Another is that, with individual viewing, three dimensional effects are obtained—and that in moving pictures, not only in stills.

INTERSTELLAR COMMUNICATION

He calculated that if the nearest civilization were 10,000 light years away, a reasonable assumption, we would need an aerial as large as the earth to catch its signals, but the consequences of success in interstellar communication would be enormous.

" It is emotionally destructive ", he observed, " to find man clinging to the thin skin of a slag heap, revolving about a middle class star, one of 200,000 million located in a remote suburb far from the centre of our own Milky Way. But more violent in emotional impact than the number and scale of things is the question of the possible presence and high evolution of life and mind."

X-RAY GALAXIES

A five-minute Aerobee rocket flight has discovered the first two " X-ray galaxies " and doubled the number of known X-ray sources in the universe. United States Navy scientists announced today.

The two galaxies—Cygnus A and M-87—are among the strongest radio sources in the sky. The X-radiation they emit was found to be 10 to 100 times greater than all the radio and light energy they send out.

The biggest solar flare for six years reached its peak today over the American city of Boulder, Colorado, the Institute of Telecommunications Sciences reported.

. . . no imagination is needed to work out the properties of superdense stars. So far no superdense star has been identified.

1966

FORMATION OF STARS OBSERVED

The intense source of infra-red radiation discovered this year in the Orion Nebula seems now to be accepted as a group of stars in the course of formation from a cloud of gas and dust. This is clear from a detailed calculation of the expected course of events carried out by Dr. William Hartmann, at Arizona University.

The discovery of this infra-red source is thus the most direct evidence so far of the process of star formation in the Galaxy. It appears to consist of about 100 newly condensed stars, embedded in an opaque cloud through which visible radiation cannot penetrate.

NEW ATOM IS HEAVIEST KNOWN

The Lawrence radiation laboratory at Berkeley, California has produced the heaviest atom so far identified—an isotope of the artificial element, mendelevium. The isotope, Mendelevium 258, has 101 protons and 157 neutrons, and the remarkably long half-life of two months.

DEATH OF COSMONAUT

Colonel Vladimir Komarov perished today on board Russia's new spaceship Union I, when the parachute strings twisted after the spacecraft had successfully reentered the atmosphere. His flight, clearly timed to coincide with the anniversary of Lenin's birthday, thus ended in tragedy.

PULSING LASERS

The discovery that the light output from ruby and neodynium lasers contains brief flashes lasting for a millionth of a millionth of a second is striking, although not by itself a great surprise. For some months it has been likely that light from lasers like those made of ruby, which seems to consist of flashes lasting a millionth of a second or even a fraction of that, might really consist of flashes of much shorter duration.

The group also says that the appearance of these rapid pulses of light is an intrinsic property of the solid lasers. It suggests that random exchanges of energy in the form of light between molecules in the laser can build up to produce these exceedingly brief pulses, and that this will happen spontaneously.

By ordinary calculations, the power of the radiation in these rapid pulses is enormous. For the brief interval for which pulses of light were obtained from the neodynium laser, no fewer than 10,000 megawatts of power were being transmitted by each square centimetre of the laser during the picosecond for which each pulse lasted.

COSMOLOGICAL X-RAYS

The United States Naval Research Laboratory has reported the first evidence that X-rays are produced as far away as the rim of the universe. An Aerobee rocket, launched on May 17, has brought back evidence that they come from quasars, one of which is thought to be 1,500m. light years from earth.

ASTRONAUTS KILLED

On the day that the three astronaut victims of last Friday's fire in an Apollo spacecraft at Cape Kennedy were buried, a similar fire today in a spacecraft simulator at San Antonio, Texas, killed two airmen.

As in the Cape Kennedy disaster, the two men were working in a 100 per cent oxygen environment. They were in an altitude chamber in which pressure simulated a height of 18,000 feet.

BIG-BANG THEORY

If the "big bang" theory that the universe came into existence some 10,000m. years ago by the explosion of a fireball of hydrogen is correct, one would expect that the oldest stars would be formed roughly of two-thirds of hydrogen and one-third of helium.

The theory is that within the larger stars nuclear reactions generated successively heavier elements up to the atomic weight of lead. The heaviest natural elements may have been formed later, in Dr. Fowler's view, by "little bangs", violent explosions while the universe was still young.

Although the original fireball could not have lasted long enough to produce heavy elements, it is hard to see why it should not have converted some of the original hydrogen into helium.

One of two possible explanations was that helium, being slightly heavier than hydrogen, had sunk into the centre of the stars and become indetectable in their light. The other was the existence of a cloud of neutrinos which would interact with neutrons to produce protons and electrons, and they in turn would combine to form hydrogen atoms and limit the formation of helium. "Neutrino astronomy", Dr Fowler observed, is just getting started.

RELIC OF THE BEGINNING OF THE UNIVERSE

A striking refinement in the study of the bath of radiation known as the cosmic microwave background has been carried out at Princeton University by a group of physicists under Dr. D. T. Wilkinson.

By the use of equipment designed for the precise measurement of this radiation, the group has been able to fix the background temperature to within roughly a fifth of a degree or so of 2.7° absolute. So far the uncertainty in the measurement of this temperature has amounted to roughly 1°.

The cosmic microwave background radiation is now considered to be a physical relic of the intense pool of energetic radiation which would have filled the universe at the beginning of its expansion. At that point, perhaps 10,000m. years ago, all the matter and energy now visible would have been concentrated into a small space.

With the passage of time, and with the continuing expansion of the universe, the primeval radiation has been degraded into a much less energetic form corresponding to a temperature only a few degrees above the absolute zero.

1967

FIVE SITES FOR LANDING ON THE MOON

Five areas on the surface of the moon have been chosen as a possible site for the landing of a manned spacecraft by the United States National Aeronautics and Space Administration. The areas have been selected by a committee of scientists who laid down stringent conditions about the choice of landing zones.

Two of the sites are in the Sea of Tranquillity, a third in the Central Bay and the fourth and fifth in the Ocean of Storms. Thirty areas were examined as possible landing zones, and a short list of five was drawn up after studies of information from the Lunar Orbiter series of satellites and the Surveyor craft that soft-landed on the moon.

THE HUNT FOR QUARKS GOES ON

A few years ago it was soberly suggested that the understanding of nuclear particles would be easier if these in turn were made up of sub-particles. The name "quark" was given to these hypothetical constituents of the nucleus, after a quotation in *Finnegan's Wake*. Inquiries yesterday failed to identify the precise sentences.

No one knows if quarks really exist, but if they do they must be massive compared with other atomic particles.

Another interesting property of quarks is their charge. Several varieties of quark have been suggested, some having an electric charge one-third that of the electron and others with a charge two-thirds the electronic charge. No other atomic particle has a fractional charge, so this should be a good method of distinguishing quarks. It has been thought that cosmic rays might have sufficient energy to give rise to heavy particles like quarks but no fractionally-charged particles have been found this way.

Many searches for fractionally-charged particles in meteorites, air, and seawater have been conducted, with no success. It is beginning to look as though the elusive quark may exist only in the minds of theoretical physicists.

MACHINES AS LORDS OF EARTH

"The more sophisticated computers become, the more heavily we shall come to depend upon them to help in human decision-making, and nations that do not invest will be at a very severe technological disadvantage."

"There is a real possibility that we may one day be able to design a machine that is more intelligent than ourselves."

The species of superior intelligence could be "morally much superior to ourselves, since there is no reason to build in the ingrained selfishness which we have needed to survive and which leads to so much human irrationality"

PLANS FOR GRAND TOUR OF THE PLANETS

American scientists are charting the course for a "grand tour" by one spacecraft to Jupiter, Saturn, Uranus and Neptune—a journey that could begin in 10 years but could not be repeated for another 179 years.

Some flight plans even provide for such data-collecting voyages to reach as far as the orbit of Pluto, the outermost planet of the solar system.

Opportunities to launch the mission will occur in 1977, 1978 and 1979 with October, 1968, considered the most suitable take-off time. At those times the big outer or Jovian planets will be aligned so that a spacecraft leaving Earth could swing by each one, ricocheting from one to the next on the energy supplied by each planet's gravitational force

EARTH'S MAGNETIC FIELD ORIGIN

Because very little is known about conditions deep inside the Earth, there has been plenty of room for speculation. Even so, it is quite difficult to show, theoretically, that the dynamo theory could work, even if the most convenient assumptions are made. One mechanism which seems to work was put forward in 1958 by Dr. A. Herzenberg.

The real importance of the most recent experiments seems to be that they show that the model can reproduce some of the characteristics of the magnetic field of the Earth. It is well known that the direction of the magnetic field is shifting, which is why magnetic north is not identical with true north. This regular oscillation of the magnetic field is known as the secular variation.

It is also known that at times in the past the Earth's magnetic field has switched right round, so that the North Pole became the South and vice versa. A successful model of the magnetic field has to explain both the secular variation and the reversals of the magnetic field. It seems possible that the laboratory model will be able to do both.

The experiment was to bombard atoms of uranium with electrically charged atoms of helium, carbon. oxygen, neon and argon. Instruments near the uranium target detected atomic fragments produced in the collisions.

Physicists believe that collisions between the nuclei of atoms may take place in two stages. The first step is the formation of what is called a compound nucleus, made of up the bombarding nucleus and the target nucleus. The compound nucleus has only a momentary existence—it almost immediately breaks up into smaller nuclei.

Dr. Sikkeland measured the directions at which atomic fragments flew off the bombarded uranium target. From the measurements, there is evidence that a compound nucleus containing 110 protons is produced in collisions between bombarding argon atoms and uranium nuclei. he says.

GRAVITY SURVEYS BY SATALLITE

PROTOTYPE of a gravitational mass sensor developed by Hughes Aircraft Co. at its research laboratories in Malibu, California, is claimed to be able to measure a gravitational change of 0.0000000000001g. Such a sensor could enable detailed gravity surveys to be made from an aircraft or satellite.

In the prototype, the sensor spins at 16 revolutions per second on a frictionless magnetic bearing within a vacuum jar. As the weighted tips spin, the torsional strains are sensed by a flexible part at the centre and transmitted to instruments which indicate the gravitational strength and direction.

STORING INFORMATION IN CRYSTALS

His research team speaks of writing and reading information into a salt crystal in much the same way as sound is recorded and played back on a tape recorder. Several ways of using crystals have been devised, but they all hinge on the fact that the materials under study can be made to change colour when a beam of electrons is shone on them.

Compounds coloured deliberately have been put in the electron beam to find ways in which the process can be reversed to get a transparent crystal again. This can be done with laser beams. Once ways of both colouring and erasing the smudge from the material had been found it was possible to develop units for storing information.

OPTICAL FIBRE

a cable consisting of several hundred hair-thin glass fibres.

Each fibre would be capable of conveying several thousand times the normal amount of information carried by a pair of telephone wires and several times as much as the larger coaxial metal conductors used in television and multi-channel telephone cables or their equivalent microwave links.

COSMIC-RAY MUONS

Since the thirties, other particles of matter called mesons have been discovered among the cosmic rays, but the muons are conspicuous among the electrically charged cosmic ray particles because they penetrate most deeply through layers of rock. So much is clear from the way in which the Utah experiment was designed to detect muons penetrating through a mountain into the tunnel of a deserted lead mine 2,000ft. below ground.

The interest of this experiment is linked with the question of how the muons come to be present in the cosmic rays. A good proportion of them come from the radioactive decay of the particles of matter called pions (or pi-mesons), but these tend to have comparatively little energy. Others arise from the radioactive decay of the particles of matter called kaons. It is most often supposed that the muons in cosmic rays arise from one or other of these sources.

With that assumption, it is possible to show by calculation that muons of very high energies should be much more often found travelling at large angles to the vertical. The authors of the Utah experiment say they have found comparatively much less than the expected variation of numbers with angle, which is why they suggest that some other process than the radioactive decay of pions and kaons must be responsible for some of the really energetic muons recorded in their experiments.

COSMIC RAYS FROM PULSARS

Recently two pulsars have been detected which probably lie in the centre of supernovae, one in the Crab Nebula and the other in the supernova known as Vela X. This has given rise to the theory that pulsars are created in supernovae explosions. If this is true, the abundance of cosmic rays, which tallies so well with the rate at which supernovae explosions occur, can be equally well accounted for by the pulsars which the explosions are supposed to create.

FOUR NEW PARTICLES DISCOVERED

Four new particles of matter have been discovered in two independent experiments using the proton accelerators at Brookhaven, in the United States, and at the European Nuclear Research Centre (Cern), near Geneva.

The new particles, known by the Greek letter Xi, are believed to fit into the pattern of subatomic matter which many consider to be the key to understanding the origins of the 200 or so nuclear particles that are known to exist.

Physicists carry out this ordering process by looking for regularities or symmetries between groups of particles. The most successful symmetry scheme is the one in which the elusive particle, the quark, is the fundamental building block from which all the heavier particles are made. Preliminary calculations which have been carried out for the new Xi particles show that at least two of them fit into a particular symmetry scheme.

ARMSTRONG SAYS 'ONE GIANT LEAP FOR MANKIND'

It was 3.56 a.m. British Standard Time) when Armstrong stepped off the ladder from Eagle and on to the moon's surface. The module's hatch had opened at 3.39 a.m.

"That's one small step for man but one giant leap for mankind" he said as he stepped on the lunar surface.

The two astronauts opened the hatch of their lunar module at 3.39 a.m. in preparation for Neil Armstrong's walk. They were obviously being ultra careful over the operation for there was a considerable time lapse before Armstrong moved backwards out of the hatch to start his descent down the ladder.

Armstrong moved on to the porch outside Eagle and prepared to switch the television cameras which showed the world his dramatic descent as he began to inch his way down the ladder.

By this time the two astronauts had spent 25 minutes of their breathing time but their oxygen packs on their backs last four hours.

When the television cameras switched on there was a spectacular shot of Armstrong as he moved down the ladder. Viewers had a clear view as they saw him stepping foot by foot down the ladder, which has nine rungs.

He reported that the lunar surface was a "very fine-grained powder"

Clutching the ladder Armstrong put his left foot on the lunar surface and reported it was like powdered charcoal and he could see his footprints on the surface. He said the L.E.M.'s engine had left a crater about a foot deep but they were on a very level place here"

Standing directly in the shadow of the lunar module Armstrong said he could see very clearly. The light was sufficiently bright for everything to be clearly visible.

The next step was for Aldrin to lower a hand camera down to Armstrong. This was the camera which Armstrong was to use to film Aldrin when he descends from Eagle.

Armstrong then spent the next few minutes taking photographs of the area in which he was standing and then prepared to take the "contingency" sample of lunar soil.

This was one of the first steps in case the astronauts had to make an emergency take-off before they could complete the whole of their activities on the moon.

Armstrong said: "It is very pretty out here."

Using the scoop to pick up the sample Armstrong said he had pushed six to eight inches into the surface. He then reported to the mission control centre that he placed the sample lunar soil in his pocket.

The first sample was in his pocket at 4.08 a.m. He said the moon "has soft beauty all its own", like some desert of the United States.

Armstrong then started to prepare to guide Aldrin out of the lunar module as he emerged backwards through the hatch on to the porch.

By this time Armstrong and Aldrin had used up 45 minutes of their oxygen supply.

Armstrong told Aldrin: "I feel very comfortable and walking is very comfortable. You've got three steps to go and then the long one."

Seconds later Aldrin dropped down on to the lunar surface and Armstrong said: "Isn't that wonderful."

It was 4.15 when Aldrin stepped on to the surface. One astronaut was heard saying "magnificent desolation".

Armstrong and Aldrin then carried out a number of exercises. Armstrong could be seen jumping up and down while Aldrin, clutching the ladder, was doing what looked like a knees bend.

Armstrong appeared to move rapidly across the moon's surface but only seeming to take short steps.

Sharp contrast between light and shadow made the television picture partly obscure, but early overall transmission was good.

One of the astronauts reported that the rocks had a powdery surface and were rather slippery.

The astronauts reported to mission control that their steps tended to sink down about quarter of an inch. All the time the two astronauts could be seen moving around in front of the lunar module. Their movements were slow and they seemed to lope.

They then unveiled the plaque which contained President Nixon's signature and with an inscription saying: " Here men from the planet earth

" First set foot upon the moon
" July, 1969, A.D.
" We came in peace for all mankind."

The two astronauts then removed the television camera from the storage compartment in the lower half of the lunar module. As they carried the camera away from the module they pointed it towards the ground to show the first view of the lunar surface.

One of the astronauts asked: " Have I got plenty of cable.". The other replied: " You have got plenty."

After Armstrong had moved about 50 feet from the module Aldrin said: " Why don't you turn round ? " The camera then swung round and for the first time viewers saw a full view of the module as it nestled on the lunar surface.

Aldrin said he was " filled with admiration " at their first-in-a-lifetime sight and experience.

Armstrong turned the camera to give a panoramic view of the Sea of Tranquillity panning slowly and picking out small craters. He then pointed it towards the sun. A long piece of angular rock could be seen in the distance. It was about 18 inches long.

There was then a shot of the shadow of the module and beyond that there were two craters. Armstrong said they were about 40ft. long and about 20ft. across.

Armstrong then swung the camera round again towards the lunar module and Aldrin could be seen erecting the solar wind instruments. Armstrong then left the TV camera on its tripod and moved into view. walking slowly towards Aldrin.

By this time Armstrong had been on the lunar surface about 70 minutes. Aldrin somewhat less.

The two astronauts then began a thorough inspection of the lunar module to check whether there had been any damage to the struts on landing or any other exterior damage.

They could be heard talking to each other by radio as they checked their fragile lunar craft.

As they then prepared to collect a larger sample of the moon dust and rocks the voice of Michael Collins could be heard reporting to mission control as he came round from behind the moon having completed an orbit in the mother ship. Columbia.

The two astronauts told him that they were just erecting the American flag, the Stars and Stripes. and Collins replied: " Gee, that's great." The flag was planted at 4.41 a.m.

It seemed at first that Armstrong and Aldrin were having difficulty in getting the staff of the flag into the ground, but after a few minutes it was standing securely.

One of the astronauts was then photographed standing by the side of the flag.

Aldrin could then be seen loping across the surface and guffaws of laughter from mission control were heard in the background. The astronaut began performing high jinks as he bounced up and down like a jack-in-the-lift off, still set for 6.50 p.m. (B.S.T.) tonight.

Armstrong and Aldrin were then told to stand by to receive a message from President Nixon. The President told them that he was speaking from the White House by telephone and " this certainly has to be the most historic telephone call ever made ". He said he wanted to tell them how proud everyone was. For every American it must be the proudest moment of their lives, perhaps for everyone in the world.

MAN ON THE MOON (APOLLO LANDER)

1970s

The decade saw a change of emphasis in the space exploration programme. Apollo 13 was a failure (leading to many a superstitious comment), yet the human drama provided by the astronauts revived popular interest in what were already being dismissed as straight-forward landings on the Moon. The large quantity of lunar rock brought back from these trips was to keep many laboratories occupied over the next few years. However, the cost of the landings, plus apparently diminishing returns in terms of political and technological impact, led to the discontinuing of such landings in 1972. Future developments followed two different lines—the quest to put manned space stations into Earth orbit and the exploration of the planets using unmanned space probes.

Jupiter's atmosphere from the Voyager fly-by

Once the Soviet Union had decided not to compete with the Americans in a race to the Moon, it took the lead in developing link-ups between manned space capsules. When the USA decided against further lunar landings, it therefore concentrated its attention on pursuing similar missions. In 1975, the first link-up between American and Russian astronauts in space occurred, but despite obvious parallels, the two countries still pursued rather different objectives in their plans for unmanned spaceflight. At the same time, other countries entered the world of space: both China and Japan launched their first satellites in 1970.

During the 1970s, the USA took a clear lead over the USSR in planetary exploration. The Russian exploration of Venus continued to prosper, but the Americans triumphed with the first soft-landers on Mars. In addition, they obtained the first close-up pictures of Mercury and, at the end of the decade, of Jupiter. The Voyager probe to Jupiter revealed that the planet possessed rings of material round it: something that had also been found not long before for Uranus as a result of ground-based observations. Whereas the rings of Saturn had previously been thought unique, it now appeared that large planets typically possessed a ring system.

As with this study of planetary rings, a number of advances in astronomy during the decade employed a mixture of satellite and ground-based observations—often using large new telescopes—to study the nature of objects in the Universe. γ-rays from sources in space were identified for the first time and, along with increasingly detailed x-ray observations, led to a more informed discussion of such exotic objects as black holes.

High-energy physics flourished on the ground, as well as in space, during the 1970s. An increasingly integrated theoretical approach was supported to some extent by new experiments. Early in the decade, the basic ideas of quantum chromodynamics were elaborated. An important idea behind this theory was that the various kinds of quarks previously specified by the theoreticians could also be distinguished via a new label called 'colour'. The quarks are bound together in this theory by particles called 'gluons', for which there were claims of experimental detection by the end of the decade. In 1974, the discovery of a new, massive particle (now usually called the 'psi' particle) suggested that there must be an additional quark characteristic, labelled 'charm'. Different types of quark were now beginning to proliferate in the way that 'fundamental' particles had done in earlier decades.

This progress encouraged new attempts to find a theory that would unify known forces into a single entity. Such theories, now called GUTs (Grand Unified Theories), attracted more and more attention as the decade progressed. Of the four basic interactions defined long since, gravity had always seemed the odd one out. Now Einstein's original ideas were extended to a new theory of 'supergravity', which involved a re-look at the likely quantum characteristics of gravity.

Both nuclear physics and space research required the application of increasingly sophisticated computers. But much the more important computing development for the future was at the other end of the scale, with the appearance of the first personal computers. Other technical developments, such as the use of optical fibre for telephone cables, similarly held promise for the future of networking. Though computers had featured in popular media reports down the years, it was only in the latter part of the 1970s that they became a regular feature of media reporting.

Big computers and weather satellite data together were steadily improving understanding of the atmosphere. the 1970s also saw the launch of the first satellites to study Earth resources and the oceans. Apart from their utility, these satellites underlined the rapidly growing popular interest in the global environment. This was reflected, too, in the increasing amount of research into climatology, and, more generally, into Earth physics. Perhaps the most widely reported physics-related impact on the environment was, however, the accident at the end of the decade to the nuclear reactor at Three-Mile Island in the United States.

Earlier in the decade, physics had had a rather better press in terms of its contributions to health. Both CAT (computerised axial tomography) and NMR (nuclear magnetic resonance) scanning of patients came into regular use during the 1970s.

1970s CHRONOLOGY

1970
- Accident to Apollo 13, the third lunar mission

1971
- US Lunar Rover used on the Moon's surface
- USSR orbits the first space station
- USA and USSR send space-probes to orbit Mars

1972
- The last manned lunar landing takes place
- The idea of quantum chromodynamics is developed

1973
- The first celestial γ-ray source is detected

1974
- Intitial development of GUTs (Grand Unified Theories)

1975
- Russian space-probes produce the first pictures of the surface of Venus

1976
- US space-probes produce the first pictures of the surface of Mars
- The idea of supergravity is introduced
- Rings are discovered around Uranus

1977
- The first successful personal computer appears

1978
- Pluto is discovered to have a satellite

1979
- Voyager 1 discovers a ring around Jupiter
- The gluon (which transmits the strong force) is detected
- There is an accident to the nuclear reactor at Three–Mile Island in the USA

1970s

LONG RANGE WEATHER FORECASTS

it will be a few years before the long-range forecasting system is ready for general use. It depends or a method of working out long-range weather conditions by analysing thousands of measurements of wind speeds, atmospheric temperature, pressures and other data from a global net work of observation stations. All this information is used to construct a mathematical model of the weather picture.

Although this type of information is already gathered for short-range forecasting, the method under development allows for the influence on the weather of evaporation from the oceans and of the mountain ranges across Europe.

Long-range forecasts require a look at the weather conditions building up far out in the Atlantic, and determining how other events in the northern hemisphere and even some in the southern hemisphere will modify the situation about a week ahead.

SPACE STATION PLANS

Colonel Andrian Nikolayev, veteran Soviet cosmonaut, was tonight piloting a Soyuz 9 spacecraft in earth orbit in what looked to western observers to be a further step forward in the Russians' declared plan to build a huge space station.

The choice of Colonel Nikolayev, who is married to Valentina Tereshkova, the first and only woman cosmonaut, and who was last in space for four days in August, 1962, indicates that another space link-up is in mind. More Soyuz craft or possibly a new space vehicle altogether could follow in a day or two

EVOLUTION OF SEAS AND CONTINENTS

The theory is based on the recent discovery that the subsurface skin of the earth, known as the lithosphere, is divided into an irregular patchwork of rigid plates.

The plates, which are layers of rock several miles thick and thousands of miles long, are driven sideways by underlying forces which may be convection currents kept in motion by the earth's internal heat.

Where two of the plates are propelled against each other, one is thrust down into the mantle of the earth from which it came. At the same time the leading edge of the other plate is crumpled up to form an arc of islands, such as Japan, or, if a continent lies at its margins, a range of mountains such as the Andes of South America.

Early in the history of the earth, when the globe was perhaps covered in water, numerous island arcs may have been formed in this way, later to coalesce into larger land masses by a shift in the pattern of the plate system.

Some of these land masses, the theory goes, formed the nuclei for the first continents. Others would have been eroded by the usual climatic processes, discharging matter into the oceans that was later deposited as sediment. Thick piles of sediment accumulated behind island arcs or at the edges of continents.

MOON ROCKS

Conflicting opinions about the interpretation of the minerals brought back from the moon by the Apollo 11 astronauts became apparent when the lunar science conference opened in Houston today. More than 1,000 scientists have joined the 142 principal investigators to discuss the findings of analyses made over the past four months.

Differences between the way minerals have been distributed over the moon's surface and the way similar compounds have been distributed over the earth are attributed to the influences of meteorite and solar wind bombardment.

In the early days of moon formation more than 3,500 million years ago, when the subsurface layers were volcanically active, large meteorites fractured the lunar mantle to tap the underlying molten material, allowing it to flow out to cover the areas known as the Seas.

Many scientists believe that the first lunar rocks are not a good representative sample for determining the age, structure and origin of the moon, because they were gathered very quickly.

Findings from the exploration of the second astronauts to set foot on the moon may be more important because the crew had more time to choose specimens and document the condition in which they were obtained.

ANTI-HELIUM

Soviet physicists have reported the creation and detection of nuclei of anti-helium, thus strengthening the hypothesis that the universe is composed symmetrically of ordinary matter and anti-matter.

As reported in Pravda, the anti-helium nuclei were created at the world's largest nuclear accelerator at Serpukhov near Moscow where energies up to 70,000m. electron volts can be generated.

THE WORLD'S MOST POWERFUL NUCLEAR REACTOR

An imposing list of experiments is being prepared to hail the opening next year of the world's most powerful nuclear reactor designed specifically for experiments in physics. It is the high flux reactor being built jointly by France and Germany at the Laue-Langevin Institute, Grenoble.

Neutrons can be produced in a reactor at speeds ideally suited to studying the atomic structure of solids and liquids. To achieve this neutrons produced by fission inside a reactor must be streamed so that they all emerge at the same speed.

The neutrons may then be scattered by the atoms inside a solid or liquid to give key information on the positions and movements of the atoms, a theory physicists first put to the test about 15 years ago.

One question that will be investigated on the reactor is how the atoms inside a solid rearrange themselves when a material becomes magnetic. This has been puzzling to physicists for some time, although recent experiments with neutrons have shed new light on the way that the little atomic magnets making up a magnetic material like iron move up and down in simple wave-like patterns.

FUSION RESEARCH

To obtain a fusion reaction of a quantity of gas containing two of the isotopes of hydrogen (deuterium and tritium), a method has to be found of producing a temperature close to 100 million degrees Centigrade.

Once fusion can be obtained for a matter of a second or more, the energy from the gas will far exceed that needed to trigger the process. The trouble has come in devising a satisfactory container in which to compress the gas in the first place. Scientists have devised a variety of special "bottles" which are made from powerful magnetic fields.

In the Tokamak-type machine the plasma is made to carry extremely high electrical currents of about 400,000 amperes. Because of this, a complex electromagnetic interaction takes place to produce what is called the pinch effect.

Experiments indicate that with this system plasma can be contained long enough for fusion temperatures to be reached and sustained.

LASER-GUIDED BOMBS

In its latest issue, Aviation Week and Space Technology said it had been found that o nearly 1,000 laser-guided bomb dropped in combat, more tha. 700 hit their targets. Many o the others probably also hit thei targets, but this could not b verified. The magazine said tha laser-guided bombs had bee introduced in Vietnam by th Air Force after very little testin in the United States.

APOLLO 13 CRASH

The cause of the mysterious explosion or impact which damaged the command ship's fuel cells in deep space was still unexplained as harassed mission controllers continuously monitored oxygen, water, and fuel supplies in the lunar module to make sure they would last until splashdown.

The first hint of trouble came at 4 30 am, when the astronauts saw a swarm of "fireflies" outside the command ship. It turned out to be venting fuel cells. Jack Swigert then noticed a sudden power drop in the command craft, brought about by the failure of two of the three fuel cells Flight controllers at Houston told the crew to reduce the electrical load so that the remaining cell would not be drained of its power.

Without the fuel cells to power oxygen, water, and the engine system, the command module was almost uninhabitable. Commander Lovell told ground control: "It looks like we're going to have to go to an LM lifeboat." Shortly after 5 am, Lovell and Haise crawled through into Aquarius, the lunar lander, which alone has the heated oxygen necessary for survival, while Swigert remained behind in the command module to make sure that the remaining fuel cell continued to operate.

ION THRUSTERS

Ion thrusters, or electric rockets, offer the best technique for long space journeys. Most probably, missions to the other planets of the solar system will soon be carried out using this form of propulsion, in which a beam of photons is used as a thruster instead of the conventional chemical rocket thrusters.

Because the photon beam—in effect a very intense beam of light—can produce only a very small thrust, it can be used only once space craft have been lifted clear of the Earth by chemical (or, in the near future, nuclear) rockets. But once in space, the ion thruster is so efficient that it can be used continuously, gradually building up enormous speeds which chemical rockets cannot achieve.

LUNAR ROVER ON THE MOON'S SURFACE

THREAT TO APOLLO 14 ENDS

The Apollo 14 crew finally managed to link their command and landing modules tonight after hours of drama in which the entire mission's success hung in the balance.

Three previous attempts had failed, and without the docking operation no Moon landing would have been possible; the three astronauts were not themselves in any danger.

For the Space Agency, though, the Apollo 14 mission has considerable importance. The chief objective is to fly a flawless mission, with a perfect Moon landing, to reestablish some of the confidence in the programme that was lost following the near disaster in space with Apollo 13 last April. The present mission is important as a test of the Moon landing system with the additional safety precautions that have been built into the spacecraft during the past few months.

BARIUM CLOUDS AND THE EARTH'S MAGNETISM

A cloud visible to the naked eye as a bright "star" will be released in the upper atmosphere today.

Launched by a Scout rocket from Wallops Island off Virginia, in the United States, this cloud of glowing barium will be used by scientists to study the structure of the Earth's magnetic field, mapping out the magnetic field lines like iron filings clustered around a bar magnet.

For the first 20 minutes or so after release the cloud will be as bright as a first magnitude star and visible, weather permitting, over most of the western hemisphere.

CLIMATE AND SOLAR ACTIVITY

Shorter cycles in world climate have been linked definitely with sunspot activity, warmer periods coinciding with increased sunspot activity. One intriguing detail which has emerged is that in Europe periods of sunspot activity may have been associated with the preponderance of warm south-westerly winds which brought warm, dry summers in 1350, 1530, 1730, and 1920.

The warm period from 1925 to 1950 is now seen as a result of increased sunspot activity. The subsequent decline in mean temperatures was indeed tentatively predicted in 1951. Although another short period of unusual warmth is unlikely this century, we can take some comfort from the prediction that another "little ice age" comparable with the period 1550 to 1900 is not due until about the year 4300.

MAJOR TELESCOPE DISTORTION

The Mark Va will pose many engineering problems. The steerable sections will weigh about 7,000 tons, but an automatic system of correction will keep distortion of the bowl due to its weight to less than 2 kilometres at any point.

This accuracy will enable Jodrell Bank to make far more sensitive observations than has been possible in the past.

APOLLO 15 MISSION

The three first moon missions—Apollo 11, Apollo 12 and Apollo 14—were largely devoted to testing the lunar landing systems. By contrast on Apollo 15 the basic systems have now been well proved, and the astronauts have been able to go forward on a real journey of scientific exploration.

Their mobility has also been dramatically increased by the Lunar Roving Vehicle—the four-wheeled Jeep that has been carrying them and their bulky equipment up to five miles from the lunar lander Falcon.

As on all previous missions, the astronauts have found it takes longer to perform their tasks on the moon than during practice on earth. Both yesterday and today some of the less important tasks had to be eliminated from their drives on the moon. They also found in their first moon excursion that it was harder to drill the deep holes in the moon than had been hoped.

These holes, which are to hold heat-flow experiments, could not be drilled more than about five feet before hard bed-rock was reached.

For geologists at Mission Control the initial surprise was the news that the landing site was much smoother and less rocky than had been anticipated. There was no immediate explanation of how this area could have become so smooth without the wind and rain that erodes features on earth.

HEATING THE IONOSPHERE

Atmosphere scientists working with the Arecibo radio dish have been able to produce a bubble of hot plasma in the ionosphere. This bubble, 100 kilometres long and 50 kilometres across, was made by beaming intense radio waves into the F-layer of the ionosphere.

The idea of using radio beams to heat the upper regions of the atmosphere dates back many years, but it was only in 1970 that successful experiments of this kind were carried out.

The earlier Boulder experiments found that the temperature of the F-layer increased when the heater was switched on, producing airglow and infrared activity. The Arecibo observations go further, however, because the advanced 430 MHz radar used makes it possible to plot contour maps of temperature variations in the F-layer.

From the radar maps it is clear that the plasma heated by the transmitter expands parallel to the magnetic force lines in the upper atmosphere. This produces a cigar-shaped volume of hot plasma. The temperature excess in this region can be as much as 300°C above the normal 1,000°C temperature of the region.

FIRST MAP OF MARS

The first map of Mars made from the pictures sent back by Mariner 9 has been completed. It improves over the best maps previously available by a factor of 100; although there are some small gaps in the new map, and not all of the north polar region is included, it provides a remarkable new insight into the structure of the planet.

As the pictures from Mariner 9 came in during the months of its primary mission, the piecemeal evidence hinted that astronomers would have to change radically their view that Mars is a "dead" planet; the complete map emphasizes that and provides the basis from which new theories may be built up.

Perhaps the most striking result of the survey, as noted by Dr H. Masursky, the leader of the Jet Propulsion Laboratory team responsible for the television mapping, is that "we are forced to no other conclusion but that we are seeing the effects of water on Mars".

THE LAST APOLLO FLIGHT

as Apollo 17, the last such flight for the foreseeable future, heads into space, it is timely to think about the motives, the technology, and some of the achievements of the United States manned space flight programme.

Technologically, the vast Apollo programme has evolved logically from the one-man Mercury flights, skimming tentatively around the earth; and the two-man Gemini missions, on which advanced earth-orbital manoeuvres were practised. But the motives for the whole Apollo development were far from logical.

The objectives of the programme were not scientific, although much new scientific information has been gained. Considerable benefits accrued, but these had not been put forward as the reason for going to the moon.

Beyond Apollo there will be many other manned and unmanned space projects. In 1975 the Americans and the Russians plan literally to get together in space, through a rendezvous in earth orbit. The space shuttle programme will be even bigger than Apollo, but is likely to be stretched out over many years and is unlikely to generate anything like the same enthusiasm.

However mixed its original motives, and whatever the total cost, the Apollo programme has been an unparalleled achievement. It is the supreme example of twentieth-century exploration.

THERMONUCLEAR RESEARCH

A novel method for extracting thermonuclear energy from heavy hydrogen has been described by a group of scientists from the Lawrence Livermore Laboratory, California. The essence of their proposal is that the energy from a powerful laser should be used to compress a droplet containing heavy hydrogen or deuterium to such a high pressure that it will explode in the way in which heavy hydrogen in a hydrogen bomb explodes.

there is, in theory at least, a chance that the light energy from a powerful laser could be focused in such a way as to create a pressure of a million tons a square inch.

In the scheme the fuel would be a mixture of deuterium and tritium in roughly equal proportions. It would be fed into a spherical chamber in the form of droplets rather less than a millimetre across. Light from the laser would be passed through reflecting mirrors and lenses to illuminate the chamber uniformly from all directions.

The first effect would be to evaporate some of the material from the surface of the droplet, but that material would then be driven towards the centre by the converging energy from the laser. It is estimated that the density of material at the centre of the chamber would reach 1,000 grams a cubic centimetre, and that its temperature might exceed 100 million °C.

BRAIN X-RAYS ASSESSED BY COMPUTER

Mr Godfrey Hounsfield, designer of the system, called computerized transverse axial tomography, explained yesterday that it X-rays two adjacent slices of the brain, each one centimetre thick, by rotating round the patient's head and taking 56,000 readings.

These readings "learn" from each other and are fed to the computer which, solving 28,000 simultaneous equations, builds up a matrix of 25,000 points giving a highly accurate valuation of the material within the slices.

The system overcame the disadvantage of the low sensitivity of X-ray machines and the confusion of information on the plate caused by three-dimensional information being superimposed on a two-dimensional photograph.

MOVEMENT OF EARTH'S CRUST

The tidal influence of the Moon may be causing the solid crust of the Earth to drift westward over the liquid inner regions. Because the crust, or lithosphere, is of uneven thickness this drift will occur at different rates in different parts of the world.

Individual plates will tend to rotate, as well as moving to the west, because of the variation of the tidal force with latitude. This will cause a clockwise rotation in the northern hemisphere and anticlockwise rotation in the southern hemisphere. Several geophysicists have found evidence for just such rotation. Further evidence supporting the theory comes from the apparent eastward movement of regions of volcanic activity

Because volcanic activity is associated with the mantle, and not with the lithosphere, each volcanic island is carried off to the west of the volcanic region where it was formed. But to anyone on the surface of the Earth being carried westward with the lithosphere it seems that the islands are stationary and that the region of volcanic activity moves to the east.

SATELLITE SCATTERS PLUTONIUM

A United States Navy satellite that burnt up in the atmosphere eight years ago scattered traces of radioactive plutonium over 12 countries in the Southern Hemisphere, the Atomic Energy Commission disclosed today.

About 2 lb of plutonium were strewn by the satellite. This equalled about one sixth of all the plutonium that fell over the Southern Hemisphere from more than 300 nuclear tests conducted in the atmosphere by the Soviet Union and the United States.

EVOLUTION OF MARS

American scientists who have studied pictures and information sent back from the Mariner 9 spacecraft orbiting Mars say that the planet was until recently in geological terms a living, breathing, evolving body. Unlike the Moon, Mars had a long life, they say.

Some think it might still be stirring internally. What they agree generally is that:

1 Fiery volcanoes once burnt over the planet, belching smoke and gas and playing a prominent role in shaping Mars. The fires, perhaps heavily banked by time, may still burn deep within Mars.

2 Strong erosional forces have been at work, shaping some regions in ways strikingly similar to Earth. There are strong indications of erosion by gas, such as winds, and by liquid, such as flowing water. But the scientists hasten to add that they have little information about what caused the erosion.

3 Permanent dust storms may rage over certain regions and the entire planet may be dustier than a half-century of careful telescope observation from Earth has indicated.

REACTORS AND HEALTH

A report from an international group of scientists, on how to minimize the effects of radioactive waste while building more nuclear reactors for power generation, contains a serious warning about developing the next generation of fast breeder reactors. This new system is believed to pose the most serious health hazard and greatest risk of proliferation of weapons material if it is introduced for routine operations.

The Pugwash group believes it is technically feasible as well as desirable to restrict the radiation exposure to members of the public to levels less than 1 per cent of the average "natural background" radiation. This level has been set for the routine emissions from reactors used in power stations.

The long-term problem of storage of waste is seen as the most difficult question of all.

GRADUAL APPROACH TO A UNIFIED FIELD THEORY

One curious feature of the many reactions associated with radioactivity is that until now whenever the uncharged particles called neutrinos were observed interacting with other particles they were transformed into charged particles, called muons. These neutrino interactions involve the weak force, and it had seemed a fundamental feature of them that a transfer of electric charge was involved.

That was discomfiting for the theoreticians, since a few years ago Professor Steven Weinberg and Professor Abdus Salam found a way to combine the electromagnetic and weak interactions into one almost perfect theory.

The obvious flaw in their theory, however, is that it allows the existence of "neutral currents", and predicts that in reactions involving these neutral currents no electric charge is transferred, and neutrinos can thus interact through the weak interaction without always being transformed into (charged) muons.

Now the CERN experiments have produced evidence of just that kind of neutral current interaction.

NEW INTERSTELLAR MOLECULES

Radio telescopes capable of detecting microwave radiation from space have enabled astronomers in the past few years to identify radiation from molecules in interstellar space. The work requires a microwave facility because whereas individual atoms produce characteristic spectral features in visible light, molecules are larger than atoms and so radiate at correspondingly longer wavelengths.

Twenty-one such molecules were found, in a flurry of discoveries up to autumn of last year ; now, after a 12-month hiatus, the discovery of the twenty-second interstellar molecule has been reported.

It seems very likely that this list of 22 is far from complete, because microwave astronomers are more concerned with understanding the discoveries they have already made than with setting out deliberately to find new molecules.

But that is not to say that the painstaking work now going on is unlikely to yield spectacular results. If organic molecules exist even in clouds of gas and dust between the stars, the argument runs, then surely any planet formed in our galaxy stands a good chance of picking up enough of these building blocks to start the laborious build-up of increasingly complex molecules until life begins. That is one of the most powerful arguments for thinking that we are not alone in the galaxy.

NEW COMPUTER MEMORY

A new type of computer memory, based on tiny magnetically charged zones or "bubbles" inside a thin film of garnet material, could be adopted commercially within three years, according to a report published today.

The magnetic-bubble memory would hold information in very concentrated form, and would be reliable, a small power consumer and relatively cheap.

BLACK HOLES

There has been much excitement among astronomers lately because of the discovery of X-ray stars in binary systems. From the way in which the two stars in a binary disturb each other it is possible to calculate their masses, and in one particular case, the X-ray star known as Cygnus X-1, it seemed that the X-rays detected by rockets and satellites such as Uhuru are coming from a star more than 10 times as massive as our Sun. Yet this star is invisible in ordinary light and cannot be detected by optical telescopes.

In a black hole matter is squeezed together under the influence of its own gravitational field until it is cut off from the visible universe altogether. The gravity field becomes increasingly powerful as the object becomes more dense, until eventually even photons of light cannot escape from it for the escape velocity becomes greater than the speed of light. Thus, the object is invisible, hence its name. That, at least, is the theory.

Indeed until very recently, there seemed no hope of ever finding a black hole because of their invisibility. But obviously such a massive, tiny and invisible object could provide exactly the intense gravity field needed to explain the X-rays coming from the dark star in the Cygnus X-1 system.

WEATHER SATELLITE

Even with satellites and huge computers for analysing the climate, meteorologists are uncertain about forecasts of more than two or three days, because their picture of conditions in remote parts of the world and in the layers of the upper atmosphere is far from complete.

But two types of new satellite will make possible new studies. One, called Tiros N, is being produced in the United States; the other, Meteosat, is a joint European venture. The American vehicle will circle the Earth in a low solar orbit.

ORIGIN OF MARTIAN CHANNELS

The discovery in Mariner 9 pictures of Mars of features that seem to be eroded stream channels has caused much discussion among planetary scientists. The problem they are faced with is that Mars as it is today has an atmosphere too thin to allow running water to exist for long, even though the temperature at the Martian equator does rise above freezing.

Although the low pressure of the Martian atmosphere would cause ice exposed on the surface to sublimate directly to vapour, without passing through the intermediate liquid phase, it is generally accepted by geologists that a substantial layer of permafrost might exist below the surface of the Martian ground. If some of this ice could be heated and exposed to the atmosphere, it might exist as water for long enough to form stream channels before evaporating into the thin atmosphere, there are at least three ways of liberating this subsurface water without invoking a change in the density of the atmosphere and a big climatic upheaval. First, volcanic activity might melt the permafrost; second, landslips might uncover ground water, which could have melted in the Martian "tropics" while remaining underground; and finally the impact of meteorites could liberate water from the permafrost.

FUSION AND FISSIONS

Energy could be extracted efficiently from very small pellets of fissile material, such as uranium-235, coated with a mixture of deuterium and tritium, the heavy isotopes of hydrogen.

The point is that the mass of fissile material required before a chain reaction can be started and energy extracted, as in a nuclear power station, is quite small if the density is much higher than the densities of solids in everyday experience (typically a few grams a cubic centimetre). That state of affairs can be achieved if the outer part of a pellet is quickly heated by means of a laser pulse, thus rapidly compressing and heating the rest of the pellet.

If a coating of deuterium and tritium is applied, the pellet has a neutron-reflecting surface which will increase the number of neutrons available for the fission reaction. The critical mass is then only a few milligrams. But once the fission reaction has been started and has been "amplified" by the coating thermonuclear fusion—in which atoms of deuterium and tritium combine and give off energy—will also commence in the coating because of the high temperatures generated. Extra neutrons are produced in this fusion process and this enhances the fission reaction, which in turn boosts the fusion reaction and so on.

The star's temperature varies with the light variation: it is, paradoxically, hotter when decreasing in temperature.

COSMIC RAYS FROM PULSARS

FIRST PICTURES OF MERCURY

WHERE DOES THE HEAT GO?

According to the best astronomical theories, the radio stars known as pulsars are created in violent explosions of stars—supernova explosions. Such explosions should produce large bursts of cosmic rays, which might leave a trace on the Earth; it now seems that evidence of such bursts of cosmic rays has been found in the sediments of the sea bed.

When cosmic rays arrive in the Earth's atmosphere, they intereact with atoms such as nitrogen and argon to produce isotopes of other elements. These include radioactive atoms (radioisotopes) of beryllium and aluminium.

Because these radioisotopes decay into stable elements only fairly slowly, they remain present in the environment long enough to provide a record of cosmic ray activity. The more cosmic rays there are, the more of these radioisotopes there are in the air at any epoch; and since some of these atoms from the air get into sediments being laid down on the sea floor, measurements of the radioactivity of samples from different layers in the sediments ought to provide an indication of how cosmic ray activity has varied over millions of years.

Unfortunately, the technique is accurate enough only to date this increase as occurring between one and a half and three and a half million years ago; but that interval does coincide with the ages of 15 known pulsars.

The planet Mercury gave up some of its secrets to the Mariner 10 spacecraft today as it flashed back television pictures showing an arid, moonlike surface covered with big and small craters.

The surprising discovery was made that the planet has an atmosphere and a magnetic field.

Mercury, half as big again as the Earth's moon, is the smallest of the nine planets of the solar system and the one nearest the Sun.

The dramatic photographs were flashed back over 96 million miles today, taking eight minutes to reach Earth, and showed Mercury's craters were much like those of the moon.

The recent wide acceptance of plate tectonics—the idea that the surface of the Earth comprises a small number of moving plates—has added enormously to our understanding of geophysical phenomena such as earthquakes. But the new insights have so far failed to reveal an explanation for the surprisingly complex pattern of heat flow.

The puzzle for geophysicists has been that it is possible to predict how much heat should be emerging near ocean ridges and the thousands of measurements made so far indicated values below expectation by a factor of three or more. Further, the heat flow values are highly variable from place to place, being almost zero in some areas and only a few kilometres away much nearer to the expected value. Where is the heat going to, and why is the pattern complex?

Conduction is one way to transport heat in rocks, but another is convection—the actual movement of hot material. Normally we think of convection in terms of fluids—the atmosphere, the air above a candle, porridge. Rocks clearly are not fluid, but they do possess interstitial water and it has gradually been becoming clearer that much of the movement of heat in the Earth's crust is not by conduction but by the rising of hot and the sinking of cold water in the interstices.

If water flow in near surface rocks truly accounts for the way that a large amount of heat escapes from the Earth, that is the second striking role water plays in geophysical processes. It is already quite widely accepted that interstitial water is centrally important to the initiation of earthquakes.

Through July the solar system has served us well, running an average of 5 hours daily.

What needs to be explained is really a considerable puzzle. The energy of some cosmic ray particles is staggering—the equivalent of, say, a million million, million million votes.

PARTICLES FASTER THAN LIGHT

Although it is widely believed that Einstein's theory of relativity precludes the existence of anything going faster than light, some scientists have pointed out that one reading of the equations allows the speed of light to be exceeded. For that to happen, however, a number of conditions have to be fulfilled. These make faster-than-light objects rather bizarre creations.

In the hope that faster-than-light atomic particles might one day turn up, physicists have reserved the name "tachyon" for them, from the Greek for "speed". So far all searches for these hypothetical tachyons have given negative results.

showers of atomic particles can be detected at ground level as a burst of radiation. About 1,300 of these showers detected by instruments at Adelaide University between February and August last year have been studied by Dr Clay and Dr Crouch.

They have made the surprising discovery that just before the arrival of each burst of radiation their detectors pick up a weak extra signal. This extra signal precedes the main shower by only a few millionths of a second, but all the same they can find no conventional way to account for it. So they suggest that it may represent a burst of tachyons created when a cosmic ray hits the atmosphere. Because these tachyons are travelling faster than light they would get to the ground before the main shower of particles.

ORBITING CITIES 'ARE NOW FEASIBLE'

Dr Gerard O'Neill, of Princeton University, airs the idea of building "a habitat capable of supporting and maintaining some 10,000 people" in space.

That kind of suggestion 's familiar to readers of science fiction But there seems to be a real possibility that the idea is feasible if advantage is taken of the curious stability of certain possible orbits for such a city in space.

There are two such stable orbits, occupying what are called the Lagrangian points of the Moon's orbit around the Earth. As Lagrange discovered, the combined gravitational pull of the Earth and Moon produces stable gravitational "islands" 120° ahead and 120° behind the Moon in its orbit around the Earth.

In most orbits, a collection of material will gradually be dispersed by the gravity of the Moon and of the Earth. But at the Lagrangian points dust, spacecraft or the components of a space city would be held together and kept from drifting apart.

ARTIFICIAL AURORA

A party of French scientists has just left Paris for the Kerguelen Islands, in the Indian Ocean, to make final preparations for a joint Franco-Soviet experiment designed to provoke an "aurora borealis" by artificial means.

Energy particles will be produced by Soviet engineers from an electronic "gun" carried by a French "Eridan" rocket launched from the Kerguelen site. The French scientists come from the National Space Research Centre in Toulouse.

It is the electrons, as they follow the inductive currents in the higher atmosphere, which will produce the aurora effects.

The aurora effects ought to be visible soon afterwards both in the Indian Ocean and at the other magnetic extremity round Archangel in the northern hemisphere.

There have been occultations of Ganymede by the star HD 186800.

1974

SILICON CHIPS

A near-magical piece of electronic technology is threatening, or promising, to revolutionize the world of computing. It will make possible a host of new computing uses and it will give to existing computing the flexibility it has never had. It will also reshape a large part of the electronics industry.

That is not bad for a package no more than an inch square. It goes under the abbreviation VLSI, standing for "very large-scale integration".

Motorola, one of the main United States semiconductor component companies, showed the scale of this development in a recent briefing. An integrated circuit in 1960 might have contained four elements (the equivalent of four transistors or diodes). An LSI calculator chip in 1971 might have had 600 elements, while an LSI chip in 1974 could carry 10,000 elements. By 1980, the company estimates, a single VSLI chip will probably be able to hold no fewer than one million elements.

By that time, another source has predicted, the cost of a transistor will approach that of a word on a printed page.

As processing and memory functions of great power become available on the tiny microprocessors—cheap in high-volume production, flexible in use via external programming—a range of posible new applications emerges.

The ones most commonly cited include process and machine control in industry, traffic light control, terminals, coin changers, ticket machines, copying machines, cash registers and other point-of-sale systems, environmental control, accounting machines, teaching machines, automotive use, household appliances and electronic games.

Here the potential is for a basic change in the nature of commercial computing—away from the rigid, predetermined programme sequence and towards an adaptable system which can respond to changing needs.

Groups of microprocessors —like groups of clerks in pre-computing days—would handle a network of parallel but dove-tailed processes which would be built up or changed as required. Like clerks, but unlike conventional computers, they would be capable of handling an awe-inspiring number of different situations.

STATE BANS AEROSOLS

A Bill to be signed tomorrow will make Oregon the first American state to ban aerosol cans that use fluorocarbons as propellants. The ban would take effect in February, 1977.

STONEHENGE: FORESIGHT SAGA

Further research on the possible prehistoric use of Stonehenge as a lunar observatory has led to the identification of some of the foresights on the horizon at which the rising and setting positions of the Moon could be observed.

Locations for four are suggested at Gibbet Knoll, near Market Lavington; inside the later hillfort of Figsbury Ring; near Hanging Langford Camp; and on Chain Hill. The first of these is the most distant, more than nine miles away, the last the nearest at 3.7 miles, while another foresight may have existed at the site of the Coneybury tumulus less than a mile from Stonehenge. During an experiment to test the visibility of Gibbet Knoll it was found that a high degree of refraction, the result of heat radiation from the ground at sunset, made it difficult to see a low foresight, a condition that would have applied to at least two of the other locations. The remaining three sight lines are now interrupted by modern building.

EARTH RESOURCES SATELLITE

A new aid for discovering oil and mineral deposits and for improving agricultural development throughout the world is promised with the latest advanced satellite of the American National Aeronautics and Space Administration (Nasa). More than 100 research teams in 40 countries are preparing for the launch of ERTS-B (Earth Resources Technology Satellite) today.

Basically, the satellite is taking pictures of the Earth so as to detect various objects on the ground simply by the way they reflect natural light and radiation. However, the instruments developed for this process of multispectral remote sensing enable the teams receiving data on the ground to explore agricultural changes, forestry, geological formations, marine resources, fresh water reserves, climatic conditions and almost any issue to be covered under the general title of natural environment.

CAUSE OF ICE AGES

What Dr J. F. Lindsay, of Texas University, and Dr L. J. Srnka, of the Lunar Science Institute, Houston, have now found are periodic variations in the composition of the lunar soil with depth. They believe that the variations may reflect the periodic passage of the moon through galactic dust lanes.

The variations consist in differences in the proportions of large to small particles. Although no direct age measurements are possible for lunar soil, Dr Lindsay and Dr Srnka show that the periodicity of these variations is consistent with their resulting from passage through a dust lane every hundred million years or so.

STILL NO GRAVITY WAVES

Over the past few years it has become increasingly clear that the claims of Professor Joseph Weber, the American astronomer, to have detected gravitational radiation were probably mistaken.

So where does all that leave the study of gravitational radiation? Has the frenzied activity encouraged by Professor Weber's claims been a waste of time? Those are natural questions, but curiously the answer is that the spur of Professor Weber's claims has encouraged a great advance in techniques which may yet produce a genuine detection of gravity waves. It looks as if the theorists were right 20 years ago, when they pointed out what a long way experimenters had to go before producing equipment capable of detecting this radiation. But in the effort to prove or disprove Professor Weber's claims, the observers have developed equipment that is half way to the target sensitivity. Having got so far, some groups at least will probably press on in the effort to detect this radiation at the levels predicted by General Relativity.

That is likely to take some time. Basically, gravity-wave detectors consist of large metal bars which are expected to be squeezed and stretched very slightly by the passage of a pulse of gravitational radiation. The squeezing is so small that the detectors can be disturbed by the small vibrations of traffic in the street outside the laboratory, or by thermal effects in the bar. But two possibilities in particular point the way for future developments.

the site is so positioned in a dip of land that, although the pyramids will be seen from the city, the city will be invisible from the pyramids.

WHY TIME GOES FORWARD

Professor David Layzer, of Harvard University, who specializes in the physics of time, has suggested a new explanation for the unsolved puzzle of why time should go in one direction only.

Professor Layzer has taken the three different aspects of the problem, known as the three "arrows of time", and has tried to explain the asymmetry for each. The first is the thermodynamic arrow, which can be seen operating when a cube of sugar is dissolved in a cup of tea : although it is perfectly feasible as far as the laws of physics are concerned, no one has ever seen a cube of sugar rematerialize from tea !

The second, Professor Layzer calls the historical arrow, which is exemplified by evolution. Starting with a single-celled organism, evolution has produced increasingly more varied and complex species, in an apparently one-way manner. Finally there is the cosmological arrow : the universe is expanding steadily from a "big bang" in the past.

MARS HAS THAT FAMILIAR LOOK

The Viking lander touched down on the Plain of Chryse at 13:57:07 BST, seven seconds later than scheduled but in perfect shape.

The whole of the long descent of the landing craft after its separation from the mother ship in orbit round the planet depended entirely on pre-programming, and the lander had been sitting quietly on the surface of Mars for some 20 minutes before the cheers went up on earth.

The landscape of sand and rock fragments which look strangely uneroded, has an earth-like quality for, unlike the moon where the sky is dark, on Mars the sky is bright. In Martian terms the landing took place at evening and the television pictures showed the glare of the setting sun.

SUPERFLUID HELIUM

A superfluid is a liquid that has lost its viscosity (that is, its internal friction).

When, in 1972, it was found that the isotope of helium containing in its nucleus two protons and one neutron (therefore called helium-three), became superfluid at extremely low temperatures, there was great excitement, because it was only the second superfluid known. When the temperature drops to within one thousandth of a degree of absolute zero, minus 273°C, the Van Der Waal's force becomes dominant and causes the helium atoms to form bound pairs.

A similar phenomenon occurs with the electrons in some metals, which also form pairs, giving rise to the remarkable property of superconductivity. There is, however, a crucial difference between the two cases : the helium atoms are much larger than the electrons and therefore cannot bind closely because they would bump into each other. The dynamic tension between the two effects, attraction and repulsion, keeps the atoms of a pair at a distance from each other.

That is the unique property of the helium system. From it theoreticians have calculated and experimenters confirmed that there should be at least two different helium-three superfluids ; two different pairings of the atoms with different stability regimes. Dependent also on the angular momentum are the different sorts of excitations that could occur in the superfluid : excitations such as breaking up the pairs, or more exotically, the formation of vortices.

WHAT IS GOING WRONG WITH THE WORLD'S WEATHER

The prospect of food shortages in the developed world and of famine in the poor countries by conditions that have been described as "a greater incidence of bad weather" has been increasingly troubling many climatologists. Weather conditions around Britain are so variable from year to year that the detection of slow changes is not easy. Nevertheless Professor H. H. Lamb, of the School of Environmental Sciences, University of East Anglia, has argued that agricultural growing seasons can be seriously affected by small changes in seasonal temperatures.

Some firm evidence should be available from the global atmospheric research programme in progress under the auspices of the World Meteorological Office. This project will provide an important source of information to the United Nations meeting. It should also throw light on other phenomena causing anxiety. For instance, the Indian monsoon has failed twice already in the 1970s, compared with only once in the 1960s.

RESEARCH AT IBM

What IBM does today, the rest of the industry reacts to tomorrow. These are some of the "today" projects at the computer giant's laboratories in Yorktown Heights, New York; San José, California; and Zurich, Switzerland:

The use of electron beams and X-rays in producing ever smaller integrated circuit patterns has led to significant off-shoots. One of these is a new X-ray microscopy technique which enables biological specimens to be observed at magnifications higher than those attainable with optical microscopes.

In extending the existing silicon-based transistor technology, IBM scientists have made experimental field-effect transistor memory elements that enable about five million binary digits to be packed into one square inch. In ordinary language, this means 30 times the density of memory units in IBM computers announced only three and a half years ago.

On the software side of computer science, which in general is still struggling to catch up with the accelerating advances in hardware performance, IBM's research includes an attempt to build up a sound theoretical understanding of programming. If this could be achieved, interactions between programs could be anticipated and the programs themselves could be made more general, easier to write, and more efficient. At present, many programs are too complicated and their real performance and potential defects are not fully known.

CHARM AND STRANGENESS

If nuclear matter possesses the elusive property that theoretical physicists have named "charm", then when it disintegrates it should produce particles possessing the similar, but uncontroversial property of "strangeness".

Two teams working independently have now found the expected "strange" particles, in nuclear reactions in bubble chambers at the Fermi National Accelerator Laboratory (FNAL), near Chicago, and at the Centre Europésnne pour la Recherche Nucléaire (CERN) at Geneva. Their reports substantiate the recent tentative evidence, in the discovery of psi particles, that charm, invented nine years ago on purely theoretical grounds, is really a fundamental property of matter.

They fired a beam of neutral particles (neutrinos) into a bubble chamber and watched for leptons (electrons or muons) to be produced. The leptons are the tell-tale signature showing that a weak interaction has occurred, possibly the "weak" decay of a charmed particle. To their delight, the experiments revealed that often associated with the production of leptons was a strange particle (which leaves an unmistakable track in a bubble chamber), making it very likely that charm exists.

SURFACE OF MARS FROM THE VIKING LANDER

RINGS AROUND URANUS

• The passage of the star behind Uranus was visible only from the region around the Indian Ocean. The most complete observations were made by a team of astronomers from a plane at 41,000ft above the South Indian Ocean.

About 40 minutes before the star was expected to disappear behind Uranus it vanished for seven seconds. During the next nine minutes it disappeared four more times for about one second each time.

Those unexpected observations have been interpreted as showing the existence of five small rings encircling Uranus. The rings are very much smaller than the broad rings around Saturn, and appear to be located in a narrow belt about 4,000 miles wide and about 11,000 miles above the planet's surface. They are very narrow—the inner rings are estimated to be only six miles wide and the width of the outer ring varies from 20 miles on one side to 50 miles on the opposite side of Uranus—and seem to be made up of fragments' less than a mile in diameter. The origin of the rings is a complete mystery so far.

ORIGIN-OF-LIFE CONTROVERSY

A discovery by an Anglo-Canadian team of scientists, disclosed on Thursday, will increase controversy among astronomers about the origin of the earliest primitive life forms. The controversy concerns a new suggestion that, long before the appearance of Earth, complex organic chemicals reacted to form proteins and nucleic acids in the dust clouds that occupy the vast space between stars. the chemical compound cyanotriacetylene was detected near the constellation of Taurus.

This large organic molecule has the essential elements that would allow its conversion in a simple chemical step to form an amino acid, or the building block of life from which proteins, nucleic acids and genes are composed.

The existence of such substances in space was considered surprising at first because scientists had assumed that the harsh conditions there, including extremely cold temperatures and strong ultraviolet radiation, would break them up.

PIANO CONCERTO TO MAKE PLANTS GROW QUICKER

The methods for increasing the fertility of seeds and the growth rates of plants include the use of sound waves to stimulate the metabolic processes of various varieties. The notion of composing a piano concerto to be played to plants is, to say the least, an eccentric proposition. Yet Dr Charnoe uses just such a composition to demonstrate his method for triggering early germination and showing that seeds of differing varieties respond to different sound frequencies. If the idea was based only on some eccentric notion, a meeting with such a technically qualified group as the Acoustical Society would be sheer folly.

The response of seeds to germination has been measured by elaborate electronic equipment as they undergo bombardment by sound and other possible stimulants.

Those processes have produced, for instance, a way of growing tomato plants that, in their fourth year of growth, yield at given notice ripe fruit on any individual truss.

THE PROTON'S SPINNING INTERACTION

One of the peculiar properties of the proton, the elementary particle found in the nuclei of all atoms, is that it behaves as if it were spinning like a top. This property is well known but a recent experiment has come up with unexpected results which show that the spin affects the way protons react with each other. In extremely energetic collisions the contribution of the spin to the total available energy is minute and was expected to be unimportant. But the recent experiments show that to be wrong; the spin refuses to be ignored and produces huge effects.

To simplify the situation as far as possible the physicists looked only at those interactions where the protons bounced off each other unchanged and at wide angles. This means that they were looking at head-on collisions To their astonishment they found that in the most violent collisions they could see, where the protons emerged at wide angles and high energies, the spin of the proton made a big difference. When the protons are spinning in the same direction they are twice as likely to collide head on than when they are spinning in opposite directions.

'JUPITER EFFECT' AND EARTHQUAKES

The theory is that the planets orbiting round the Sun create tides in the same way that the orbiting Moon creates tides on Earth. Jupiter produces the largest force because of its enormous mass, which is why the theory has been called the Jupiter effect. The other planets also exert tidal forces, but usually they are positioned so that the different forces cannot reinforce one another.

But at regular intervals several planets line up ; and in particular, approximately every 179 years Jupiter, Saturn, Uranus and Neptune are co-linear. It has been suggested that those "superconjunctions" may be the cause of the increase in earthquake activity, and as the superconjunction is expected again in 1982, it is of some practical interest to know whether the "Jupiter effect" should be taken seriously.

According to the Chinese records, none of the 11 big earthquakes since AD 1,000 has coincided with a planetary superconjunction, and since 780 BC only one out of 125 significant earthquakes has coincided. So it seems that the "Jupiter effect" does not exist, at least in northern China. Californians living above the San Andreas fault need have no special fear of 1982

ATOMS CONFRONT UNIFIED THEORY

The most attractive theory today in elementary particle physics attempts to unify the weak and electromagnetic interactions. It claims that the force responsible for radioactive decay and the force between charged particles that holds atoms together are but different aspects of the same fundamental force. But doubts are now raised by the results of two independent atomic physics experiments, both of which disagree with the predictions of the unified theory.

The weak interactions are able to distinguish between right and left ; in physicists' jargon they do not conserve parity. The electromagnetic forces does conserve parity but if the two forces are aspects of the same force, then there should be very small electromagnetic effects that distinguish between right and left. As electromagnetism is the force that constrains electrons in atoms to remain around the nucleus, then there should be tiny parity nonconserving effects in atoms.

According to the predictions, the direction of vibration of the light should rotate by a tiny amount, less than a millionth of a degree, but the experiments are so sensitive and carefully controlled that they could see a rotation ten times smaller than that predicted. Both experiments see nothing : the rotation predicted by the theory is not there.

UNIVERSE MAY CONTRACT

It is nearly 15 years since quasars were discovered and, as they are the most distant objects ever found, it was hoped that they might provide clues to the evolution of the universe. In that respect they have, until now, been rather disappointing.

If there is enough mass in the universe its gravitational attraction will overcome the expansion and eventually the universe will begin to contract. Otherwise the universe could expand for ever. Which of those will happen can be predicted if we know the rate at which the expansion of the universe is slowing. In principle that can be determined from measurements of quasars of different distances but it is necessary to know how bright the quasars actually are. How bright a quasar appears to be depends, of course, on its distance from us.

It has been shown that the true brightness of a certain class of quasars is related to the details of light emitted at one particular wavelength. That light is given out in the ultraviolet region by hot ionized hydrogen and the velocity of the quasars moves the light towards the visible range through the redshift effect. Ultraviolet light is absorbed by the atmosphere, and so from the Earth it is possible to measure quasars with very high redshifts only where the light is shifted into the visible part of the spectrum. To check on the rate of expansion needs a measurement from a quasar with a lower redshift, but here the light is still in the ultraviolet and can be observed only from space.

The measurement has now been made by physicists from Johns Hopkins University, Maryland, who sent an ultraviolet telescope above the atmosphere on a rocket. Although the telescope was above the atmosphere for only four minutes they found the quasar they were interested in, pointed the telescope at it and measured its ultraviolet light spectrum. The quasar used is the brightest one seen from the Earth and has a relatively low redshift. Combining their observation with previous measurements on high red shift quasars, the physicists show that the data strongly suggest a model of the universe in which expansion will eventually stop and the universe will begin to fall back on itself.

Temperature in the centre of the planet reaches probably 300,000 degrees on the Calvin scale.

1978

Two scientists proposed this year that certain disease-causing viruses and bacteria fall to earth from cemetery dust in space.

LANDING ON VENUS

WESTERN space scientists were full of praise yesterday for the Soviet Venus probe which continued to send back messages for 110 minutes after it had landed by parachute on the planet's 900F surface.

The American and Russian probes this month have uncovered a sensational amount of information about the Evening Star, as Venus has long been called. We have now learned that:

There is nearly 40 times more of the rare gas argon in its atmosphere than on the Earth. There is so far no explanation of this, except that Venus must have evolved differently from Earth.

The clouds of Venus not only contain unexpectedly large quantities of sulphuric acid, but particles of free sulphur.

Yet the dense carbon dioxide atmosphere of Venus is not only of interest to pure scientists and future astronauts, it is an unpleasant foreboding of what might eventually happen to the Earth if mankind keeps burning fossil fuels.

SPACE COMMUNICATION BY TV

The 300 star systems nearest the Earth could detect the presence of intelligent life here from our television signals. the most intense radio emissions from Earth come from the United States ballistic missile early warning system and its Soviet counterpart. They could be detected 250 light-years away by an observer with our present technology who built an antenna system like the array of a thousand 100-metre dishes proposed for the American Project Cyclops. a receiver on a distant planet would detect the signals twice every Earth-day, when it rose above the transmitter's horizon and again when it set.

The University of Washington scientists calculate that a strong (five megawatt) UHF television station could be detected (by a receiver of the Cyclops type) up to 25 light-years away. About 300 stars and their orbiting plants lie within that range.

INTER-UNIVERSE TRIPS IN DOUBT

Scientists believe that black holes are formed when certain stars of very large mass collapse. The gravitational attraction of such objects would be so high that not even light could escape and anything falling into the black hole would be smashed out of existence at its centre.'

If a black hole is created, then it will most likely be rotating and may carry an electric charge. In that situation the conditions inside the black hole become more complicated. It turns out that the geometry of space-time inside a rotating black hole is such that a connexion is created with another universe. An object falling into a rotating black hole would not necessarily be crushed out of existence but might escape along a space bridge linking a black hole in our universe to a white hole in another universe.

The geometry of space-time inside a rotating black hole was calculated under idealized conditions; otherwise the equations cannot be solved. One fact that was neglected was the presence of material in the universe, and it has been shown that matter falling into the black hole might smash the internal geometry and destroy the space bridge.

In 1923, the Committee on Measurement of Geologic Time by Atomic Disintegration of the National Research Council was appointed.

The Royal Observatory on Blackford Hill in Edinburgh are planning a £160,000 extension. . . . There are a total Royal Observatory staff complement of 10^8

CONFIRMATION OF UNIFIED THEORY

Physicists in America announced early this week an experiment that confirms one of the most important theories in physics since Newton's theory of gravitation. The theory, originally proposed by Professor Steven Weinberg, of Harvard University, and Professor Abdus Salam, of Imperial College, London, states that the force that drives radioactive decay and the force that holds atoms together are two different aspects of the same force. If that unifying theory is correct, its importance is comparable to that of Newton's theory of gravitation, which showed that the force responsible for a falling object is the same force that keeps the Earth in orbit round the Sun.

But the Weinberg-Salam theory has run into trouble recently because of an apparent fundamental difference in the properties of the two forces it seeks to unify.

But now an experiment has been done that provides convincing confirmation of the unified theory. The experiment did not use atoms, which are complicated, but instead the physicists observed what happens when aligned electrons are scattered off protons—the nuclei of hydrogen atoms. According to the unified theory there should be a small left-right difference in that experiment.

The results, announced by Dr Dick Taylor, of Stanford University, are that the left-right difference is seen in exactly the amount predicted by the unified theory.

IS THE EARTH EXPANDING?

The surface features of the Earth change so rapidly, on a geological time scale, that it is difficult to be sure that the Earth is the same size now as it was in the past. The question is important for cosmology, the study of the overall behaviour of the universe, because some cosmological theories predict that the Earth is slowly expanding.

An accurate answer has been obtained by scientists at the Australian National University, in Canberra, who estimate that the size of the Earth has not increased by more than 0.8 per cent over the past 400 million years. That presents difficulties for some cosmological theories, but the scientists place even stricter constraints on the theories with arguments based on the appearance of the planet Mercury.

Rocks brought back from the Moon have been dated and show that the bombardment happened 3,500 million years ago. Photographs of the surface of Mercury show no evidence of any of the geological changes that would be expected if the planet had expanded since the bombardment. The radius of Mercury is the same now, give or take a kilometre, as it was 3,500 million years ago.

PLUTO'S MOON

Astronomers at the United States Naval Observatory today announced the discovery of a new body in the solar system which they believe is a moon circling the distant planet Pluto.

The satellite was spotted as an elongation of Pluto on a telescope photograph which had previously been regarded as a defect in the image.

Astronomers have now recalculated Pluto's diameter as only about 1,500 miles. Previously it was thought to be about 4,000 miles.

PROGRESS IN NUCLEAR FUSION

A cautious congratulation was given yesterday by the Atomic Energy Authority to an American team at Princeton University which has produced temperatures of more than 60 million °C in a machine that fuses together atoms of hydrogen to form helium. The reaction, which produced the highest recorded temperature in a man-made experiment, comes from the energy released as a by-product of nuclear fusion.

The long-term aim of the research is to develop a new type of nuclear reactor for generating power, but the experimental machines cost more than £100m.

1978

MICROWAVE OVENS ARE JAMMING

Sir Bernard Lovell, director of the radio telescope base, says British housewives are being sold microwave ovens which are so badly constructed that they play havoc with reception from space.

The ovens are licensed to operate outside protected bands, but are so poorly constructed they emit radio frequency energy far outside their assigned bands.

If used within a few miles of the telescope, they cause havoc to the reception of the very low intensity signals from space.

ANIMALS AND EARTHQUAKES

ANIMALS in a safari park near the epicentre of the moderately strong earthquake which shook San Francisco and northern California behaved erratically for hours before the tremors

Their behaviour, observed by many full-time keepers and trainers, supports the belief that animals are especially sensitive to seismic activity.

While some scientists are doubtful, others believe that many animals can hear either high or low frequency sound waves or booms in the earth's crust which are inaudible to the human ear.

The safari park is monitoring the behaviour of its 200 species of animals as part of the United States Geological Survey. When the earthquake was over all the animals returned to their regular behaviour patterns.

ENDLESS JOURNEY OF PIONEER SPACECRAFT

Pioneer Eleven established that Saturn's concentric rings of rock or ice chunks are extremely cold (minus 350 deg F), that the rings absorb charged particles from the Saturnian stratosphere, and that Titan has a smoggy atmosphere.

Most importantly, however, the pathfinding spacecraft showed that the outer edge of the rings can be navigated safely.

Radio contact is expected to be lost with the sturdy, 570lb craft in 1985 when it will be on the very outskirts of the solar system.

Just in case it should eventually encounter intelligent beings elsewhere in the universe, it carries a 6in by 9in plaque symbolically explaining its origins on Earth.

A naked man and woman are portrayed on the plaque, with the man's hand raised in a gesture of goodwill.

Originally the couple were holding hands, but this idea was discarded because it was thought a non-human civilisation might think the couple were one creature.

ANTI-MATTER FOUND

Instruments sent aloft in a 60-storey high balloon by scientists at New Mexico State University have found the first evidence in nature of anti-matter—particles that are the exact opposite of all matter and for which scientists have been searching for 50 years. The instruments carried to a height of 120,000 ft detected 29 particles of anti-matter.

WATER IS FOUND ON MARS

LARGE areas of water have been discovered on Mars, it was announced yesterday, increasing the likelihood that life — perhaps in bacterial form — exists on the planet or beneath its surface.

Two wet regions have been located by a water-detection device on a Viking spacecraft orbiting the planet and by an earth-based telescope using a spectrograph.

The two Martian sites, called Solis Lacus and Noachis-Hellespontus, are 400 miles and 720 miles in diameter respectively.

A STORMY WEEKEND ON JUPITER

BOLTS of lightning on the dark side of Jupiter were observed over the weekend from a distance of more than four million miles, making it the third planet so far where lightning has been seen.

Flashes nearly 20,000 miles long were seen by the American spacecraft Voyager I as it left Jupiter on its long journey towards the ringed planet Saturn.

The only other planets known to have lightning in their atmospheres are Earth and Venus.

GRAVITATIONAL LENS HINT

A few months ago, astronomers detected two quasars very close together and with almost identical spectra. Several possible explanations were proposed.

The more conventional explanations assume that the quasars really are two distinct objects ; the similarities in the spectra can then be ascribed to coincidence or to identical evolutionary histories of the objects. Several difficulties arise with these explanations however.

Certain features in the two spectra indicate that the light from the quasars is being absorbed—again, identically in both cases. Such absorption features can arise in various ways, but if the two objects really are distinct, further coincidences or unusual physical mechanisms have to be involved.

The final possibility raised by the astronomers was that there is only one quasar and that some object (probably a galaxy) near the quasar's line of sight is bending the light by means of its gravitational field, in a manner similar to a spectacle lens, to form two images. the new measurements indicate the presence of a very faint object between the two images which the astronomers suggest is the galaxy, or gravitational lens, itself. These tests do not constitute proof, but until some conclusive difference is discovered between the two quasar images the gravitational lens hypothesis must be considered the most likely alternative.

Since the estimated number of stars in the universe is 1 to the 20th power.

Telescopes orbiting other planets will have another immense advantage. It will be possible not only to vary the distance between the antennae but to also change their direction of orientation. Up to now, observers on Earth have only had a flat two-dimensional representation of the Universe. Cosmic telescopes will enable us to view everything from a new angle. With their aid, radio astronomers will for the first time have the opportunity to view distant objects from their previously "blind side" and thus obtain a three-dimensional picture of the Universe.

As NASA (National Aeronautics and Space Administration) in the United States increasingly concentrated on its manned flight programme, so its programme of planetary exploration declined. The two major encounters during the decade—with Saturn and Uranus—had been planned long before. The Russians weighed in with the first soft-landers on Venus, but the most original space investigation of the decade was not of a planet at all. Comet Halley returned to the Sun's vicinity in 1986, and a series of probes were dispatched for a close-up examination of its structure and composition. The most important of these was the European Giotto probe which passed close enough to the cometary nucleus to examine it in detail.

Challenger space shuttle disaster

Shortly before, there had been another important 'first' for space astronomy—the launching of IRAS, the first satellite capable of observing the Universe in the infrared. Besides detecting comets, asteroids and dust within the solar system, this provided extensive new data on dust clouds round other stars, as well as new information on distant galaxies.

Both the USSR and the USA extended their programmes of human activity in near-Earth orbit. In the mid-1980s, the Soviet Union established the first permanently manned space station, with individual astronauts visiting it for varying periods of time. The Americans were more concerned with developing a reusable space shuttle. The utility of this approach for carrying out a variety of missions in orbit seemed to be establishing itself; but disaster struck the Challenger launch in 1986, killing all the crew. The investigation of what had gone wrong, and the recriminations, delayed NASA's plans for the rest of the decade.

Research on fundamental particles and on the nature of the early Universe explored increasingly esoteric possibilities during the 1980s. The idea of an 'inflationary' Universe, which could give an initial push to the 'Big Bang', provided one much-discussed model, and was linked to contemporary discussions of GUTs (Grand Unified Theories of fundamental forces). The detection of W particles and neutral z particles (together with previous observation of neutral currents) gave strong support to the theory, developed in the previous decade, which linked electromagnetic and weak nuclear forces. Theoretical attention continued to concentrate on quarks. The new interest of the 1980s was in 'string' models, in which, effectively, quarks are envisaged as the ends of a 'string', rather than as point particles. It became apparent that such models could simplify the equations, but only so long as space was assumed to have eleven dimensions. This approach was soon extended into 'superstring' theory, which similarly required an assumption of multi-dimensional space (though not necessarily of eleven dimensions).

The nuclear physics that hit all the front pages, however, was the catastrophic explosion of one of the reactors at Chernobyl in the Soviet Union. For days, Europe watched apprehensively as the radioactive products from the explosion circulated round much of Eastern and Western Europe. The impact on the environment was obvious in this case. Many other environmental concerns of the decade were less easily pinned down. An example was the continuing debate about the 'greenhouse' effect. Was the average temperature at the Earth's surface increasing appreciably because of atmospheric pollution, or not? One new concern related to ozone. It was found in 1985 that there was a hole in the ozone layer over Antarctica. Was this also due to atmospheric pollution, and, if so, what effect might the ultraviolet radiation allowed through have on human beings?

The 1980s was a period of unexpected announcements in physics. One which spawned a rush both of research and of publicity was the discovery that superconductivity could occur at quite high temperatures—up to that of liquid nitrogen. Alongside such an accepted breakthrough came a number of claimed fundamental advances which proved harder to substantiate. One was the claim that neutrinos have a finite mass (whereas they had always been supposed to have zero mass). Such a discovery would have greatly helped astrophysicists, who have long speculated that there is more mass in the Universe than they can observe directly. Other claimed observations were of the long-sought magnetic monopole and of a fifth fundamental force. However, most controversy centred on the claim at the end of the decade that it was possible to produce nuclear fusion at low temperatures. The occurrence of such 'cold fusion' was thought to have been detected as the result of chemical experiments. But subsequent studies by physicists failed to confirm that nuclear reactions were involved.

One fascinating feature of the cold fusion debate was that it tended to by-pass the scientific journals—the traditional channels of scientific communication. Instead, the researchers typically used electronic means of communication to speed the flow of information. By the end of the 1980s, information technology was making an impact on all areas of physics. The number-crunching of new supercomputers carried on the long tradition of complex calculation in physics. What was new during the decade was the rapidly growing power of personal computers (reducing the need for physicists to access mainframe computers) and the joining together of these microcomputers via networks. As the decade came to a close, the mutual dependence of physics on computers and computers on physics was growing even stronger.

1980s CHRONOLOGY

1980
- There are claims that the neutrino may have a mass
- The quantised Hall effect is discovered
- Voyager space-probes fly by Saturn
- The idea of an inflationary universe is proposed

1981
- The cosmic background radiation is found not to be entirely isotropic
- The AIDS disease is identified

1982
- The US Space Shuttle becomes operational.

1983
- The first infrared satellite, IRAS, is launched

1984
- Optical disk storage of computer data developed

1985
- First quasicrystal discovered
- A hole is detected in the ozone layer over Antarctica
- The technique of genetic fingerprinting is developed

1986
- US Space Shuttle blows up after launch
- There is a nuclear reactor accident at Chernobyl in the USSR
- The first 'high temperature' superconductor is discovered
- Comet Halley returns to the Sun's vicinity
- Voyager 2 flies by Uranus
- The idea of a 'Great Attractor' in the Universe is proposed

1988
- US patent is issued on a genetically engineered mouse

1989
- Major controversy over the claim that cold fusion is possible

CURRENTS IN ALASKAN PIPELINE

The Alaskan pipeline was built over the past decade to transport oil from Prudhoe Bay in North Alaska to Valdez on Alaska's south coast, a distance of more than 700 miles. The pipeline is made of steel with a very low electrical resistance; it could be described as a giant electric wire.

Electric current will flow within a wire only if a voltage is applied, and under normal circumstances the pipeline, which is buried underground for a considerable proportion of its length, would not experience electrical voltages from any earthbound source. However, electrical disturbances in the earth's upper atmosphere can indirectly cause voltages at the earth's surface, and particularly so in Alaska. Those voltages cause the electric currents to flow in the walls of the pipeline.

Dr Wallace Campbell, of the United States Geological Survey in Denver, Colorado, has measured those electric currents and obtained values of over 50 amps during periods of moderate atmospheric disturbance. (By comparison, the average household plug is designed to carry up to 13 amps.) During the periods of greatest disturbance, about every five years, Dr Campbell expects currents of over 1,000 amps to occur.

NEW APPROACH TO FUSION

The council has funded the project, which started three years ago, into a concept known as inertial confinement fusion (ICF). It is estimated to have cost less than 1 per cent of the amounts spent in the world in the past 22 years to try to perfect what is called the Zeta system.

ICF involves deuterium and tritium being inserted under pressure into a glass sphere no larger than a small pea and frozen to about four degrees above absolute zero.

The pellet is dropped into the reactor chamber, where it is bombarded by a beam of high energy ions, causing rapid evaporation of the glass shell. Evaporation acts like a rocket motor compressing the plasma inwards and raising it to a temperature of 200 million °C. The sheer inertia of the compressed plasma is enough to provide the required containment time, which is no more than a 1,000 millionth of a second, for fusion to take place.

GIANT GAS BUBBLE

A very hot bubble of gas larger than anything else found in earth's galaxy has been discovered 6,000 light years away.

Th gas, apparently in the form of a gigantic glowing ring or shell 1,200 light years in diameter, is centred on the star constellation Cygnus, the Northern Cross. Its energy is 10 times greater than all the energy emitted by the Sun since it was formed 5,000 million years ago.

"There is no known phenomenon in the galaxy that could be responsible for releasing that much energy", Dr Cash said. "The only thing we can think of is a chain reaction of exploding stars, one going off every 50,000 or 100,000 years."

Optical telescopes differ from radio telescopes, which gather sound waves that can be translated into pictures.

EVIDENCE OF NEUTRINO'S MASS

Dr Frederick Reines told the meeting that his research group at the University of California, Irvine, had discovered that neutrinos oscillate regularly between different states. His picture of the oscillating neutrino contrasts with the conventional idea that the particle is stable.

If neutrinos oscillate, then theory requires them to have mass. Dr Reines's results do not give an exact figure for the mass, but they would be consistent with something of the order of 10 electronvolts. That would be 50,000 times less than the electron and 100 million times less than the proton or neutron.

Even such a tiny mass for the neutrino could have revolutionary cosmological implications, because the universe is so full of neutrinos, believed to be left over from the big bang that started it. Every cubic metre of space is thought to contain 100 million neutrinos. Together they could more than double the present estimate of the total mass of the universe.

HYDROGEN CLOUDS IN SPACE

The universe still contains vast quantities of original material left over from the big bang that started it, according to new observations by a group of British and American astronomers. They reported that intergalactic space is filled by a medium of extremely diffuse ionized hydrogen, with slightly denser clouds of hydrogen gas embedded in it.

Their observations indicated that the total quantity of material in intergalactic space is roughly equal to the amount that has condensed to form galaxies. This is ten times too little to " close the universe ", giving it enough mass eventually to pull it together by force of gravity.

Although it is possible astronomers will eventually find enough matter, perhaps in the form of some exotic and currently undetectable particle, all the recent evidence points towards an " open " universe that will go on expanding for ever.

THE SUN'S NUCLEAR FIRES

The Sun's energy is generated by thermonuclear reactions in its core. Those reactions, by which hydrogen is converted to helium, are the same in principle as those used in the hydrogen bomb, and in the future perhaps they will be used in nuclear fusion reactors to generate electricity.

If we detect the neutrinos, we see particles that left the Sun's centre only about eight minutes previously. The light and heat, on the other hand, result from nuclear reactions that took place in the distant past.

In the early 1960s astronomers realized that neutrinos provided an excellent means of testing the then highly successful theories of stellar structure. They found that the Sun's temperature, composition, diameter and radiation were all consistent with a particular set of central nuclear reactions. But it soon became clear, and remains so today, that the numbers of neutrinos observed at the Earth's surface are well below those that had been expected.

TIMEKEEPING BY TEMPERATURE

The American-based Bulova watch company has perfected a miniature cell capable of generating electricity caused by differences in temperature.

The device, developed at the company's Bienne Plant in Switzerland, is the result of three years' research and development by the watch manufacturers and although the unit could have many applications it will be used at first to power quartz watches.

The cell is called a thermoelectric generator and is able to harness the temperature of the body to provide between 8 and 12 microwatts which is at least a factor of three times the power necessary to drive a watch.

When the nature of a black hole has been explained several times, there's nothing more to say about it. What we need is action.

LASERS, LIGHT & LENGTH

Precise standards of measurement have long been of both intellectual fascination and practical value to scientists; and as measuring techniques have been refined, so it has become obvious that the motion of the Earth is not sufficiently regular to maintain a definition of the second as merely one 86,400th part of the day.

The proposal is that, if the speed of light at which the laser beam moves is taken as a fixed value, and since its frequency can now be measured with a high degree of accuracy (the same laser beam can be produced by anyone using the same design and materials), then the metre can be safely defined as the distance that light travels in a vacuum in a particular fraction of a second.

Lasers could now be sufficiently stabilized so that the frequency need only be measured once. After that any other similarly constructed laser would also emit that standard wavelength, and could therefore be used as a standard of length without further measurement.

NUCLEAR STRUCTURE

As conventionally understood, the nucleus of the atom is made up from smaller particles called baryons, usually protons and neutrons. Theoretical physicists have developed laws that predict the behaviour of such particles.

Those laws may have to be revised after experiments carried out at the University of California's Lawrence Berkeley Laboratory, which involved studying the interactions between nuclei when they collide at energies approaching the speed of light. The experiments showed fast-moving secondary fragments from the collision that appeared to be far more ready to collide with nuclei of other atoms than the particles in the primary beam itself.

QUARKS OBSERVED

Quarks certainly exist in the atom, where they are so tiny they have no discernible size. They combine together in triplets to make the protons and neutrons of the atomic nucleus. Quarks also have an exotic electric charge —one third or two thirds that of a proton—and should be easily detected if they escaped singly.

Most experiments to find single quarks have failed: but one, run by Professor William Fairbank, of Stanford University, continues to claim their occasional detection. Mr Fairbank is a renowned experimenter and, although greater men have been proved false before, the steady drip, drip of his results has begun to shake the now established, but unproven, theoretical convention that quarks are eternally confined in the atom.

MATTER & ANTI-MATTER

European physicists reported today that they had collided anti-matter with normal matter for the first time. The European Centre for Nuclear Research, whose experimental site straddles the Swiss-French border near Geneva, said the achievement constituted "the opening of a new window" on the basic structure of the universe.

The 12-nation research organization recently developed a technique to obtain and store dense beams of anti-protons, the anti-particles of the proton, which is the nucleus of the hydrogen atom. One week ago the scientists accelerated anti-protons to collide with a proton beam in intersecting experimental tunnels known as storage rings.

The planet Uranus is a very inactive "dormant" world. . . . The most obvious difference between Uranus and the other planets is that Uranus dies on its side

THE ONSET OF CHAOS

A theory of turbulence has been sought for many decades, and at last one is beginning to emerge. In the latest step, two independent groups of theoretical physicists at the universities of Harvard and California have confirmed earlier speculations that the best mathematical models of turbulence show properties very like those of the soundly based theories of melting, boiling, magnetization and similar phase-changes.

Thus many disparate phenomena appear to be coming under the same mathematical roof: the theory of " non-linear difference equations ·with random noise ". Those relate a property of some system (the velocity of a fluid, the number of mosquitoes in a population, the amount of money in an economy) at one time to the property at a later time, throwing in a. bit of random variation for good measure.

The solutions of those apparently innocent equations are full of fascinating variations, which can connect ordered behaviour with disordered behaviour, like smooth fluid flow with turbulence.

DOES MATTER DECAY?

Protons and neutrons until recently have been considered completely stable. In fact if they did decay at a substantial rate the whole basis of matter—not to say the gold standard—would crumble, and the Universe would be left merely as a vapour of electrons, neutrinos, and light.

The Universe is clearly not like that, so if protons and neutrons decay they must do so very slowly compared with the present age of the Universe.

The Indian experiment, done in collaboration with Japan, monitors a 100-ton mass of iron, which contains about 60,000 billion billion billion protons plus neutrons, for decays. Six decays in a year would correspond to an average lifetime for each individual proton or neutron of 10,000 billion billion billion (10^{31}) years, atomic decays being random in time. By comparison, the age of the Universe is only 15 billion (1.5×10^9) years.

FIRST STEPS IN NUCLEAR FUSION

Sixty watts of nuclear fusion reactions, the reactions that make the Sun shine and which one day may provide unlimited energy from the sea, have been created in an experiment at Princeton University, United States. Sixty watts, if it is all converted to electricity, is only enough to light a light bulb ; but the excitement is over how the reactions were generated. collided with other nuclei in the plasma, in particular deuterium (heavy hydrogen). Deuterium and helium-3 nuclei fuse to make a highly stable helium-4 nucleus, and expel an energetic proton. This proton is too fast to be retained by the confining magnetic fields, and shoots out of the plasma to be detected by equipment outside the ring.

The Princeton team detected sufficient protons to calculate that 20 million million such reactions were taking place in the plasma each second, enough to yield 60 watts if the energy released was all converted to electricity.

THE END OF THE UNIVERSE

They have taken to its logical conclusion the theory current among high energy physicists that ultimately (after some 1,000 billion billion billion years) the nuclear constituents of atoms will decay, leaving only leptons : light electrons, positrons, and massless neutrinos.

That would mean that if the universe expands for ever, which it will do if its density is less than a critical value, all structures will disappear into a lepton gas. Even black holes will ultimately radiate themselves into particles, in the famous process discovered by the Cambridge physicist Stephen Hawking a few years ago, and whatever those particles are they will eventually decay into the lightest leptons.

So it is now thought that Titan could have temperatures as low as ·2 degrees Centigrade. In those conditions, it's unlikely that any life has developed there.

ALL IS NOT WELL FOR EINSTEIN'S THEORY

if the Sun is not a perfect sphere, Einstein's prediction of the exact orbit of the panet Mercury (the nearest planet to the Sun) would be affected.

Since the Sun spins, it might be expected to be slightly fatter at the equator than at the poles, as centrifugal force flings out its equatorial mass. However, calculations show that if the Sun were spinning as a whole only as fast as it appears to spin on the surface (about once ever 25 days) the distortion would not be enough to upset Einstein.

Nevertheless, the core of the Sun might be spinning faster than the exterior — which is slowed down by the solar wind and the magnetic fields which link the Sun to interstellar space. Professor Hill's measurements show that this is the case, with the core spinning about four times as fast as the exterior.

And the calculated distortion of the Sun puts Mercury's orbit out of upset Einstein.

Nevertheless, the core of the Sun might be spinning faster than the exterior — which is slowed down by the solar wind and the magnetic fields which link the Sun to interstellar space. Professor Hill's measurements show that this is the case, with the core spinning about four times as fast as the exterior.

And the calculated distortion of the Sun puts Mercury's orbit out of reach of Einstein.

FUZZY QUARKS

Quarks, the particles that compose atomic nuclei, are not the clean, pointlike entities they first appeared to be; they are fuzzy. That is the result of an experiment at the Stanford Linear Accelerator Centre, in California, where quarks were first revealed in 1969.

However, in the interval between the two experiments scientists have developed a theory of quark interactions in which the smearing can be understood, even though the underlying quarks in the theory remain pointlike.

The theory is misleadingly called "colour", and it assumes, on good grounds, that the quarks attract one another by the exchange of another kind of particle, called gluons. Thus a quark is never alone, it is always attended by its gluon cloud.

The quark "fuzziness" is thus really a measure of the fuzziness of the gluon cloud, rather than of the quark itself.

EARLY EXERGY

The world is concerned about a shortage of energy, but really it should be exergy, not energy, that concerns us. Exergy is energy which can be "expressed", which is available to do work.

The problem is that the very early universe had no exergy at all. The universe was then uniformly hot, with no differences from one part to another. With no thermal differences, no work could be done by one part of the universe on another because heat engines cannot function without a difference in temperature somewhere in the system.

Exergy was created essentially because the universe was expanding too fast for thermal processes to keep up with it.

The main creation of nuclear exergy began 10 seconds after the beginning of the "big bang" the researchers calculate, and it was essentially complete within 24 hours. The stars, with the Sun among them, are now using up and radiating this first day's exergy.

TWINKLE, TWINKLE LITTLE NEUTRINO

Observations of ultraviolet light from deep space may indicate that one of the tiniest and slightest pieces of matter — the neutrino — has a composite structure.

Neutrinos are produced in abundance by stars and were produced even more abundantly in the Big Bang about 10,000 million years ago. There are now about 1,000 neutrinos to every cubic inch of space and matter; but they have been undetectable because they interact hardly at all with the rest of matter and carry very little, if any, mass. Recently, however, experiment and theory have combined to create speculation that the neutrinos may be detectable after all: they may decay, causing a faint emission of ultraviolet light.

The satellite observations do indeed show a "background" of diffuse ultraviolet emission which might be produced by neutrino decay.

MAKING NEW ELEMENTS

The basis of the new alchemy is to make two atoms of different, naturally occurring elements collide with sufficient energy for their nuclei to fuse and so form a much heavier atom. In their experiment, a particle accelerator was used to fire a beam of iron nuclei, each containing 26 protons, into a target of bismuth atoms, each with 83 protons in their nuclei.

After bombarding the bismuth target with billions of iron nuclei just a single atom of element 109 was detected, and five-thousandths of a second after it was formed it began to decay, releasing first an alpha particle (composed of two protons and two neutrons) with an energy of more than eleven million electron volts. The remaining nucleus continued to break down by radioactive decay and the world's only atom of the new element was gone.

OBSERVING FATIGUE

NMR spectroscopy makes use of the fact that the nuclei of certain atoms behave as tiny magnets, aligning their magnetic moments either with or against an applied magnetic field. The nuclei can absorb radiowave energy to flip from one alignment to another. Analysis provides detailed information on the nature and amount of the chemical species with magnetic nuclei.

The decision to use NMR to investigate the two sisters was taken after one sister suffered her worst attack of muscular weakness and sickness. Hospital tests after the attack showed abnormally high blood levels of lactic acid, a chemical that normally accumulates during exercise and is thought to be a cause of muscle fatigue. A central role in such reactions is played by molecules which contain phosphorus. Since the common isotope of phosphorus, ^{31}P, has a magnetic nucleus, it was feasible to use NMR spectroscopy to measure the levels of molecules containing phosphorus in the muscular tissue of the sisters before and after exercise.

Compared to spectra from healthy individuals, those of the sisters clearly showed an abnormal response.

Some curious Britishisms, such as the use of Greek letters to designate specific stars, mar an otherwise interesting text.

THE GREENHOUSE EFFECT

The greenhouse effect is caused by an increase of carbon dioxide in the atmosphere from the burning of fossil fuels in power stations and from the emissions of motor vehicles.

Warming of the Earth's surface occurs because sunlight penetrates the atmosphere but the carbon dioxide acts as a screen to prevent the escape of heat radiation into space.

Estimates suggest that the global average temperature will rise by 2 deg C (3.6 deg F), by the year 2040, and by 5 deg C (9 deg F) by 2100.

These changes would cause a profound alteration to the world's climate. The most devastating damage would by caused by the transformation of temperate agricultural regions into arid areas.

SATELLITE STUDY OF THE OZONE LAYER

The first results are coming from a scientific satellite, the Solar Mesosphere Explorer, designed specifically to study the ozone layer in the upper atmosphere. Its findings are particularly important in trying to resolve controversies about possible destruction of the ozone layer, which protects life on the earth's surface by filtering out the high levels of ultra-violet radiation from the Sun.

The anxiety is that oxides of nitrogen and chlorine from industry and agricultural processes will disturb a delicate series of chemical reactions which occur at different levels in the upper atmosphere. The satellite's instruments observed depletion of 70 per cent of the ozone at 65 degrees latitude, at a height of 78 kilometres, on the morning of the day that the proton burst was detected. In the afternoon the ozone level was depleted by only 10 to 20 per cent.

THE INFRA-RED ASTRONOMICAL SATELLITE (IRAS)

INFRA-RED VIEW OPENS UP HEAVENS

the helium-fuelled infra-red astronomy satellite (Iras), has been giving scientists the first glimpse of some of the hottest material in the universe, hidden hitherto from ground or space-borne telescopes.

Five new comets, a ring of dust around the Andromeda galaxy, and a newly-forming star the size of the Sun in a sector of the Milky Way nearby are among the findings.

About 2,000 objects have been viewed each day and the accumulated data may take 10 years to sift.

SUPERSYMMETRY

there are reasons why supersymmetry is so attractive – and these are largely that it offers more conceptual unity than its competitors, the so-called "grand unified theories" or GUTs.

The GUTs attempt only to unify three of the four fundamental forces of nature (electromagnetism, the weak nuclear force and the strong nuclear force; the fourth force is gravity). Supersymmetry links the first three forces, and all the particles in one scheme, and may be neatly extended in a theory called "supergravity" to include gravity also. The result is some very beautiful mathematics which bears a tantalising similarity to reality. Unfortunately, it predicts whole classes of particles which do not appear to exist, a problem known as the supersymmetry (credibility) "gap".

The GUTS, on the other hand, are not so ambitious, but are full of arbitrary, unexplained numbers.

A TUNNEL PURPOSE BUILT FOR COLLISIONS

The machine on which the W and Z particles were uncovered was used to accelerate protons. But proton machines do not allow an extensive study of the weak force because of unavoidable interference resulting from the strong force which manifest themselves with protons, which are complicated particles in comparison with electrons.

Therefore an electron-positron machine provides better experimental conditions. It will cost the 16 member countries of Cern £300m to build. Much of that cost lies in the civil engineering work for a tunnel. It could have been a smaller circumference, but the eventual choice was a compromise between many technical factors governimg the behaviour of particles in an accelerator.

COLLIDING BEAMS MAY SOLVE SCHRODINGER EQUATION

Scientists in the United States, Britain and France are collaborating on a low-energy physics experiment which involves colliding electron and hydrogen beams. They expect their first full set of results to be available for analysis next spring.

The object of the project is to provide more precise data for atomic theorists. For the first time the Schrodinger Equation, which defines the interaction between atomic particles in collision, will be able to be solved for three particles.

It has been solved in the past for two bodies but not for three.

The three particles consist of the electrons in the electron beam and the electron and proton contained in the basic hydrogen atom. An important part of the experiment is monitoring the behaviour of the electron spin during the reaction of the beams, which is a crucial factor in solving the Schrodinger Equation for three bodies.

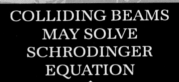

EVIDENCE OF APPAREN

We are able to give below the substanc yesterday at the London School of E Science by Sir William 11. Beveridge, School), setting out a new theory of per and the crops:-

Many attempts have been made in th world-wide fluctuations of trade a recurrence of commercial crises with s meteorological cycle affected the weath this periodic fluctuation of the ec connected with anything in meteorology? I thin From the Repeat of the Solar Ph

1983

A newly discovered comet will skim past earth next week at the "extremely" close range of 3 miles .

1984

NEW MICROSCOPE COULD SEE SINGLE ATOM

Work has begun on building the world's most powerful microscope, which could allow scientists to observe directly individual atoms of almost any solid material.

The instrument will need such delicate movements to bring objects into visual focus that it will have to be operated under computer control.

The most powerful instruments can resolve objects down to about two angstroms in size. Most atoms are spaced about one angstrom apart.

The high resolution which is expected with the new instrument depends on an invention made at Chicago University to correct the inherent distortion created by the magnetic lenses used in electron microscopes.

HIGGS' BOSON

Higgs's boson, if indeed it exists, is one of those tiniest of fragments of matter called subatomic particles. Although every object, living or inanimate, is made up of them, they are only seen for fractions of a second when scientists cause slightly larger particles to collide in powerful laboratory accelerators.

The current excitement stems from a report that something very much like a Higgs's boson has at last been observed. But this new particle did not behave in quite the way that Peter Higgs, a theoretical physicist at Edinburgh University, predicted for such a fragment thrown like a spark from a collision in an accelerator.

An international team of 78 physicists were involved in the discovery of this elusive object, which has been given the name Zeta.

GRAND UNIFIED THEORIES

According to schemes for the grand unification of theories about the cosmos, particle physics and cosmology were clearly one and the same thing in the cataclysmic conditions of the first few minutes after the big bang that formed the universe.

During the subsequent 20,000 million years or so nature has been much less spectacular and clues to the big bang have mellowed with age.

The astronomers unravel those tell-tale traces, nevertheless, with their optical, radio, X-ray, ultraviolet, and infrared telescopes. The particle physicists search for clues by mimicking in accelerators the sort of conditions involving immensely energetic reactions between matter that must have occurred if the big bang theory is correct.

Hence, the astronomers are engaged in the physics of the extremely large, and the particle physicists are probing the behaviour of the extremely small by searching for the tiniest subatomic fragments from which all matter was originally derived.

The excitement caused by the discovery last year of the so-called W and Z particles at CERN was due to the encouragement given to grand unified theories.

Soviet plans have concentrated on drilling of several very deep boreholes (12,000–15,000 km).

SIXTH QUARK COMPLETES FAMILY

One of the most exhaustive experiments by particle physicists has ended with the discovery of one of the tiniest building blocks of matter. It is called the sixth quark.

The discovery of the new object, now called the top quark, was made by analysing immense amounts of data gathered by a team of 151 scientists from nine countries who collaborated in an investigation at Cern, the European Laboratory for Particle Physics, near Geneva.

The everyday world can now be explained in terms of the family of quarks in one group of basic building blocks and a family of leptons in another.

The sixth quark is produced when a W boson undergoes a transition, but its existence was predicted. In fact, its particles table by Top quark.

existence was necessary if theories about the forces of nature were to make sense.

CARBON COATING HELPS NUCLEAR FUSION

The Joint European Tokamak (Jet), the world's largest fusion energy machine, designed to emulate the interior of the sun and test the possibility of making energy by the nuclear fusion of atoms of pure hydrogen, has made a big leap forward, right at the end of its recent experimental run.

The discovery appears to remove the "impurity problem" from which the £100 million machine had been suffering. Jet has been bothered by atoms of heavy metal, chromium and nickel, finding their way into the ring of superhot hydrogen plasma in which the solar reactions are planned to take place. These impurities radiate energy from the plasma and make it difficult to heat further to cause atoms of hydrogen to fuse together.

But in the last two weeks of a recent run, at the centre near Oxford, physicists found that the problem could be dramatically reduced by giving a coat of carbon to the inner surface of the doughnut-shaped vacuum chamber in which the plasma circulates.

ANIMAL 'MAGNET COMPASS' FOUND

Evidence has been growing to suggest that those avian species which migrate over vast distances have a tiny magnetic compass to guide them. Support for the idea of such an inbuilt inertial navigation system has come from research showing that birds are not the only species to rely on the magnetic field of the Earth.

Fish also depend on the variations of the magnetic field to travel particular routes.

Moreover the navigation signals used by one aquatic family, the Yellowfin tuna called *Thunnus albacares*, are obtained from microscopic crystals of magnetite (the natural black oxide of iron, with formula Fe_3O_4) in tissue in the skull of the fish.

Each crystal is about one-millionth of a millimetre in size, arranged apparently in a chain or cluster which is enough to give the fish a sensitive system for following the magnetic lines of the Earth's field.

The other day, landing at Heathrow, the pilot told us that although they were watching like hawks the plane had been landed by electronics.

I am glad to say that before this he did not ask for a show of consenting hands.

Scientist finds proof of plant life on star

The strongest evidence yet that plants surround a nearby star has renewed speculation about possible "life on other worlds",

JAPAN AND FIFTH-GENERATION COMPUTERS

European industry will find it harder than it believes to catch up with the Japanese in its efforts to develop the so-called fifth generation of computers. The prediction carries some weight because it is from the West German Institute of Economic Affairs.

Two key reasons lay behind the assessment. In Europe, the efforts to produce the new generation of equipment and systems were seen as "too splintered to achieve the necessary programme of consolidation". And second, even in the US, there was not thought to be an obvious programme of research and development that was "all-encompassing and sufficiently capital-intensive in direction" to make up the leeway.

The current fourth generation is evolving into large-scale integration with several thousand devices on a chip.

The most commonly discussed feature of the forthcoming generation is the ability to allow "knowledge-based" or artificial-intelligence systems to be developed.

In conceiving the Esprit approach, advisers to the European Commission prepared a list of the key technologies which they believed were needed, then set about devising a plan of how those technologies could be acquired.

In that planning process a number of broad fields of application were identified from a market point of view. These sectors included telecommunications and satellites, integration of communication and data-processing technologies for purposes such as office automation, home and personal electronics, and industrial measuring and control equipment.

RANGE OF CAT'S HEARING EMBRACES 'UNCERTAINTY PRINCIPLE'

According to Dr William Bialek and Dr Allan Schweitzer, of the University of California in Santa Barbara and San Francisco, experimental measurements show that cats can respond to sounds of extremely low intensities, lower than a few ten thousandths of a femtowatt (ten followed by nineteen noughts).

Such intensities move the tiny hair cells in the inner ear of the cat by less than ten trillionths (ten followed by eleven noughts) of a metre, less than a tenth of an atomic diameter, Bialek and Schweitzer calculate. These distances are so small that Heisenberg's uncertainty principle must take effect, the two researchers believe.

Bialek and Schweitzer calculate that the quantum vibrations of the hair cells (the fundamental sound detectors of the ear) are of the same order of magnitude as are the minimum sounds that a cat can detect.

Therefore, it seems a cat can hear sound right down to the minimum noise level implied by quantum mechanics, far fainter than the much larger interfering noises of heat vibrations in the ear or even the noises of turbulence in blood flow.

SHUTTLE TO TEST STAR WARS LASER

The space shuttle Discovery will blast off today on a commercial, scientific and military mission which will include the first space test of laser technology for President Reagan's controversial Star Wars programme.

In the test of a laser tracking system, a low-power laser beam will be fired at a mirror, eight inches in diameter, attached to the shuttle's window. The object of the test is to see whether a ground-based laser could be aimed accurately enough to destroy a Soviet nuclear missile.

ATMOSPHERIC CHANGE AND THE POLAR ICE CAPS

Conditions essential to life depend on a certain level of carbon dioxide in the atmosphere, providing an insulating layer. But indirect records going back to the beginning of the industrial revolution suggest that the natural balance of carbon dioxide in the atmosphere would have been concentration of about 270 parts per million.

Measurements made at remote observatories during the past 40 to 50 years show a steady increase. The latest results come from scientists at the Physics Institute of the University of Bern, showing that during that period the concentration has risen by more than 25 per cent.

The Polar ice caps provide a unique record because annual surface melting and compression has created layers, akin to the growth rings of a tree, which trap air samples going back hundreds of years.

The data locked in the ice core gives a record going back to 1750, when the atmospheric carbon dioxide was 260 parts per million. Through the eighteenth century, carbon dioxide in the air stayed below 280 parts per million, and the curve plotted by the Bern team rises to 290 ppm in Victorian times and up to 330 ppm in the late 1970s.

Today the figure is about 345 ppm.

STAMPS COMMEMORATING HALLEY'S COMET ISSUED FROM HALLEY'S STATION
IN THE ANTARCTIC

HAZARDS OF VDUS

The purpose of a VDU is to emit visible light, which allows the text to be seen. But the worry is that the box emits other and noxious rays. The prime suspects are X-rays, ultra-violet light, micro-waves and magnetic fields.

Given a sufficient intensity and exposure time, each of these can produce adverse reactions in man or in animals in special circumstances. It is not surprising, therefore, that anything that radiates and is near our bodies should arouse doubt, suspicion, and even fear.

This concern has been fuelled by a number of user reactions, such as eye-strain, back-ache, abnormal pregnancies and skin problems. Their causes are not always known but several potential explanations are beginning to emerge.

A strong point appears to be posture. If our spine is distorted for any length of time it will rebel. This may be the result merely of the user's desire to avoid a plaguing glare reflection that interferes with the legibility of a text.

Even though the answer may be wrong, in every case it was the same. While the presence of traces of such radiations could sometimes be detected, their intensity was far too low to cause concern and much lower than in other appliances such as micro-wave ovens.

There have been several detailed studies undertaken in about six different countries to find out how much of the potentially noxious X-rays and ultra-violent light is emitted by a variety of VDUs.

GIOTTO, the £40m spacecraft . . . will intercept Halley's Comet about 92 miles from Earth, and will pass about 300 miles from its nucleus. By then, Giotto will have travelled 450 miles from Earth .

It has been financed jointly by the British (80 per cent) and Dutch (40 per cent) science research councils.

RADIOACTIVE RUSSIAN DUST CLOUD ESCAPES

A major nuclear power accident in the Soviet Union yesterday sent a cloud of radioactivity drifting across much of Scandinavia.

The alarm was first raised at a Swedish nuclear plant at Forsmark, on the Baltic coast, where staff at first thought their own reactor was leaking. But as further reports of unusually high radioactivity began to come in from Stockholm, Helsinki and elsewhere the Soviet authorities admitted than an accident had occurred at one of their plants in the Ukraine. A nuclear power reactor had been damaged and there were casualties.

Although the Tass news agency report was slightly ambiguous, the Chernobyl plant which has suffered the accident is believed to be one of a complex of four light water reactors built by a lake just north of Kiev — about 800 miles from the Scandinavian coastlines where the radioactive plume was being monitored.

Helsinki reported six times the normal background level, Stockholm twice the usual level and Oslo 50 per cent higher than normal. Swedish scientists compared the abnormal readings with those recorded during the 1970s, when China was testing nuclear bombs in the atmosphere. But analysis of the isotopic content of the Scandinavian dust samples confirmed the Tass statement that a nuclear power reactor was the source, probably as a result of cooling failure and severe overheating.

For the Russians to admit to an accident of any kind sets a precedent, which may be a reflection of Mr Gorbachev's policy of greater openness. But Tass sought immediately to put the incident in perspective by listing several foreign nuclear disasters, including the near

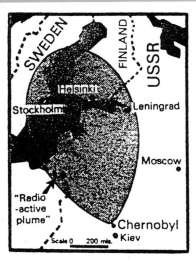

meltdown of the American Three Mile Island reactor in 1979. The Chernobyl accident, the Soviet news agency said, was the first to have occurred in the Soviet Union, ignoring reports reaching the West of a serious accident in the Urals in 1958 which is believed to have killed many people and contaminated a vast area when a nuclear waste dump exploded.

Yesterday's announcement said simply : " An accident has occurred at the Chernobyl nuclear power plant as one of the reactors was damaged. Measures are being taken to eliminate the consequences of the accident. Aid is being given to those affected. A government commission has been set up."

In Sweden, 700 workers at the Forsmark plant north of Stockholm were evacuated and radio warnings were broadcast for other people to stay away, before it became clear that the radioactive plume was drifting in across the Baltic with the south-easterly winds.

At first, suspicion was directd towards the Soviet Baltic republic of Lithuania, where the Ignalina power station is reckoned to be the world's largest nuclear plant. Finally, the Soviet announcement put the source in the Ukraine, whose distance from the Baltic suggests that the Russians have a major disaster on their hands.

Three Mile Island was a di-saster as far as the American nuclear industry was concerned, but it did not release large amounts of radioactivity as this accident has evidently done.

Accidents like this are extremely difficult to disguise or hide because even the smallest sample of wind-blown dust, when put through a gamma ray spectrometer, will indicate that the level of radioactivity is unusually high, and that the source is either a nuclear explosion, a leaking reactor, or an accident involving a reprocessing plant for spent power station fuel.

The Chinese nuclear tests conducted in the atmosphere during the 1970s, for example, were readily identified within a couple of weeks by laboratories attached to British nuclear power stations as well as Swedish ones. All that was needed was a scrap of sticky muslin stretched on the roof and a spectrometer — admittedly a highly sophisticated piece of scientific equipment — to analyse the gamma rays given off by the minute particles of dust it collected.

If the Soviet plume begins to drift towards Britain — and there have been easterly winds — the National Radiological Protection Board will quickly pick up the signs from its fall-out monitoring stations in Glasgow, London, Belfast, Bridgend, Shrivenham, Glos, and Chilton in Oxfordshire.

Sweden's Environment Minister, Birgitta Dahl, said yesterday that while the contamination reaching her country was not considered harmful to health, her government was concerned that it had not been given any warning by the Soviet authorities and had still not been given any real information about the source, or whether the leakage must be expected to continue. Despite her reassurance, however, it was clear last night that many Scandinavians intend to stay indoors until they are certain it is safe to go outside.

Britain has no need to fear the radiation, the Protection Board said. The fall-out reaching Scandinavia was only twice the natural background level of radiation. This was so low that it should not cause any harm if it was ever carried to Britain. There was no need for any special precautions to be taken in this country.

Anthony Tucker adds: The two nuclear reactors at Chernobyl, near Kiev, are each of 1,000 megawatts electrical—comparable with the largest stations in Britain—but are of a water-cooled and graphite moderated design that is used only in Russia.

The graphite cores have no secondary containment. Both reactors have been in use for several years and each would contain about 10,000 million curies of radioactivity in fission products and about 3,000 million curies of what are called heavy actinides—which is the uranium fuel and its derivitives such as plutonium.

Accident planning always assumes that, at worst, about 10 per cent of any radioactive core will be vaporised and that the volatile elements — such as iodine and the rare gasses — will dominate releases of radioactivity when accidents are small.

But reports already suggest that caesium is a measurable component of the radioactive cloud coming from Russia. This suggests that a substantial proportion of a reactor core may have vaporised with massive releases of radioactivity.

Close to the plant the radiation dose levels may be lethally high, but the amount of radioactivity so far detected in the west presents no immediate hazard especially as the isotopes are apparently of a kind which do not accumulate in bone or other tissues.

The caesium isotopes which have half-lives of 2½ and 30 years respectively behave like salts in the body and therefore are excreted fairly rapidly. However, when they fall out on land they may remain for years raising the local background radiation levels. If fallout is heavy as it will be in Russia, areas may remain too hazardous to enter for many years.

PROGRAMME IN TURMOIL AFTER SHUTTLE TRAGEDY

The future of America's ambitious space shuttle programme was in complete turmoil last night after a devastating mid-air explosion, just one minute after takeoff, which killed the crew of seven, including the first ordinary citizen on the programme, a New Hampshire school teacher, Mrs Christa McAuliffe.

Nasa said last night that it had suspended operations of the shuttle indefinitely, pending the findings of a board of inquiry. But in a televised address to the American people, President Reagan stressed that the space programme would go ahead.

Some 60-seconds into the launch, as the elegant spacecraft soared nine miles into the deep blue sky above Cape Canaveral, it turned into a ferocious fireball as a bolt on the main fuel tank exploded, detonating an eruption with the power of what was described by Nasa scientists as a "small nuclear weapon."

Last night, after examining video film of the takeoff, Nasa experts said that the first sign of trouble came when small flames began to leak out from one of the two giant solid rocket boosters which propel the Challenger orbiter into space. One of the solid rocket boosters fell away and the liquid fuel of hydrogen — which is highly volatile — exploded.

The commander and crew of the space shuttle may have been able to see that something was wrong from the computers and equipment in the small crew capsule at the front of the shuttle. But the explosion came so quickly that it was unlikely they had time to take any precautionary action. This shuttle mission, unlike some of the earlier craft, carried no ejector seats. They had been considered an unnecessary safety precaution.

SATELLITE PHOTOGRAPH OF THE CHERNOBYL NUCLEAR POWER STATION AFTER THE EXPLOSION

PROVING EINSTEIN WRONG?

Einstein made this apparent equality between "inertial" mass and "gravitational" mass of an object a cornerstone of his theory of gravity.

Detailed calculations find that the acceleration of an object under gravity depends on both the mass and temperature of an object in a way that makes heavier, or cooler, objects fall faster than lighter, or hotter, ones.

The effect, however, is extremely small. Delicate experiments carried out by American and Russian physicists in the 1960s and 1970s have shown that the inertial and gravitational masses of objects are the same to one part in a million million. The effect predicted by the two theoreticians is about 100 thousand times smaller still.

NEW SUPERCONDUTORS

The fast-breaking research on superconductors, materials which carry electricity without any loss of energy, has overwhelmed the ability of scientific journals to keep up with it.

The reason for the excitement is the discovery of materials which carry current with no loss of energy whatsoever at record-high temperatures, as high as 92 kelvins (a degree equivalent to the Celsius degree, with zero at absolute zero)

Before the breakthroughs, superconductors were a multibillion-dollar business, but they were limited to applications which justify the enormous expense of cooling the materials almost to the physical limit. Now, with a host of everyday applications in sight, scores of laboratories around the world have joined the effort to understand the materials, to discover new ones and to turn them into shapes which can be used in technology.

BITES

Sharks are literally biting into the new fibre optics telecommunications business at an enormous cost per bite. The cost of a disruption in such lines across the Atlantic could run into hundreds of thousands of pounds a minute.

Repairs on a single bite into a fibre optic cable can cost as much as £150,000 and a week's worth of time to send a ship across the ocean to find the cable, haul it to the surface and resplice the glass strands inside.

THE FIRST PLASTIC MAGNET

Scientists in Russia have succeeded in making the world's first plastic magnet.

Made from material more than six times lighter than magnetized iron, the new magnet opens the way to lightweight motors in everything from toys to aircraft.

Theories that magnetic plastics might exist date back to the late 1970s. They predicted long, thin chains of molecules might behave like iron.

CHEAPER CABLING

Compared with copper-cabled networks, fibre optic LANs have a number of important advantages.

Being non-electrical, they are immune from electrical interference and therefore offer much greater flexibility in routing cables during network installation.

Unlike copper cables which have to be routed away from power cables and electrically-noisy equipment, optical cables can be run alongside them in existing cable runs previously unsuitable for data cable.

Fibre optic networks are also safer. Their glass fibre cables eliminate the possibility of electrical equipment faults being relayed through the network, damaging equipment or indeed people.

Glass fibre also offers infinitely better data security because it is almost impossible to tap into a fibre network. In addition fibre cables have a much greater traffic-handling capacity.

TALKING TO COMPUTERS

IBM showed off a personal computer in New York last week said to be able to recognize the sound of 20,000 words and print them instantly on a video screen.

A person using the newest computer speaks into a microphone and pauses between each word briefly. Once the whole text is on screen, **a verbal order causes the computer to print it on paper. A person dictating to the computer must first read aloud to it for 20 minutes so the machine accustoms itself to the person's pronunciation.**

COMPUTERS AND BRAINS

Like their counterparts in nature, neural networks are endowed with a sophisticated system of "nerves" that can transmit messages. The network can "learn" any information needed to perform tasks and make decisions.

Scientists, however, are quick to caution that neural-network technology, which grew out of research into bionics in the 1960s, is nothing like computer science as it is known today. Scientists are designing systems with special functions — particularly an ability to learn — with the hope of one day merging that capability with sophisticated robotics.

SOLITARY WAVES

Scientists at British Telecom are experimenting with a new means of sending laser signals over long distances that could lead to a revolution in communication.

The research team at BT's Martlesham research centre near Ipswich hopes to achieve its aim by exploiting the properties of one of the most bizarre phenomena in Nature: the soliton, a single pulse of energy that can, in principle, travel forever utterly unchanged.

This unique property of solitons would allow tens of millions of simultaneous telephone calls to be carried on a single optical fibre. It could also speed the introduction of videophone communication, the development of which has been held up by the relatively small amount of data current fibres can handle.

Mrs. Thatcher rejected Mr. Gibson's plans to treble the centre's £100 budget and ruled out switching money from other research projects to space.

RAINFALL CHANGES

The Norwegian Prime Minister, Dr Gro Harlem Brundtland, has said that the impact of world climatic change over the next decades "may be more drastic for mankind than any other challenges except for nuclear war".

She was speaking in particular of the so-called "greenhouse effect", the warming of the Earth's atmosphere thought to be caused by the increased emission of man-made gases into the air. Some scientists think this could be linked to changes in rainfall.

It is known that the past 40 years have seen slowly but steadily rising average annual rainfall in the upper latitudes of the northern hemisphere — roughly above 35 degrees north, with considerable local variations — and a corresponding decline in rainfall in more southerly latitudes, as the "rain belts" move north.

The change is not very great; approximately 5 per cent since 1950. Moreover, it is obscured by the fact that individual years may in any case see very wide variations in temperature and rainfall,

Is this global warming, linked to the "greenhouse effect", responsible for the shift in rainfall?

Scientific "models" established by research teams in Britain and America suggest this is very likely the case and there is strong evidence that the warming at least is definitely occurring, and may well be speeding up. The four hottest years of the past century, on world average temperatures, have all been in the 1980s and 1988 may well make a fifth, with 1987 the hottest year since regular scientific records began.

Dr Jim Hansen, of the US space agency Nasa, told a news conference in June that it is time to "stop waffling and admit that there is considerable evidence that the greenhouse effect is here already". He said that for the purposes of political decisions, you shouldn't need 99.9 per cent proof. "The fact is that the case is now clear that we have to anticipate big climatic changes over the next decades. The policy makers have to know that."

STRINGS AND THE UNIVERSE

Some physicists pointed out that a theory developed by an Italian physicist, Dr Gabriele Veneziano, to understand the strong "fundamental force" could make sense if subatomic particles are not just infinitesimal points, but have a definite size, like a length of string.

The idea was not taken seriously by many. But a handful of renegades, including Green and Schwarz, saw that strings were far more than a curiosity, they had an astonishing potential to explain not just one, but perhaps all, the "fundamental forces".

In 1984, after 10 years of gruelling effort, Green and Schwarz proved that if all particles and forces, including gravity, are really manifestations of the writhings of these inconceivably small strings, then all the mathematical reasons that had killed off the other attempts at "theories of everything" vanished.

CAPTIVE GALAXY

The astonishing observation of a galaxy whose central region is spinning rapidly in a direction opposite to that of its main body has been reported by two astronomers based at the US Space Telescope Science Institute near Baltimore.

The observers say the galaxy was probably formed by the merger of two pre-existing galaxies, and that a search for other such objects may show how often galaxies collide with one another.

MOLLUSCS AND COMPUTERS

the complexity of the human mind has greatly hampered efforts to copy its abilities. Scientists at Mitsubishi's central research laboratories in Tokyo have decided to attack the problem by working on the thought processes of sea molluscs, creatures with only about 100,000 brain cells, compared to the 10,000 million in the human mind.

The researchers are working on an radically new type of memory for an optical computer, which has the human-like ability to recall large amounts of information after being given small clues.

LOOKING AT ATOMS

Individual atoms and molecules can be seen and handled — and now rearranged — using a device called a scanning tunnelling microscope (STM).

The STM depends on a needle only a few atoms wide at the tip. Held with its point just 10 ångströms (a millionth of a millimetre) away from a plane of graphite, and with the help of a small voltage difference, electrons can "tunnel" between point and plane. The needle scans the surface at high speed, using variations in current to build up such a detailed picture of the surface as to show individual atoms.

While single bacterial cells are just discernible in light microscopes, and the structure of viruses can be made out using electron microscopes, with the STM it should be possible to examine the texture of virus coat proteins.

SPIN CONTROL

Earthquakes, by enabling the Earth to become more compact, make it spin faster and thus shorten the length of the terrestrial day; the effect is like that of ice-skaters spinning faster by folding their arms. But the acceleration of the Earth's rotation caused by earthquakes seems to be small compared with other forces, chiefly tidal, working in the other direction.

A further puzzle concerns the effect of earthquakes on the position of the Earth's rotation pole on the geographical globe. The effect of most recorded earthquakes, it appears from the calculations, would be to nudge the position of the rotation pole in the direction of 150 degrees east. But, it seems, the rotation pole has actually moved in the opposite direction during the past 15 years.

CLUSTER PHYSICS

Cluster physics, a new field of study, is driven by the question "How small can a grain of material be and still be counted as a solid?" The question may help define the limits that can be reached in making electronic devices smaller (and therefore faster), but deliberately designed clusters may also be powerful chemical catalysts.

Several techniques have been developed for making clusters containing only tens or hundreds of atoms, usually by looking for aggregations of atoms in a vapour.

When there are 100 atoms or fewer in a cluster, metals cease to be metallic and crystals lose their characteristic form. By ordinary standards, these clusters are submicroscopic: the smallest grains of material visible to the human eye contain thousands of billions of atoms.

1988

1989

NEW US SPACE PROJECTS

President Bush today marks the 20th anniversary of the first moon walk with a speech which is expected to set out bold new space initiatives for a colony on the moon and a mission to Mars

Apart from these cursory details, little is know about the missions. However, space exploration planners and engineers are starting a technological and logistical list of questions needing to be answered, centring on whether to lift off from earth, a low earth orbit, or the moon.

The favourite is a lunar launch but a decision would then have to be made over whether it is cheaper and more technologically practical to accommodate workers in a self-sufficient environment or maintain an umbilical cord of life support from earth.

And should the space vessel be shipped from earth in pieces, then be assembled on the moon? A further question concerns which currently available technologies will still be of use in 2010;

SCIENTISTS PURSUE ENDLESS POWER SOURCE

Scientists in the United States will announce today what they maintain is a breakthrough in the quest for an endless source of cheap power.

Professor Martin Fleischmann, of Southampton University, and Professor Stanley Pons, working at the University of Utah, Salt Lake City, claim to have found a radical way of achieving controllable nuclear fusion reactions.

Professor Fleischmann called his discovery "a shot in the dark". He said: "We thought it was a chance in a billion that it would work".

The professor said the research had achieved "a fairly sizeable release of energy".

The new research has concentrated on ways of starting the fusion reactions in a small reactor vessel using an isotope of hydrogen derived from seawater, known as deuterium.

COLD FUSION

Professor Martin Fleischmann, of Southampton University, and Professor Stanley Pons of Utah University, show now they appear to have observed nuclear fusion in a small glass vessel filled with heavy water and the chemical lithium.

Results given in the report show that in some cases the vessel gave out in excess of twelve times more power than was fed into it to start the reactions.

In one experiment, a power output equivalent to a one-bar electric fire was obtained from a volume no larger than a teacup.

The researchers conclude: "It is inconceivable that this could be due to anything but nuclear processes". They believe that the large heat output may be the result of the ignition of nuclear fusion reactions on one of the metal electrodes inside the glass vessel.

Professors Fleischmann and Pons admit, however, that they cannot explain the reactions. They speculate that they are the result of "a hitherto unknown nuclear process". Until now, experts in fusion physics have had virtually no details with which to study the claims. But as details do emerge there is increasing doubt over their validity.

—— and —— are well-respected observers and their infrared excess is undoubtedly real.

SINKING OF THE FIFTH FORCE

Researchers looking for a proposed "fifth force" of nature are finding the trail growing cold.

Although the search for the fifth force may ultimately leave physics unscathed, it has spawned some useful and ingenious experiments. Bennett set up his apparatus in a truck parked beside a rural canal lock in Washington state, looking for changes caused by the filling and emptying of the lock whenever a barge passed through.

Knowing the shape of the lock, Bennett could calculate the gravitational effect of the water in it. Ideally, he would have liked to empty and fill the lock at will, so that differences could be measured when he knew his apparatus to be working properly.

Unfortunately, last summer saw widespread drought across north America, and to conserve water the lock was only operated when a barge was passing through. Bennett had to keep his experiment ready, constantly prepared for an approaching barge. His best day was Sunday August 21, when an armada of pleasure craft went up and down the river.

As with many other experiments, no unaccountable fifth force was found.

US DROUGHT 'NOT CAUSED BY GREENHOUSE EFFECT'

A new study published today confirms suspicions that last summer's severe drought in the United States was more likely to have resulted from natural causes than the so-called greenhouse effect.

Many researchers suspect that responsibility for the drought lies with a mysterious weather system that emerges every two to four years in the equatorial Pacific Ocean, known as El Nino.

This starts as an abnormal warming of oceanic surface waters, with consequences felt much further afield. The displacement of air masses as a result of El Nino may be enough to affect the Earth's rotation

This does not mean that the greenhouse effect is unimportant, far from it, but an informed knowledge of the greenhouse effect can only be gained if it is studied in the context of global weather systems, rather than as an isolated phenomenon.

CAUSES OF MAN-MADE EARTHQUAKES

Researchers have recognized for some time that pumping fluids at high pressure into deep mines can set off earthquakes. Disaster was narrowly averted in January, 1986, when an earthquake registering 4.9 on the Richter scale rocked a nuclear power station near Cleveland, Ohio, just as it was about to start commercial electricity generation.

Researchers traced the source of the earthquake, at first thought to have been perfectly natural, to a waste disposal plant near by where liquid residue from agrochemical factories had been pumped down deep wells for 12 years, at pressures of 1,400 pounds per square inch.

But the implications of the reverse process - removing fluids have posed problems for theorists. It is easy to see how injecting fluids into underground cracks can distort rocks, but somewhat harder to envisage what goes on when fluids are removed. Nevertheless, the evidence that earthquakes occur near drilling rigs is overwhelming:

In France, almost 800 earthquakes have been recorded around a gas field in the western Pyrenees. These earthquakes, quite distinct from the regular small tremors associated with the mountains themselves, started in 1969 after a substantial drop in pressure in the gas reservoir, a 500-metre-thick bed of limestones.

1989

Illustrations on every page are razor-sharp and the text so clear that even scientists will be able to understand it.